OWMC—The *"Wasted"* Years
Volume One: The Early Days

☆ by Marilyn J. Gracey ☆

A Story of The
OWMC
The "Wasted" Years

Volume One — the early days

Belleville, Ontario, Canada

A Story of the OWMC—The *"Wasted"* Years
Volume One: The Early Days
Copyright © 2005, Marilyn J. Gracey

All Rights Reserved. No part of this publication may be reproduced, stored in a retrieval system or transmitted in any form or by any means—electronic, mechanical, photocopy, recording or any other—except for brief quotations in printed reviews, without the prior permission of the author.

All Scripture quotations are from *The Holy Bible, King James Version.* Copyright © 1977, 1984, Thomas Nelson Inc., Publishers.

Library and Archives Canada Cataloguing in Publication

Gracey, Marilyn J. (Marilyn June), 1935-
  A story of the OWMC : the wasted years / Marilyn J. Gracey.

Includes bibliographical references.
Contents: v. 1. The early days
ISBN 1-55306-866-1 (v. 1.)
ISBN 1-55452-118-1 (LSI ed.)

  1. Hazardous wastes--Ontario. 2. Hazardous wastes--Ontario--Management. 3. Hazardous waste treatment facilities--Location--Ontario--Niagara Penisula. 4. Hazardous waste treatment facilities--Ontario--Niagara Penisula--Planning--Citizen participation. 5. Ontario Waste Management Corporation. I. Title.

TD1045.C32O58 2005   363.72'87'0971338   C2005-901287-0

**For more information or
to order additional copies, please contact:**

Marilyn J. Gracey
4996 Elcho Rd. E., R.R. #3
Wellandport, ON L0R 2J0
Or order on-line at:
www.essencebookstore.com

Epic Press is an imprint of *Essence Publishing.* For more information, contact:
20 Hanna Court, Belleville, Ontario, Canada K8P 5J2
Phone: 1-800-238-6376
Fax: (613) 962-3055
E-mail: publishing@essencegroup.com
Internet: www.essencegroup.com

To those, past and present, who carried the torch before Clifford and I realized there was a torch to be carried.

– *In Memory* –

Mary Dykstra (November 14, 1947 – August 11, 2003)

Donald Harold Austin (July 27, 1932 – February 25, 2003)

Stanley Kszan (March 26, 1929 – March 4, 2002)

Edith Mabel Hallas (November 18, 1947 – February 6, 2002)

June Marie Austin (July 6, 1932 – April 29, 1999)

Zygmunt Soijka (In his 74th year: November 5, 1992)

Rudolph G. Ohler (In his 69th year: October 14, 1988)

– 1 –

Marilyn held the neck of her housecoat closed as the young man at the door handed her a large brown envelope. "Do...you know what the decision is?" she asked.

The young man's eyes wavered, then shifted nervously to look beyond her to where her husband stood. "I'm sorry," he said, edging toward the steps. "They've chosen this area as the preferred site." He turned and hurried toward his vehicle.

Marilyn struggled to hold back her tears until she had closed the door. Then, her eyes streaming, she leaned against the door casing of the century-old farmhouse and sobbed. "Oh, Clifford, I *knew* they'd pick our area, I knew it!"

For more than a year, she and Clifford had kept out of the controversy sweeping the community like a fierce storm. They were convinced that common sense alone would keep the Ontario government from siting a toxic waste facility in the middle of a farming community, directly upwind from the Niagara Peninsula's famed tender fruit belt. But Marilyn and Clifford had been wrong, and now, in one shattering instant, their lives had become a shambles.

He handed her a rumpled tissue from his pocket and drew her into his arms. "There's no use crying," he told her. "We didn't help Helen and the others try to fight it when we had the chance. Now it's too late."

The reminder that they had done nothing to stop the facility brought on a fresh flood of tears. For a few moments, they held each other. Then he slowly released her. "Honey, I have to go to the barn."

"I know," she murmured, hating the defeat in his voice and the unaccustomed droop to his shoulders. She blew her nose and made an effort to compose herself. He had at least two hours of hard, heavy chores ahead of him: filling the hog feeders, cleaning the pens, and throwing down straw for fresh bedding.

He kissed her cheek and headed down the back steps. Tall and hard-muscled, with his peaked cap pushed back to reveal a lock of unruly fair hair, he made her heart beat a little faster as she watched him jog across the lawn toward the barn. Behind him, gold and brown fields of ripening corn and soybeans glowed in the early morning sunshine. Thursday, September 26, 1985. Marilyn doubted she would ever forget this date.

With heavy steps, she started up the stairs to dress and begin her day. She put her hand on the varnished handrail, where once there had been no handrail, and she remembered what a horror the house had been in their early years. Plank floors, no plumbing, and a cantankerous, old, broken-down coal furnace that either left them to freeze or roasted them out. She stopped on the landing at the top of the stairs. Whole walls had been missing then, and sections of the ceilings had been torn away, revealing the dark, shadowy attic hung with cobwebs and crawling with spiders and flies. Tin cans, pails, and even a pig trough had been strategically placed on the slanting bedroom floors to catch the water dripping from the leaky roof during rainstorms.

Barely in their twenties when they married, Marilyn and Clifford had been as penniless as a young couple could be. Clifford had taken a job in a steel mill and, a year later, when they moved from their one-bedroom city apartment, Marilyn had been a new mother with a six-week-old baby. But they had scrimped and struggled and put in years of backbreaking labour until board by board, nail by nail, field by field, they had turned the rundown farm and the dump of a house into something to be proud of. They had raised two daughters and had seen both girls through university. They had done everything they were supposed to do. Yet, now...

Tears filled her eyes and rolled down her cheeks. Now, just when she and Clifford had finally seen their way clear to take things a little easier, spend more time together, and even go off sometimes for a day, a government edict had changed everything.

She stood at the bedroom window, gazing into the yard below. Bright morning sunshine filtered through the huge maple trees on the front lawn, dappling leaves that already showed hints of the autumn colours to come.

A long, drawn-out sigh escaped her lips. She and Clifford had paid no attention when, four years earlier in 1981, the Ontario government set up a new corporation—the Ontario Waste Management Corporation—with instructions to build an industrial waste-disposal facility somewhere in the province. A little over a year ago, the OWMC had slimmed the list of hundreds of potential sites to sixteen, and then to eight. "Candidate sites," they had called them. Four of the eight had been in the Niagara Peninsula; and of those four, two had been in the Township of West Lincoln, near Clifford and Marilyn's farm.

With the announcement of Elcho and Bismark/Schram Road as two of the eight possible locations, some in the community had begun to stir in opposition. Neighbours who had previously lived in peaceful obscurity began rushing around, trying to rally the community to action.

At the time, Clifford and Marilyn had felt too much was being made of the matter. Yet there had lurked, within the recesses of Marilyn's mind, a dread similar to the one that had begun to grip the community. She was torn between a longing to be left in peace and an inability to rid herself of the memory of U.S. television documentaries showing toxic waste dumped into rivers and streams or dribbled onto side roads and highways. She had no reason to expect the practices in Ontario to be different.

At the time, after weeks of private struggle, she had murmured, "Oh Lord, if this is the only way to clean up Your world, so be it." She had said the words. She had truly tried to mean them. Yet in her heart, she had wanted the OWMC to take its ugliness and noise to some industrial site far from the peaceful farmlands of West Lincoln.

Now, more than a year later, the thing she had secretly dreaded had come to pass.

Marilyn turned from the window and pulled on clean blue jeans, a tailored white blouse, and a long-sleeved, dark green sweater. She crossed the hall to what had been her elder daughter's bedroom and sat down at her father's old rolltop desk. But, try as she might, she couldn't stop crying long enough to start work on her Sunday-school lesson. Her head ached, and her eyes had swollen nearly shut.

Then, from beneath the window, came the unexpected sound of singing—Clifford, in his coveralls and barn boots, was serenading her from the yard below.

"Gloom, despair, and agony on me..."

It was a popular comedy song about depression, misery, and bad luck, and she laughed through her tears. But they were some of the saddest tears she had ever shed. For she recognized, behind Clifford's boyish grin, a bleakness that matched her own and it nearly broke her heart to see it.

*   *   *

The next day, the local newspaper put out a special edition. As Marilyn stared at the headline in the *West Lincoln Review*, reality settled like a dead weight inside her chest: "Residents' fears confirmed; Bismark/Schram Road site selected."

She studied the photograph of Donald Chant, the Chairman of the Ontario Waste Management Corporation, with his greying hair and his carefully clipped beard. Beside him sat a younger man, in his late thirties or early forties, identified in the caption as Michael Scott, information director for the OWMC. Caught on camera as he reached across the table to adjust the microphone, Mr. Scott looked every bit the suave and upcoming young executive. Behind the two men hung a map of southern Ontario and a close-up of the Bismark/Schram Road site.

Marilyn felt a desperate sense of helplessness. Sixty-five kilometres away, in the city of Hamilton, two Crown corporation executives had held a press conference and decreed that a parcel of farmland less than two kilometres from her home would become the site of a rotary kiln incinerator, a physical/chemical treatment

OWMC—The "Wasted" Years: The Early Days

plant, and an engineered landfill for hazardous waste. The chairman told the press, "I regret that this is affecting the lives of the people of West Lincoln." Marilyn studied the face, but could discern no recognizable sign of regret.

The *West Lincoln Review* had printed the lengthy official announcement, in its entirety, on the inside page. Marilyn read it slowly, carefully, making notations. When she came to the reference to supposed financial gain, she marked it with blue marker and a measure of scepticism.

Not until the chairman spoke of residents' fears did Marilyn feel the stirrings of personal offence. "Industrial waste treatment does not mean dumping raw toxic waste into a hole in the ground," he assured the people of the province. "A facility consisting of a physical/chemical treatment plant and a high temperature incinerator has never been built in Ontario before. It is no wonder most people fear these technologies."

Marilyn resented the word "fear." Born in Welland, a highly industrialized town, she knew all about smokestacks and pollution. She had awakened more than one morning to see that the water of the Welland River, which flowed behind her home, had turned the colour of rust. To her, the reference to fear made those living near the site sound like children afraid of the dark.

But her greatest outrage came as Dr. Chant explained that the OWMC placed great importance upon "minimizing risk to human health," then, three paragraphs later, revealed one of West Lincoln's most favourable features: "It has fewer households and fewer residents than any of the other candidate sites."

With furious slashes of red pen, Marilyn linked the two statements together. Apparently, to the OWMC, minimizing the risk to human health meant picking an area with fewer people to suffer the consequences. Too bad about the few! she scrawled in the margin.

Once the proposed site had been announced, a reporter from the *Review* had interviewed Alan MacKinnon, chairman of the West Lincoln Task Force Against Toxic Waste, one of the local citizens' groups. In the accompanying photo, Mr. MacKinnon looked heart-rendingly downcast. "The selection of this site was inevitable," he stated, "given the weighting that the OWMC placed on clay soil in the site selection process."

Marilyn did not know what "weighting" meant, but she had a better idea after she read the Quick Fact Summary put out by the Crown corporation. Location, size, access route, residents on site, and agricultural activity—all had been presented as concise facts.

Then came the part about the clay. Given its own heading, larger than the rest, the seventeen-word description flowed with superlatives: "At least 35 meters of deep, consistent, and uniform clay soils…will provide excellent natural containment."

Marilyn stared at the page. The words deep, consistent, uniform, excellent, and natural stood out in her mind, stirring a sense of uneasiness. In her experience, whenever someone went into overkill on the virtues of something, it usually signalled a problem. So, why had the OWMC made such a noticeable sales pitch about the landfill?

* * *

Five days after the announcement, Marilyn drove past the site on her way to the shopping mall in St. Catharines. In those few short days, Bell Canada had poured a cement slab near the road and put in an oversized telephone junction box. At the western end of the property, the OWMC had stoned a long driveway. Steel chain-link gates, complete with padlocks, barred entry. No fence, just gates.

At the end of the new roadway stood a portable air-monitoring station. The red caution light atop the tower had already become the bane of Marilyn's life. No matter where she walked in the yard around their house, no matter what time of day or night she looked out the kitchen window, there shone the red light, alien, intrusive, a constant reminder that their lives would never be the same.

The media seemed insatiable. Night after night, television cameras flashed clips of the proposed site onto the TV screen in their living room and conducted interviews that heralded the coming of the facility. Articles and photographs flooded the newspapers. For Marilyn, sleep became elusive as, daily, the OWMC presence in their lives became more and more pervasive.

OWMC—The "Wasted" Years: The Early Days

Rightly or wrongly, she and Clifford, like most residents, attached a spirit of deviousness to the actions of the Crown corporation. They had received no advance warning to soften the blow, nothing to prepare them for the announcement that came at eight a.m. on September 26. The question arose: What other surprise was the OWMC planning behind closed doors?

As a result, even the most routine procedures gave rise to suspicion. Because the OWMC had placed the oversized telephone junction box so far from the air monitor, some residents were convinced that the Corporation had installed a wiretap. Marilyn pooh-poohed the idea, insisting it sounded like something out of a James Bond movie. Nevertheless, although using the private line that had been installed some years before, she found herself increasingly guarded in her conversations. She knew she was probably imagining it, but thought she heard a clicking on the line lately that she hadn't noticed before, a sound reminiscent of a receiver being lifted on a party line, and later put down.

- 2 -

A few weeks later, Marilyn sifted idly through the mail, then stopped: another newsletter from the Ontario Waste Management Corporation had arrived.

"No sleep again tonight," she muttered irritably, tempted to toss the letter into the garbage can and pretend it had not come. But it *had* come, and she felt an obligation to read it. "Maybe later," she told herself, dropping it onto the shelf in the kitchen beside the telephone book.

That evening, seated in her dilapidated orange recliner, she forced herself to read the newsletter. The assurance that emissions from the stack would dissipate over Lake Ontario and beyond did not rest easily with her. She marched into the kitchen, where Clifford was peeling an apple, and waved the newsletter in his face.

"I don't want that stuff coming down into the lake or landing on another community!" For the second time in a matter of days, she felt an unaccustomed violence welling up. The last time she had passed the site, she had seen men along the perimeter of the field using drilling rigs. Overcome by instant and impotent rage, she had later been horrified by her reaction. But at the time, she had wanted to slam on the brakes and jump out of the car, flailing a piece of two-by-four to drive them off.

She shook the newsletter under Clifford's nose again. "It's like sweeping dirt from the kitchen floor under the living-room rug. You shift it from one place to another, but you don't solve the problem!"

When Clifford didn't answer, she crumpled the newsletter and dropped it into the garbage can. She wanted nothing lying around to remind her of the OWMC.

Several days later, notice of a public meeting arrived in the mail. "Are you going to the meeting?" Marilyn asked Clifford when he came dashing in the back door an hour later.

"I don't think so."

She hesitated, then pressed the issue. "Don't you think we should go?"

"I don't like meetings."

"Neither do I, but—"

"I'm not going, so quit asking."

"Well," she said after a moment's thought, "I think I should go. One of us needs to know what's going on."

Clifford shrugged as he grabbed his wallet. "Do as you like. I'm going to the hardware store. The pigs have broken the valve on one of the water bowls, and I've got water everywhere." With a glance at the kitchen clock, he hurried down the back steps.

As Marilyn turned away, she caught sight of herself in the mirror of the small bathroom just inside the back door. Hazel-coloured eyes stared back at her, broadcasting anxiety. I don't want to go to the meeting alone, she thought.

She had been five-foot-eleven ever since her mid-teens, and still felt self-conscious in public about her height. When she stood beside Clifford, who gave no thought to his six-foot-two frame, she felt self-assured and out-going. Alone, she felt vulnerable. Small children called out in the grocery store, "Mommy, look at the big lady!" Without Clifford's presence, shorter men invariably made thoughtless remarks about her height.

I do wish Clifford would come to the meeting with me, she thought, knowing he would not. With a sigh, she tucked back a strand of softly curled brown hair. Today, her eyes had taken on the dark green of her sweater.

* * *

Two weeks later, Marilyn nodded to neighbours and acquaintances as she sought a chair near the rear of the Wellandport

Community Hall. The large turnout surprised her. Across the front, a panel of four men in dark suits sat at a table, facing the rows of chairs rapidly filling with residents. The only woman at the table had been seated at the end, where she was obliged to sit half-turned in order to see the audience. Wearing flamboyant, sixties-style clothing and a remarkable wide-brimmed, dark-coloured floppy felt hat, she made a great contrast to the men in their suits.

Almost immediately, another dark-suited man seated in the front row stood and turned to face the audience. He held up small sheets of paper and announced, "We'll take a minute for you to write your questions on one of these slips of paper. Then we'll go though the questions, and select which ones to ask the panel."

Marilyn frowned as he passed slips of paper to those who raised their hands. To the stranger at her side, she murmured, "How do I know what to ask until I've heard what they're going to say?"

"I do not know," the lady responded, with a heavy accent. "I no like that they pick and choose the questions. I like to ask my question myself."

A man seated in the row ahead of them turned around and grinned amiably. "It's a form of crowd control. Originally they picked a site in South Cayuga, you know, on a tract of land the province had bought up."

Slowly shaking her head, Marilyn confessed, "No, I didn't know."

"Yeah," he continued. "They held a meeting, and I heard that close to a thousand people packed the Dunnville High School auditorium. They say that some guys threatened to kill Dr. Chant." A hint of amusement touched his features. "I heard he took the threat real serious, considering the meeting took place just a couple of days after John Lennon got shot."

Marilyn looked beyond the man to search the faces at the table. "If you're looking for Chant, he's not here tonight," the man told her. "They've sent his sidekick instead, Michael Scott. He's the guy talking to the woman."

Upon closer scrutiny, Marilyn recognized the OWMC's information director from the photo taken at the press conference the day he and Dr. Chant had made the announcement.

OWMC—The "Wasted" Years: The Early Days

The woman seated next to Marilyn wrote her question, then leaned closer and asserted, "I still no like them to ask my question for me." She wagged her head unhappily. "Or maybe they no ask my question at all."

Marilyn felt a sense of grievance. It's not right to muzzle all these people or to censor their questions, she thought, as she scanned the sizable crowd.

The dark-suited man who had explained the format for asking questions now called the meeting to order and collected the slips of paper. Then he introduced the panel. The woman, Pat Hayes, was a local environmentalist from Port Maitland, near Dunnville. The men all worked for the OWMC. To Marilyn, they looked remarkably alike—dark business suits, carefully knotted ties, and not a hair out of place. With nothing to distinguish one man from another, she found that their names, except for Michael Scott's, quickly slipped from her memory.

"We have done the most extensive research of any similar project anywhere," Michael Scott assured those present. "If the same process had been used earlier, the South Cayuga site would never have been proposed."

He fielded selected written questions on such things as scrubber effluent. There were serious restraints involved with sending the effluents by pipeline to Lake Ontario, he said. One solution would be to solidify the effluent and residues from the plant in concrete; then, the end product could be stored in a warehouse until suitable uses could be found for it, such as road foundation or patio stone. When asked whether OWMC would accept wastes from the United States, he spoke of a possible reciprocal arrangement, since Canadian liquid waste was already going to the States because American waste-treaters were asking for it for business purposes.

Marilyn felt her uncertainties begin to fade. Mr. Scott exuded confidence and authority as, over and over, he assured the audience that everything would be "state-of-the-art" technology.

Only Pat Hayes, the woman in the flamboyant clothing, seemed less than impressed. She took obvious delight in poking holes in the statements made by the previous speakers. Slowly, any rose-coloured hue that had been cast over Marilyn's concerns began to dissipate.

"You've got to fight for yourselves," Pat Hayes told the audience. "Demand a moratorium on landfill. Landfill is only an excuse for having done nothing. Your only beauty, your great redeeming quality for a waste facility site, is your clay. You have to start working for yourselves. Nobody else can do it for you." She spoke with such passion that Marilyn could almost believe her. "Start reading the documents," she urged. "I used the OWMC documents to learn about hazardous waste, and anyone else can do the same."

Marilyn slid lower in her seat. Not me, she thought. I won't be reading any documents. She remembered only too clearly what a blockhead she had been in science all through school.

"But don't take them as gospel. If you read something you don't agree with or you don't understand, call the OWMC," Ms. Hayes insisted. "Suffering in silence won't do any good. Make them explain it. I call the head office in Toronto, and I don't hang up until I know what it says. I don't care if they have to explain it to me six times."

Marilyn admired the woman's grit, but she knew herself. She would never have the courage to call up and demand an explanation. Yet, the fervent urgings to read the documents lodged in her mind.

As the spokesperson continued to read out selected questions from the public, Marilyn found that, without the proper background of information, she understood only parts of what she heard. Yet, she understood enough to find questions had begun to buzz around in her mind like nagging insects. She fidgeted in her chair, frustrated because the audience had been informed there would be no questions from the floor.

One of the last questions expressed concern over air emissions and accidental releases from the incinerator's stacks.

"Air safety regulations would prohibit such releases," Michael Scott assured them.

Dissatisfied with the answer, Marilyn drew her eyebrows into a puzzled frown. It doesn't work that way, she thought. Meanwhile, those at the front table began shuffling papers and stuffing briefcases. This isn't right! she thought, becoming increasingly agitated. People will go home believing what they've just been told because none of us can challenge it.

Not one to break the rules, certainly not one to speak up in public, Marilyn felt her heart begin to thud as her sense of injustice built. Suddenly she could hold back no longer. Her heart thundering in her chest until she could hardly breathe, she forced herself to her feet.

"Ex—excuse me—" she said, in a small voice over the rustle of the crowd stirring to leave.

Several voices hissed, "Quiet! This lady wants to speak."

Before those in charge could ask Marilyn to be seated, she blurted out, "That business about the stacks is a lot of rubbish. My husband works in industry. Air safety regulations don't mean a thing. They aren't allowed to clear the stacks where he works, so they do it at three in the morning when no one's around to notice. This facility will be no different. People are people, and—"

Before she could finish, the auditorium broke into thunderous applause. She sank to her chair, her knees too weak to hold her. Michael Scott attempted to play down Marilyn's remarks. "As an organization, if we don't have credibility, we have nothing. We walked away from the site in South Cayuga, and we've also fired a consultant that Pat Hayes brought to our attention at a public meeting for misusing reports."

Shakily, Marilyn rose to her feet a second time. "You only have to blow your stack once and the hazardous material is in my well and in my body, and I don't know how to trust Dr. Chant's figures [for emissions that say it would be] 99.9999 percent safe."

Noise from the audience drowned out any further assurances, and the meeting concluded shortly afterward. Amid the ensuing hubbub, Marilyn slipped unnoticed from the hall and hurried to her car. She was still shaking when she arrived home. Clifford shook his head in disapproval when he heard what she had done.

"Clifford," she declared, "I only said what people were thinking, but were too scared to say." She held out her hands. "Look at me. I'm still trembling."

"You'll get yourself into trouble," Clifford warned.

"I—I can't help it. I couldn't let it go by. It just wasn't right!"

*   *   *

As the weeks went by, Dr. Donald Chant's face and name became almost as familiar to Marilyn as her own. Radio talk shows introduced him as one of the founders of Pollution Probe, an environmental activist group formed in 1969 in conjunction with students from the University of Toronto, where Dr. Chant had been chairman of the zoology program.

Not certain which side to take on the OWMC issue, environmental groups, including Pollution Probe, pointed to Dr. Chant's long involvement with the environment. Even before he was appointed chairman and president of the Ontario Waste Management Corporation, he had served as chairman of the Canadian Environmental Advisory Council in Ottawa. Currently, he was serving as chairman of the wildlife toxicology program funded by Environment Canada and the World Wildlife Fund.

In some circles, he was touted as one of the Niagara Peninsula's local celebrities, having spent four years during the early sixties as director of the research laboratory at the Canada Department of Agriculture in Vineland.

Detractors, however, pointed to his PhD in zoology and scoffed, "What does a zoologist know about building and operating a toxic waste facility?"

But Marilyn found the lengthy list of his accomplishments both impressive and intimidating. How could a small rural community hope to oppose someone as knowledgeable and illustrious as Dr. Donald A. Chant? Or, more importantly, should they even try?

Although Clifford and Marilyn tried to keep the OWMC from intruding upon their lives, rarely a day passed without some reminder of the proposed toxic waste facility. Each edition of the *West Lincoln Review* carried accounts of the latest developments. Letters to the editor crowded page five, written by members of either the West Lincoln Task Force or the Citizens for Modern Waste Management, two citizens' protest groups that had formed in the area.

OWMC—The "Wasted" Years: The Early Days

One morning, after an OWMC pamphlet arrived in the mail, Clifford and Marilyn pored over the tiny map with an oval marked on it. Using a magnifying glass, they determined which properties, besides their own, fell within the "magic circle," as Marilyn dubbed the two-kilometre study area.

Another morning, Marilyn found a thick report from the OWMC crammed into the mailbox. She flipped through the pages, pausing now and again to glance at diagrams or maps. She stared at the title: *Facilities Development Process, Phase 3 Report*. She could hear the flamboyant woman from the meeting several weeks ago urging them to read the documents.

Reluctantly, Marilyn waded through the pages, discovering she understood more than she had expected. Again, her sense of injustice troubled her. She had assured herself the government would bend over backward in its efforts to be fair toward the residents. Now, after reading the report, she felt less certain.

She pitched the report into the garbage can in the back entrance, then fished it out again. Inside the front cover, she jotted down her questions as well as her angry feelings, and shoved it under the telephone book on the shelf by the back door.

"It's like being under siege," she told Clifford over supper that night. "We woke up one morning and there they were, with their cannons primed and ready to fire. It's just not fair!"

"Aw c'mon," Clifford growled. "Don't get carried away."

Marilyn's mouth set in a straight line. "Don't you know what they call us?" She didn't wait for an answer. "The *target* area."

* * *

Opponents of the toxic waste facility scheduled a meeting for seven-thirty p.m. in Vineland for the last day in October. When the night rolled around, the struggle to get supper over early and the kitchen cleaned up made Marilyn reluctant to make the twenty-five-kilometre trip. Yet, the need to learn all she could, and not just from the OWMC, overcame her reluctance.

As she pulled out of the driveway onto the crushed-stone road, Clifford roared past, pulling two empty grain buggies behind the tractor. That was quick, she thought. He had made the trip to the nearby grain elevator since supper, unloaded two bins of soybeans, and returned. The empty buggies rattled and banged as he wheeled into the field west of the house. Beyond him, Marilyn saw a cloud of dust in the distance. It meant that Ernest, Clifford's brother, was still running the combine. She knew from experience that they would work until ten or eleven that night if there was no dew.

As Marilyn zigzagged over stone roads toward Highway 20, she passed the proposed toxic waste site. Scattered across the countryside around the site stood beef, dairy, hog, and poultry barns. Most of the soybeans had been harvested, and the fields lay bare and brown. Beside them, fields of corn, still showing hints of green, crowded the roadsides. Here and there, fields of winter wheat or new-crop clover broke the monotonous brown with expanses of spring-like green.

Within minutes, she turned left onto Regional Road #24 toward Vineland. Soon the flat clay fields gave way to rolling hills and lighter soil. Vineyards and orchards began to appear along the roadside, their dark shapes sculpted against the rosy hue of the evening sky.

Marilyn came to the brow of the escarpment. As always, she thrilled like a child at the wide, unbroken view of Lake Ontario. On the opposite shore, the misty outline of Toronto, lit by the waning sunset, took on the ethereal aspect of a city floating above the skyline.

Below the escarpment, Marilyn drove past orchards planted to sweet black cherries and sour cherries, peaches, and pears. Eastward, beyond the spreading subdivisions, she knew that vineyards covered several thousand acres.

Minutes later, she pulled into the crowded parking lot of Vineland's elementary school. She recognized no one. Not sure what to expect at a meeting of this kind, she felt ill at ease as she climbed the steps and made her way into the auditorium. Spotting a seat near the middle of the room, she stepped over feet as she inched toward it. Just as she sat down, the room darkened and the organizers rolled the video, commissioned jointly by the North Niagara and the South Niagara Federations of Agriculture.

OWMC—The "Wasted" Years: The Early Days

Marilyn had long known that agriculture, the tender fruit especially, held great importance for the Niagara Peninsula. But until she watched the video, she had not realized how much. Nor had she known the large amount of revenue generated for the area by farming and related industries.

On the screen, aerial views taken periodically throughout one crop season fanned out over thousands of acres of Niagara's vineyards, orchards, pasture, and croplands. Fields, lush and fertile, and planted with strawberries, corn, wheat, hay, and tomatoes, stretched like a patchwork quilt beneath the airplane. How could anyone even *think* of putting a toxic waste facility in the middle of that? she wondered. After the video came speeches, and then the organizers opened the floor for questions. Mostly, Marilyn had heard it all before—including the ugliness of name-calling.

"You're just a bunch of *NIMBYs!*" one man shouted as he struck out with passion at the opponents of the toxic waste dump. "You don't care about the environment. You just don't want OWMC's toxic waste facility in your backyard."

Marilyn thought the label, NIMBY, which stands for "Not-In-My-Backyard," grossly unfair. She shifted uneasily; in part, he was right. She did not want it in her backyard. Nor, however, did she want it in any other person's backyard. From everything she had read and heard, she knew that the opponents were as dead set as she was against visiting the dump on anyone, anywhere.

"Aw, sit down!" called another man's voice. "Before they picked West Lincoln, every one of the eight candidate sites had protest groups. I didn't hear anybody calling the people of Milton or Niagara Falls or Brampton NIMBYs! And where are those groups now?"

For the second time in as many weeks, Marilyn's heart pumped wildly with the fear of speaking in public as she forced herself to her feet. Unable to reach the microphone without climbing over a dozen people, and certain she would lose her courage before she ever reached it, she said, "Excuse me—"

The babble of voices gradually died away. Although her voice was, by nature, too soft and light to be heard easily, she willed it to carry throughout the auditorium. "We're farmers in the area close to the site. Perhaps you should remember that what lands in my backyard will end up on your dinner plate." She almost fell into her seat amid enthusiastic applause.

People continued to stream to the microphone. Often, the questions, opinions, and possible ways to fight the OWMC proposal showed considerable thought and concern. One young woman in particular impressed Marilyn. Quiet-spoken, refined in her speech and her manner, she urged people to write letters to the editor.

"You have no idea how much power a letter to the editor has," she told them. "You can reach more people—including politicians—with a good letter than you can in any other way. Members of provincial parliament have people in their offices that clip and file letters to the editor. Don't underestimate the value of a strong, well-reasoned letter."

Marilyn watched the young woman take her seat. Again she had been challenged to do something totally out of character, and felt "discombobulated," as her father would have said, as well as somewhat annoyed. She didn't want her complacent little niche in life set in disarray.

The next day, Marilyn read an account of the meeting in Vineland. She was astonished to see her statement in print. She was even more astonished to see it attributed to some man of whom she had never heard.

*   *   *

Early in November, the West Lincoln Task Force Against Toxic Waste organized a meeting with Phil Andrewes, the local member of the provincial parliament. Questions raised at the meeting were incorporated into a letter from Mr. Andrewes and sent to the OWMC. Several weeks later, the *West Lincoln Review* published both the questions and Dr. Chant's responses.

"Listen to this," said Marilyn, as she read excerpts to Clifford later that night. "Question: Does an agency such as the OWMC have the right to override existing bylaws? Answer: As a Schedule 2 Crown agency, the OWMC is not bound by existing—or future—municipal bylaws."

Marilyn peered over the page at Clifford. "Well, that shoots down the citizens' ideas to pass bylaws to keep the trucks off local roads, and to refuse permits for an industrial facility."

Clifford shook his head, anger darkening his expression. "I don't see how they get the right to come in here and do as they please."

Marilyn made no comment. She shook out the paper and held it under the reading lamp as she read on. "Question: Would the proposed facility be used to treat toxic waste imported from other provinces or the States? Answer: the OWMC is sizing and building to treat only Ontario's wastes...Having said this, there may be opportunities for the reciprocal exchange of wastes with other jurisdictions."

Marilyn looked over at Clifford. "Dr. Chant gives an example of how it would work, but..."

"Waste will be shipped across the border, no buts about it," Clifford stated firmly.

Marilyn read the last question. "It's about recycling. Dr. Chant doesn't sound as if he thinks too highly of recycling. He says it won't significantly affect the need for the OWMC's facility, because recycling produces toxic residues of its own." She stopped reading, a stunned look on her face. "Clifford, he says they'll double the size of the facility if the market justifies it."

She hastily turned the page. "He says, 'The single most important principle that we have learned from Western Europe is that when waste treatment facilities are commissioned, there must be a tightening of the regulations for waste disposal, and more effective policing, if all the wastes requiring...more appropriate treatment are in fact to move to the facility. With the tightening of regulations...the OWMC will always be assured of a very substantial market.'"

Marilyn fell silent under the weight of the circumstances over which she and Clifford had no control. With a heavy sigh, she folded the paper and tossed it onto the chair. "Have you noticed the way the rest of the world says 'proposed,' but the OWMC says 'when' and 'will,' as if the matter has already been decided?"

Clifford said, "It probably has been."

"But what about the Environmental Assessment Hearings?"

"What about them?"

– 3 –

For Marilyn, the long, arduous days of harvest passed in a haze of weariness. Exhaustion dogged her steps as she carried a dishpan loaded with pint jars of pickles down to the dark stone cellar. Icicle pickles spent fourteen days in the crock and required attention every day. This year, she vowed silently, is the last time. She vowed the same thing every year. As she opened the door to the fruit cupboard, she let out a cry of dismay. A quart jar of peaches had spoiled, and sticky black ooze had run down the sides and seeped under other jars. Grumbling at the mess, she put the pickles aside and pried the jar from the bottom shelf. Then, she fetched a cloth and soapy water, and set to work.

Minutes later, she reached for a newspaper to spread on the freshly washed and dried shelf. As she spread the pages, an article caught her eye: "New Mayor Claims Difference in Philosophy— Disbands Previous Committee As Too Radical." Marilyn snatched up the newspaper. Crouched uncomfortably on her haunches in front of the fruit cupboard, she skimmed the highlights, her indignation mounting with every word.

She had attended the public meeting at the community hall with Helen Kszan the night the residents had elected the committee. Now the new mayor, voted in two weeks later, had overruled the residents' choice. Political puppets! she decided, viewing the faces in the photograph with distrust. They were nothing more than hand-picked puppets!

Throughout the winter months, Marilyn and Clifford became avid readers of the *West Lincoln Review* and the *Pelham Herald*.

OWMC—The "Wasted" Years: The Early Days

When cartoons began appearing in both papers, Marilyn's pack-rat instincts took over. She started saving them—especially the ones using a mouse in the corner as the signature.

Meanwhile, one by one, members of the unpopular puppet committee resigned, until the organization folded into oblivion.

In April, the OWMC held information nights at four of the previous candidate sites, far removed from the West Lincoln site. Finally, after these trial runs, the displays and the consultants arrived at the Wellandport Hall. About the same time, the OWMC consultants released research balloons from the air-monitoring station with its detested red warning light.

In mid-June, Helen Kszan called to ask Marilyn if she could spare her a few minutes.

"I'm going to town shortly," Marilyn told her.

"I won't take much of your time."

Marilyn rolled her eyes heavenward and made a face. "Sure, come on over."

Helen drove in the lane minutes later. Pretty and vivacious, her long dark hair pulled back from her face, she perched on the edge of the couch and came straight to the point. "We're planning a rally for July on the corner of the site that our family owns. I was wondering if you could help."

Marilyn's reluctance felt like an object lodged in her throat. The slow shake of her head became more pronounced. "I really don't think—"

"We need people to go door to door with our petition and urge everyone to attend the rally. If we're going to show our opposition, we need a big crowd."

Paralyzingly shy, Marilyn had always said she would rather starve than make her living as a door-to-door salesman. Now, Helen had just asked her to do something infinitely worse. Besides, she thought, it's too late. The decision's been made. "I can't. I really can't."

"We're asking people to bring a picnic lunch and lawn chairs. We figure if it's a family picnic, with children and grandparents present as well, there won't be any rowdiness." Helen's shoulders sagged when Marilyn again refused.

"We're having Lois Gibbs as one of our speakers," Helen added.

Marilyn stared at the floor. She'd heard of Lois Gibbs from some chemical mess in New York State, at a place called the Love Canal. The hullabaloo had even made the Canadian press, cropping up night after night on the television news. All the more reason to steer clear of the conflict, she thought.

"Do you think you and Clifford could come?" Helen persisted.

"I—I don't know…" Marilyn said, as she shook her head again. "I'll speak to Clifford, but he's not keen on these sorts of things."

Helen rose to her feet. "I'd better let you get to town." She turned to leave, then turned back. "Can I leave these flyers and petition sheets with you?"

"Well—okay, I guess. But it seems a shame to waste them."

"Maybe Clifford—"

"I wouldn't count on it." Two minutes later, Marilyn waved politely as Helen backed out of the driveway.

* * *

When Marilyn came home from town, she had a quick supper and settled down to watch television. Clifford, she knew, would not be in from the field until nine-thirty or ten. She glanced at the flyers and petitions Helen had left, then quickly looked away. She had made the mistake of leaving them in plain view. Now she could not get Helen's plea out of her thoughts.

She kept her eyes on the television, but in her mind, she wrestled with the events of the past months. Although not 100 percent in favour of the opposition to the facility, she admitted to herself how disillusioned she had begun to feel. From deep within arose a strong sense of duty, urging her to action. Resist as she might, the excuses she raised did not satisfy her conscience.

She jumped up from the recliner and snapped off the television set. "You can't just sit here and let everyone else fight your battles," she told herself out loud, her stomach in a turmoil. Snatching the pamphlets and petition sheets from the arm of the sofa, she took a step, then turned back. "I don't have to do this," she argued. She almost sat down again. As she fought down her terror and the urge to back out, she exclaimed, "Marilyn Gracey, you can at least talk to your next-door neighbours!"

Thus began her daily nightmare. She canvassed her road. She canvassed the side road. Nancy, a new neighbour, offered to help her cover the village of Wellandport.

It took every ounce of courage Marilyn could muster to go, day after day, knocking on doors. Some received her warmly, most with polite reservation. Others challenged her efforts to persuade them and, on a rare occasion, a householder became downright hostile. But each night, as she and Nancy tallied up the signatures on the petition sheets, Marilyn had to admit they had made appreciable progress.

The day she completed her task and no longer had to force herself to knock on doors, she felt as though she had been given a reprieve from the rack.

* * *

Marilyn snatched the phone from the wall just as the timer went off. "Hi, can you hold the line 'til I take the beans from the stove?" She drained the kettleful of yellow wax beans, dumped them into a dishpan of ice water, and picked up the receiver. "Sorry about that."

The voice on the other end was pleasant and unruffled. "That's fine. I'm Rose[1] from the OWMC's local office in Smithville."

Marilyn suddenly felt nauseated. She had been so busy, she had actually forgotten for a few hours about the OWMC and the office that had opened a year ago.

"We're conducting interviews with residents in the area. We were wondering when would be a good time to meet with you and your husband."

"Oh dear. July is such a busy time," Marilyn replied.

"It should only take an hour or two."

Marilyn's mind raced. How could she spare an hour or two? She made no effort to hide her reluctance. "Well—all right."

"How about Friday, three weeks from today, at two o'clock?"

"Okay, I guess," Marilyn agreed. "But on the farm—"

---

[1] Not her real name.

"We'll arrange for another day if you find it's not convenient."

Marilyn slowly replaced the receiver. It's never convenient, she thought, as a troubled feeling settled over her. She drained and bagged the beans for the freezer, the bounce gone from her step. Her mind went back two years to an earlier questionnaire. At the time, she recalled, she had developed a gnawing uneasiness while answering the questions.

"I don't know enough about the proposal to answer these questions," she had told the young interviewer as she handed her a piece of fresh cherry pie. "My answers could affect the rest of our lives."

Her concerns had been brushed aside lightly, and Marilyn had attributed the girl's indifference to her youth. After all, to her it meant a summer job, no more. But when Marilyn found several of the questions unclear, the girl had neither the background nor the authority to clarify them.

Although Marilyn had blamed the OWMC for sending an unqualified person, she had been forced to swallow her annoyance. After all, venting her anger on this young person would be like kicking a puppy.

She wondered if the OWMC would send a puppy this time, too.

"I don't like answering questions in the dark," she grumbled to herself as she carried the packages of beans down the cellar stairs to the freezer.

Over the weekend, she mentioned her concerns to her brother, Jerry. As a chemist, he worked in industry. "Don't be taken in just because the interviewers are young," he warned Marilyn. "Do they collect a paycheque at the end of the week?"

"Yes—"

"Fifteen years from now, they'll be in the same line of work. They aren't innocent puppies, as you call them," Jerry said. "They work for your enemy."

* * *

Four days later, a woman named Edith Hallas called. "I was just talking to Helen Kszan," she began. "She says you did a great

OWMC—The "Wasted" Years: The Early Days

job of collecting names on the petition for the rally. So, I was wondering if you could stuff some flyers in mailboxes for me?"

Marilyn felt instant reluctance. She had neither the time nor the inclination to become further involved. "For the rally on Saturday?" she asked, her mind made up to refuse.

"No. It's about the OWMC's questionnaire. Some people who've had their interview complained to our citizens' group about it."

Marilyn remained noncommittal.

"I've talked to John Jackson," Edith went on. "The—"

"Who's John Jackson?"

"He's with the Group of Eight." She sounded as though Marilyn ought to know Mr. Jackson. When Marilyn said nothing, she explained, "Before the OWMC picked West Lincoln as the preferred site, they had eight candidate sites—"

"Yes, I know," Marilyn interrupted with impatience.

Edith scarcely seemed to notice the interruption. "The Group of Eight is a coalition of the citizen's groups from around those eight sites. They've been studying the OWMC's proposal since 1980 or 1981. John's been involved from the beginning." When Marilyn said nothing, a note of admonishment crept into Edith's voice. "We're very lucky to have John to help us out."

Marilyn drummed her fingers on the kitchen counter and glanced at the clock. Edith cleared her throat. "Anyway, I phoned you because our group, the Citizens for Modern Waste Management, have called an emergency meeting at Silverdale church for Thursday night. That only gives us two days to notify the community. Most of our members are busy helping the Task Force to get ready for the rally on Saturday, so we're asking others in the community to help. Could you canvass your road and the Canboro Road as far as Wellandport Hall?"

Marilyn released a silent sigh. "I guess we could."

"I'll be right over with the flyers."

*   *   *

Marilyn still held out hope that, in the end, the Crown corporation would do the right thing and bend over backwards on behalf of the residents. Clifford did not.

"I didn't feel right about the questionnaire two years ago," she admitted to Clifford as they drove from one mailbox to the next. "But these citizens' groups are never happy with anything the OWMC does."

When Clifford did no more than grunt, she declared, "Common sense says the OWMC must do *something* right once in a while."

*   *   *

The next day, the front page of the *West Lincoln Review* carried a picture of opponents taping posters for the People Power Rally to a van.

"For two and a half years," the article said, "the West Lincoln Task Force Against Toxic Waste has been involved in the OWMC's public consultation process. Of late, they have spent two months or more in preparing for the rally on Saturday."

"During the past two and half years," the spokesperson for the Task Force told the reporter, "our concerns have not been addressed and our questions have not been answered satisfactorily. We are tired of being insulted and ignored. We have therefore decided to organize the People Power Rally, to be held on Saturday on the corner of the proposed site which is owned by Stan and Helen Kszan." He ended his remarks with the Task Force slogan, "There *is* a better way."

Two and a half years! Marilyn thought, appalled by the idea that Helen and the others had been involved that long. Helen and Stan milked a dairy herd morning and night, did field work, and in between, Stan worked shifts at the General Motors plant in St. Catharines. Being closest to the proposed site, they, more than anyone else, were constantly being overrun with OWMC personnel. And now, this rally.

No wonder Helen looks so tired! Marilyn thought, a wave of sympathy for her neighbours sweeping over her. They had fought for so long, and she felt keenly their sense of injustice. Yet the name, People Power, made her uncomfortable. It sounded as though the answer rested solely in numbers and strength.

The rally made good sense, she reasoned. Large numbers gathered in a peaceful show of opposition would make a strong statement. I've agreed to sell tickets, she reminded herself. Clifford's going to help with parking. Why do I feel so uncomfortable with it?

Because, her inner voice answered, no one has mentioned the name of God or called upon Him for help.

On the same day, Richard Szudy, an environmental planner with the OWMC, told the *Standard* that testing at the proposed site showed the clay was deeper and even more uniform than previously estimated.

Marilyn felt her mouth go dry. It's hopeless, she thought. Utterly hopeless. And all the rallies in the world can't change the fact.

- 4 -

On Thursday night, two days before the rally, Clifford and Marilyn set off for the meeting at Silverdale. Topic? The questionable questionnaire. They drove past the proposed site and turned onto Schram Road, and ten minutes later neared Silverdale's one-room country church. Cars filled the parking lot and lined both sides of the road. Clifford dropped Marilyn off in front of the church while he found a place to park the car. As she waited on the lawn, she watched people arrive, stopping to exchange greetings, and then inching through the groups congregating on the sidewalk.

When Clifford came, he took her arm, nodding to acquaintances as he guided her toward the door. The church overflowed with residents crowding into the tiny front entrance and standing along the walls inside. Marilyn, uneasy amid unfamiliar faces and voices raised in heated discussion, sank into one of the last available pews.

Edith Hallas called the meeting to order and quickly introduced John Jackson: "John is the co-author of the book, *Chemical Nightmare*. He has been a professor at the University of Windsor. He did graduate work in political science and sociology at the University of California, Berkeley Campus, and is a gold medallist in both fields."

The name Berkeley grabbed Marilyn's attention. Hadn't Berkeley been the site of student riots, a seedbed for rabble-rousers? As Edith paused to search her notes, Marilyn searched John Jackson's face. He didn't look like a rioter, but then, what did she know about hotheads and dissidents?

OWMC—The "Wasted" Years: The Early Days

Edith gave a perfunctory smile which she quickly turned on and then off. "John's had twelve years' experience in dealing with waste management at all three levels of government. He is vice-president of Great Lakes United, which is an international organization of two hundred groups from Ontario, Quebec, and the eight states bordering the Great Lakes. Most importantly," Edith gave Mr. Jackson a mischievous sideways glance, "at least from our standpoint, he's chairman of the Group of Eight and has been studying the OWMC issue almost from its inception." She flashed a proud smile. "We're very pleased to have John with us tonight. He has come, at his own expense, to offer any advice we might want."

Mr. Jackson, dressed in casual clothes, his shirt open at the neck, had turned deep red with embarrassment. He shed his weighty-looking knapsack and placed it to one side. From the enthusiastic applause, Marilyn realized he was a recognized and respected figure.

"Who has already answered the questionnaire?" Edith asked.

Several hands went up. A lanky young man rose to his feet. "I did—part of it, at least. But I didn't like the whole deal, so I quit. What has my income or my education got to do with anything?" A rumble of indignation circled the room.

"The interviewer read off the questions," he went on. "He wouldn't give me a copy to look at or let me take time to think before I answered. He wanted snap answers. He wouldn't clarify anything or explain the ramifications of my answers."

It's the same as the last time, thought Marilyn. Her concerns, which had seemed silly an hour ago, took on substance.

The young man held up a small file card. "For some questions, the interviewer held up cards about this size. I was told to pick one of the multiple-choice answers. Finally, I'd had enough. I told him to leave. Then I called Edith." He made an all-encompassing gesture. "And here we are."

John Jackson pushed his hands into his pockets. "Without a copy of the questions," he told them, "it's difficult for me to advise you fully. However, I've spoken to several people who mentioned questions that troubled them. I'm concerned, for example, that you have no guarantee of confidentiality."

Lack of confidentiality had not crossed Marilyn's mind.

"I'm also very concerned about the 'yes, no, don't-know' format," he continued. "Before answering, you need to know how OWMC intends to interpret the information."

John raised his eyes above the sea of heads as though marshalling his thoughts. "Take, for example, the question: If the OWMC's proposed facility is built here, will you move? The choices are yes, no, don't know. If you answer, 'yes I will move,' does it mean you feel the facility is a threat to your health? Or does it mean you have no strong ties to the area?

"If you answer 'no, I won't move,' does it mean your ties to the community are too strong to leave? Or are you saying you don't feel the facility is a threat?

"If you answer 'don't know,' does it mean you don't want to stay, but you can't bring yourself to leave? Or you can't afford to move? Or does it mean you don't care, one way or the other?

"If you refuse to answer a question," he said firmly, "don't let the interviewer mark it as 'don't know'. Make sure he or she writes in that you refused to answer. 'Don't know' could be interpreted as having no opinion."

The more Marilyn listened, the more concerned she felt. Answering the questionnaire would be far trickier than she had supposed.

John paced a few steps. "I'm also concerned because the factual questions take up about the first fifty questions. What kind of livestock, crops, and so on. How many acres you own, how many pigs, chickens, or cows. Then, when you are getting tired, the questions that require judgment begin. By this time, most people are not alert and they just want to get the interview over with."

As John continued to touch on points raised by residents, Marilyn whispered to Clifford, "This survey sounds more and more like a minefield."

"Some questions require a premature judgment," John went on. "The questionnaire makes no serious effort to deal with the real

OWMC—The "Wasted" Years: The Early Days

impacts on the community—stress, effects on business decisions, the change in the nature of your community that such a facility would bring.

"If you decide to answer it, I think you should insist on getting a copy of the survey in advance of the interview," John told them at the end. "Any questions or comments?"

Helen Kszan jumped to her feet. "I don't think we should co-operate with the OWMC on this. They're just extorting information from us that they can use against us."

An older woman agreed. "If we answer questions about what impact we feel the facility would have once it is built, it's like admitting defeat."

John nodded in agreement. "The OWMC wants to get us discussing petty details, such as the size of the proposed smokestack, and landscaping, and especially compensation. They want to condition us to accept the proposed facility by referring to it as a done deal."

"But if we don't answer the questionnaire," another resident argued, "we won't have any input."

Eric Waldron, grey-haired and irate, sprang to his feet. "Many of these questions are based on the assumption OWMC is going to steamroll over us," he stated. His gaze swept the audience. "*Don't* assume they are going to build that toxic waste facility—because they're *not*!"

Cheers and applause shook the church timbers.

Edith Hallas, who moderated the meeting, closed by saying, "I will not personally answer the questionnaire until my concerns are addressed."

Outside the church, people stood in groups, hashing over the issue of the questionnaire. "I'll get the car," Clifford told Marilyn, and strode off into the night, a black shape against the long, crooked line of creeping headlights.

"Excuse me," said a young man with a pad and pen. "I'm from the *West Lincoln Review*. Could I ask you a few questions?—"

*   *   *

Both the *St. Catharines Standard* and the *West Lincoln Review* carried detailed accounts of the meeting in Silverdale United Church. "Residents Buck Survey, Say Questions 'Obscure,'" headlined the *Review*.

Marilyn read excerpts aloud to Clifford as she scanned the article. "It seems like an accurate account of the meeting," she remarked, then gasped with surprise. "Oh! They've quoted what I said!"

> One resident wondered how a computer could handle opinions. For instance, interviewers asked 'how people feel' about the impacts of a waste facility on community life and how they 'feel' about compensation.
> "Can computers tabulate residents' feelings?" Marilyn Gracey asked. "Or would an OWMC employee simply slot the answer into one of the little boxes at the bottom of the page?"

Marilyn switched to the article in the *Standard* written by Doug Draper. "Humph! Looks like he called to get OWMC's views." She pushed her glasses up on her nose. "Murray Creed says here the questionnaire is part of a 'socio-economic impact study'—whatever that is. He says the corporation is 'required' to carry out the study under the province's Environmental Assessment Act."

As she spoke the words, an unexpected sense of sadness swept over her. Despite her growing disillusionment, she secretly had clung to the hope that the Crown corporation cared about the welfare of the community. But, according to Murray Creed, OWMC was not conducting the survey out of the kindness of its heart or out of concern for this community. It was "required."

Marilyn thought back to the meeting at Silverdale church, and some of the questions mentioned on the survey. Indeed, what did anyone's income, education or ethnic origins have to do with building a toxic waste plant? How could such information assess the effect of her sleepless nights? How did Helen's ancestry account for the look of strain around her eyes?

By the time Marilyn drummed up the courage to call the local OWMC office, the receptionist interrupted impatiently. "I know, I know. You want a copy of the questionnaire in advance of your interview."

*   *   *

Saturday, the day set for the rally, turned out overcast and rainy. Conflicting reports made for confusion. No one knew whether the drizzle would cancel the event or not.

"Let's go over and see what's happening," Clifford suggested, grabbing his jacket and the car keys. Within minutes, Clifford and Marilyn approached the site. A few cars had parked along the road. People stood in small, miserable groups, looking up into the sky and shaking their heads. Marilyn had never attended a rally before. Uncomfortable with the idea of public protest, she hunched her shoulders against the wet and wished she could be invisible.

Suddenly, the sound of singing voices caught everyone's attention. As Marilyn looked toward Schram Road, fittingly pronounced "scram", first one, then three, then a whole group of bedraggled protesters marched up the road in the rain, waving signs and singing, "This Land Is Your Land."

"Where'd they come from?" Marilyn wondered aloud as cheers arose for the marchers.

"The post office at Vineland Station," said a lady standing ahead of her.

"But that's twenty-five kilometres away!" Marilyn exclaimed.

"There's Joyce McEwan," someone pointed out. "She organized the walk. And Doris Migus, I think."

"Doris has got to be sixty-five if she's a day," said someone else.

Marilyn whispered to Clifford, "I couldn't walk that far when I was half that lady's age."

Clifford shrugged, his cap cocked to one side. "Me neither."

Although three hundred showed up for the rally, including CBC television cameramen from Toronto, the organizers finally called it off. Even so, a few residents decided to release their balloons. They hoped, when the balloons came to earth with their messages attached to them, that they would be discovered and would alert people to the possibility of emissions reaching their area.

A strengthening wind carried the balloons aloft against a backdrop of scudding grey clouds. As they rose, they headed straight for the highly populated city of St. Catharines.

A cheer broke out, and then faded.

One by one, the balloons abruptly changed course when they reached a certain height, and headed due south, toward Lake Erie. Marilyn watched in amazement. She had not known that winds travelled in layers. While the ground wind continued to blow toward St. Catharines, the wind at the upper level carried the balloons in the opposite direction.

A woman who lived on the next road tipped her head back to follow the flight of the balloons. "You know," she said, "my neighbour's daughter made a pen pal from a name inside of a balloon that she found. It had been released, believe it or not, by some school kids in Detroit."

"Detroit!" Marilyn exclaimed. "That's at least four hundred kilometres from here."

* * *

The large brown envelope containing the questionnaire arrived a few days later. Marilyn sat up, night after night, until three in the morning, pondering the questions. As she worked her way through the last half of the survey, her indignation mounted. Question 75 wanted to know if anyone in the household went to church and how often. Number 82 asked if they had attended any meetings organized by the West Lincoln Task Force Against Toxic Waste, the most visible and vocal of the citizens' protest groups. Neither Clifford nor Marilyn had attended any meetings except for the rained-out rally on Saturday. But the Crown corporation had no right to ask such a question.

Outraged, she noted the response, which to her was the most telling of all. Printed at the bottom of the question, in bold print and italics, separate from the other choices and intended for the eyes of the interviewer only, she read, "No action taken." It confirmed her opinion that the OWMC was trying to pinpoint potential opponents.

The next morning, she tossed the questionnaire onto the kitchen table in front of Clifford. "They want to know about our water supply, but they've conveniently left out flowing wells and springs. I guess if you don't ask about them, they don't show up in any reports." She swept her hair back from her forehead in a gesture of frustration. "And the OWMC certainly doesn't understand the social structure out here at all! We don't borrow sugar. We borrow machinery worth tens of thousands of dollars!" She waved her arm angrily toward the kitchen window. "And that vegetable garden out there is not my idea of a 'social and recreational activity.'"

She snatched the questionnaire from the table and flipped through the pages. "Listen to this. They want us to pick out the three things we think are most important." She furrowed her brow as she thought back to the meeting at Silverdale. "Didn't John Jackson say that if we pinpoint some choices as more important, we automatically relegate others to a place of lesser importance?"

She challenged Clifford over her half-glasses with a defiant no-nonsense narrowing of her eyes. "I don't know about you, but that's not the impression I want to give."

"You're the one giving the interview," Clifford reminded her. "If it was up to me, I wouldn't let them on the place."

Marilyn experienced a thrust of fear. "Won't you be here when the interviewer comes? I don't want to do this alone."

A hint of irritation darkened his face. "I said I'd be here."

"Good. Because whatever we do, we need to do it together."

- 5 -

Saturday, July 19, 1986, dawned hot and sunny. From north, south, east, and west, vehicles choked the roads leading to the rally. Marilyn gazed at the cloudless blue sky overhead and breathed a word of thanksgiving. Despite fears voiced by the organizers that the false start caused by the rain the week before would result in a loss of momentum in the public spirit, it had not happened.

People in lawn chairs had already packed the area around the hay wagon set up as the speakers' platform. Picket signs took every form and colour: *I'm Your New Neighbour, Don Dioxin,* said one. *Dump the Dump*, said another. As three local bands played music to suit the occasion, the group of protesters from Vineland once again marched up Schram Road singing, "This Land Is Your Land."

A smaller group of marchers from Fonthill, fifteen kilometres distant, arrived at about the same time. Near the head of the weary marchers, Kenny Crowe waved and cheered as he let his signboard do the talking: *CHANT: Schram from God's Farm Country—We Don't Need Chemical Burgers, Smoke Stack Stew, or Fall-out Hot Cakes.*

Volunteers selling hot dogs, soft drinks, and grape juice donated by Wiley Brothers, local grape growers, could hardly keep up with the demand. Children, screaming with laughter and excitement, raced over the rough-cut hay field, clutching helium-filled balloons destined later for release. Marilyn gave up trying to count heads and took her position at the ticket booth.

The band sounded a sudden fanfare to quiet the crowd. Young and old alike stopped what they were doing, and drew closer to the speakers' wagon. Ted Morely, chairman of the West

OWMC—The "Wasted" Years: The Early Days

Lincoln Task Force Against Toxic Waste, acted as master of ceremonies. Holding his two-year-old son in his arms, he urged the people to unite. "This rally is for him and all the little ones. This rally is for the future!"

Protest banners waved and cheers rose as the television cameras rolled. Politicians, one and all, heartily declared themselves opposed to the proposed waste dump. Environmentalist John Jackson stated, "Statistics can be misleading. The OWMC says the incinerator is efficient up to 99.99 percent. What they *don't* tell you is that the remaining .01 percent means, each year, a quarter of a tonne of dangerous emissions will come out of that stack." He then pressed for a move toward a reduction of wastes at the manufacturing level and an increase in recycling.

Lois Gibbs, renowned for her outcry against the toxic waste buried under the homes in the Love Canal, passionately called the protesters to even greater action. "If you wait until after the dump is here," she told the two to three thousand spectators, "it will be too late."

Frank Giovannone shook his fist. "This [rally] is only the beginning!" he shouted.

The band broke into the song, "This Land Is Your Land," and voices all over the field joined the clarion call. At a given signal, thousands of residents released the balloons they had purchased. In a great cloud representing poisonous emissions, they rose almost straight up into the sky. When nearly out of sight, they slowly turned and this time, they drifted toward the city of St. Catharines.

* * *

Six days later, Marilyn watched nervously for the OWMC interviewer to arrive. Feeling queasy, she nibbled soda biscuits and wished she had refused to answer the questionnaire.

Shortly before two, Clifford came in from the field. "I'm not changing my clothes," he told her in an irritable tone. "I don't have time for this nonsense." He glanced in agitation outside at the bright sunshine. "A day like today, I should be combining wheat, not sitting in the house."

He glanced at the clock. "I should put in a bill for my time. The interviewer's getting paid, so why shouldn't I? It's like Stan Kszan says, everyone gets paid but us."

On the stroke of two, a car drove in the laneway and a young man came to the door. "My name is Jim. I believe you're expecting me." Marilyn held open the door and Clifford shook Jim's hand. Once again, the OWMC had sent a puppy.

The interview began immediately. Marilyn refused to give one-word, cut-and-dried answers. She insisted the young man recorded her opinions and Clifford's in the margin, and the interview stretched to three hours. Then Clifford called a halt.

"But we're only half done," the young man snapped.

"My wife's too tired," Clifford replied.

Marilyn could feel herself trembling with exhaustion and stress. "My husband's right. I can't go on."

"If you'd answer the questions as intended, we could finish quickly."

"No, these questions are too important to rush," said Marilyn.

"But—"

"No buts," Clifford told him. "You can just come back another day."

The young man slapped his case shut. "I'll have to drive all the way down again from Guelph." Despite his continued protests, Clifford ushered him to the door.

Shaking visibly and close to tears, Marilyn pushed back in her old orange recliner. "We're banging our heads against a wall. He wasn't interested in any of the points I raised. He'll just go back to the office and mark yes or no in his little boxes at the bottom of the page and ignore the rest."

"Don't answer any more of it!" Clifford snapped. "You're going to make yourself sick."

Marilyn could see he had turned downright bad-tempered over it. She blew her nose and sipped the scalding tea he brought to her. "We have to put on record what we think and know," she insisted. "Then, if we raise complaints, the OWMC can't make us look bad by telling the world we didn't answer the questionnaire."

OWMC—The "Wasted" Years: The Early Days

Several days later, Marilyn telephoned the OWMC office to make another appointment. The voice on the other end sounded cold and unco-operative.

Weeks passed before a second interviewer, a young woman, came to complete the survey. She introduced herself as Frances, saying to Marilyn in a pleasant tone of voice, "This seems like a waste of time. They've already tabulated the results."

"They knew I intended to finish answering the questions," Marilyn objected. "Why did they make us wait so long to complete it?"

The young woman shrugged, making a display of her patience.

Determined to finish the questionnaire, Marilyn and Clifford spent another three hours working their way through the remaining questions. Neither of them could gain a clear sense of how the answers would be used—if they were used at all.

* * *

Despite the farm work piling up, and her natural reluctance to become involved in controversy or public situations, Marilyn found herself once again sucked into "this OWMC mess," as she called it, like a bobbing pop can caught in the vortex of a huge whirlpool.

Never in her wildest imaginings had she expected to find herself picketing an OWMC Open House. As she drove down the highway toward Smithville, she repeated, over and over in her mind, I can't believe I'm doing this! I can't *believe* I'm doing this!

At every driveway, every side road, she fought the urge to turn the car around and go home, dreading the embarrassment of being the only picketer. She swept past a tiny roadside park, over a bridge, and around a huge curve. Darn! she thought, I could have turned around in the park.

As she approached Smithville, she reduced her speed. No one will be there except me! she worried, then consoled herself with the reminder that no one would know if she high-tailed it for home.

In front of the Old Farm Inn, a dozen or so people milled about on the sidewalk. Wishing she could drive on past, she turned into the parking lot and parked the car. As she took the picket sign out of the trunk, she did her best to hide it from sight. She carried it upside down and pressed against the leg of her navy slacks, as though, without her noticing it, it had stuck to her clothing like lint.

The other picketers huddled close to one another, looking sheepish and ill at ease.

"What are we supposed to do?" Marilyn asked in a low voice.

"I don't know," the lady beside her answered. She looked as scared as Marilyn felt. The woman gave an embarrassed laugh. "But given enough time, we're *bound* to get the hang of it!"

A couple of passing motorists honked in support and the picketers cheered. Kenny Crowe, bolder than some and wearing his signboard, stood at the edge of the road and motioned to passing truckers to yank on their air horns.

Several large, shiny cars swept into the driveway and pulled up to the doors of the banquet hall of the Old Farm Inn. OWMC consultants and their underlings climbed from the cars and hurried inside.

Right behind came a television news camera.

"Walk in circles," the cameraman told them. When no one moved, he demanded, "Go on! March in a circle!"

The picketers made a tentative circle, then stood motionless.

"Now yell, chant, sing, whatever!" he ordered, as he hoisted his camera to his shoulder. He waved his arm impatiently at them. "C'mon," he yelled. "What's the matter with you people? You can't just stand there!"

Several took a few steps, and some of the others slouched along behind them. They could not have made a sorrier-looking group if they had been standing in a downpour. Someone started to sing, "This Land Is Your Land," and one by one, the voices joined in.

Marilyn kept silent as she tried to shrink into anonymity. She had not counted on cameras to broadcast to the world her radical and unseemly behaviour. But before an hour had passed, the picketers were shouting and waving their signs and cheering those who tooted their support. Sheila Ayers pushed her little girl back and forth in her stroller. Kenny paraded up and down along the edge of the highway, his signboard still declaring, *We Don't Need Chemical Burgers, Smoke Stack Stew, or Fall-out Hot Cakes.*

"Well," Marilyn said, with a laugh to the lady she had spoken to when she first arrived, "I guess you could say we've finally gotten the hang of it!"

*   *   *

The next day, Marilyn received an envelope in the mail from the mayor's office. The envelope contained a brief letter offering to discuss the OWMC, and some photocopied pages from Hansard, the official report of parliamentary debates, dated June 26, 1986. The pages from Hansard contained a report given by Dr. Chant to the Standing Committee on Resources Development, as well as questions from committee members and answers by Dr. Chant regarding the progress of the OWMC project.

On Wednesday, July 30, 1986, the *Pelham Herald* printed a position paper more than half a page long, put out by Al DiRamio and John Migus from Citizens for Modern Waste Management, and John Jackson, a founding member of the Group of Eight. Marilyn read it several times, marking in orange the parts that seemed significant to her.

The Ontario Waste Management Corporation had now spent forty million dollars. She could not imagine where it had gone. Over the past months, as the message of reducing, reusing, reclaiming, and recycling wastes had gained public support, it had made increasingly good sense to her. Thrifty by nature and by upbringing, she could not abide waste. Yet, according to the position paper, the OWMC had allotted only one percent of its budget to the Four Rs.

Nor did the Crown corporation appear to have a good attitude toward the Four Rs. While advocates predicted forty to eighty percent would be possible, the OWMC stubbornly claimed no more than seven and a half percent of wastes could be put into the Four Rs.

The wide difference of opinion puzzled Marilyn until she read that the recycling of waste oils, for example, was not to the OWMC's advantage. Waste oils, along with other hazardous materials that could be burned, would be used by the OWMC as fuel, thus reducing operating costs.

But it defeats the whole purpose, Marilyn thought. Again she was besieged by doubts. Wouldn't a Crown corporation seek out the best solution? Were the opponents twisting the facts to suit their own purposes? How could there be such a broad difference in the figures?

She clipped out the article and stuffed it into an old file folder that bulged with anything and everything connected to the OWMC.

* * *

Two days later, armed with reports, a tape recorder, and accompanied by the younger of their two daughters, Clifford and Marilyn attended the OWMC Open House at Wellandport Community Hall. Kathy had taken a day off from work to go with them. With Clifford in a suit and tie and Marilyn in high heels, they intended to make an impression.

The flyer, which had come in the mail, stated in large black letters, "OWMC wants to hear your views."

"I've got some questions," Marilyn explained to her fellow picketers of two days before.

"Be careful they don't talk you into changing sides," Kenny joked over his signboard. Beneath the grin, Marilyn sensed his unease.

"We won't be long," she told him. But despite the bravado, she felt less than confident. The OWMC consultants, with their professional air and their strings of degrees, invariably siphoned off her self-assurance.

Inside, huge displays packed the hall. Everywhere, men in suits manned the booths. Clifford, Marilyn, and Kathy started with the display dealing with air dispersion. When the consultant saw the reports in Marilyn's arms, her pages of notes, and the tape recorder, he became agitated.

"Do you mind if I tape this?" Marilyn asked.

The consultant glanced nervously at his fellow consultants, who had drifted into his booth. "Why should I?"

Marilyn began by asking questions about stack emissions. When the consultant's answers failed to satisfy her, she pulled out the article printed the night before on the front page of the *West Lincoln Review* under the heading, "'Love Canal analogy unfair,'

Chant says after rally." She traced her finger under a highlighted sentence. "Dr. Chant says here that there will be '42 pounds per day for the total volume of trace organics, including dioxins.'" She slid her finger farther down the page. "And here he says, 'Emissions of heavy metals, including lead, mercury, chromium, and zinc, would be no more than one-half tonne per year.'"

"I don't know anything about those figures," the consultant told her. Those from other booths who had come to rescue him shoved their hands in their pockets and shook their heads.

"But when these emissions fall—" Marilyn started to say.

"The emissions won't 'fall,' as you put it," he interrupted. Marilyn noticed he had begun to perspire freely.

"Well, they certainly don't float up into never-never land!" she retorted.

"Because of air dispersion in the area," he explained with exaggerated patience, "stack emissions do not *fall*." The condescension in his tone angered Marilyn.

"I simply want a clear answer as to where the heavy metals and all that will land," Marilyn told him for the third time.

"I've told you, they don't *land*, as you put it," said the consultant in exasperation.

Kathy, who looked like a teenager, had been studying the maps and charts in the display. Silent until now, she asked, "What's that dark spot on the map?" and pointed to an area northeast of the site.

"That's the point of maximum impingement," he replied.

"What does that mean?" Kathy asked with studied innocence.

"That's the point where the heaviest concentration of—of—" His voice trailed away as his face reddened. "The quantity won't be as great as that chart makes it look," he said in a defiant tone.

"Who lives on that farm?" Marilyn asked.

"Ed Comfort."

"Lucky Ed. He lives right where most of what you say won't come down to earth will fall." Marilyn shook her head.

"Maybe you should talk to engineering," suggested one of the other men who had sauntered over to try to bail out his colleague. He pointed to a man several booths away.

Marilyn and Kathy moved on to engineering. "Okay if I tape this?" Marilyn asked as she flicked the switch on.

"Sure, why not?" he commented with an offhanded shrug. His nametag read Dick Griffiths.

However, when Marilyn began to ask the same questions as before, Mr. Griffiths stopped her. "I think you should be talking to air dispersion."

"They just sent us here," Marilyn told him.

"Oh—well—" He gave a short laugh. "I guess I should try to rise to the occasion."

Marilyn again quoted the figures printed on the front page of the newspaper the previous night.

The heavy-set older man had the confidence of age. He waved away the quotes. "There are all sorts of figures flying around out there," he told her. "You should know better than to put much trust in what was printed in the newspapers."

"Well!" Marilyn remarked with an amused lift to her eyebrows. "So much for Dr. Chant's own press release."

The engineering consultant embarked on a lengthy dissertation about various aspects of the plant. None of it interested Marilyn until he began to explain, in considerable detail, the method for dealing with fugitive emissions.

"Fugitive emissions are vapours which have an odour because they are contaminated with organics. They would escape during on-site handling of the wastes."

Marilyn gave him a puzzled look. "On-site handling?—"

"During transfer procedures," he explained, "such as coupling hoses or pumping wastes."

"I didn't see any figures in here for fugitive emissions," she told him, tapping the thick reports nearly breaking her arm.

"Figures for fugitive emissions aren't included in any figures released by the OWMC regarding emissions." He held up his hand to silence her. "Let me finish," he insisted. "When pumping wastes into the receiving tanks, the air within those tanks would be expelled. This air would be heavily contaminated with organic-laden vapours. The figures don't show up because the fugitive emissions would be sent to the incinerator where the organics would be destroyed by the incineration process."

Marilyn pulled out the photocopied pages from Hansard, which had come from the Mayor's office. She turned to page R-93. "According to this, Dr. Chant says it'll take eighteen months to get the plant up and running. Yet, he also says they could start accepting wastes in about the tenth month of construction." Her manner suddenly turned sharp and probing. "Tell me, how is it possible to burn your fugitive emissions eight months before the incinerator is in operation?"

Before he could answer, the sound of loud voices distracted Marilyn. She wheeled abruptly and bumped into Kenny Crowe.

"You've been in here for over an hour," he said, grinning at Kathy and then Marilyn. "I came inside to see what was happening." He gave Marilyn a broad wink. "Keep going. You're doing great!"

Again, a burst of loud voices made Marilyn look around. Just then, she recognized Clifford's voice raised in anger. "Don't you swear at me!"

"Well I'm angry!" the consultant shouted.

"I don't care how angry you are," Clifford informed him.

"I'm telling you," the young man insisted through clenched teeth, "the geology of this area doesn't allow for springs or flowing wells."

"And *I'm* telling *you*, the well at our barn flowed until they put through the new bypass to the Welland canal."

"It *can't* flow."

"It *did* flow. And there are springs all around here." Clifford turned on his heel and came over to Marilyn and Kathy. The crowd slowly dispersed and Kenny went back to his picketing.

From then on, as Marilyn, Clifford, and Kathy wandered to other displays, either singly or together, they received a cool reception from the consultants manning the tables. If they stopped at the table on the soil, consultants from other booths would drift over to help their colleague answer the questions.

As they were about to leave, Kathy got into a heated discussion with Dick Griffiths. In exasperation, he told her, "I know you can't understand it. And it is unfortunate. But sometimes a few have to suffer for the good of society as a whole."

Furious, Kathy turned away and started toward Clifford and Marilyn, who waited beside the door. Griffiths called after her. "Did you sign our visitor's book?"

Kathy stopped, turned, and marched over to the book. Marilyn followed to sign in for Clifford and herself as well. When she bent to write her name under Kathy's, she had to stifle a laugh. After her name, Kathy, who looked like a schoolgirl, had written the letters for her two degrees in chemistry.

- 6 -

Five days after the Open House at Wellandport, Marilyn told Clifford over breakfast, "I want to go to OWMC's Open House tonight in Vineland." Before he could object, she added, "Kathy and I spent most of our time on air dispersion. Then you got into a hassle with the guy over springs and flowing wells—" She raised her hand to silence his objection to her choice of the word hassle. "All I'm saying is, I didn't get to see all of the displays Thursday night at Wellandport hall."

The remark generated a silence that ended with a drawn-out sigh from Clifford. "I can't promise," he said. "I've got to finish baling the wheat straw." He stared at the clock. "If nothing goes wrong, I should be finished by five."

Marilyn nodded. Already she had begun to frame, in her mind, the questions she wanted to ask. With a wooden six-quart basket in her hand, she followed Clifford down the back steps and out the door.

Rays of morning sunshine fell through the trees in golden patches as she hurried across the deeply shaded lawn to the vegetable garden. A light breeze stirred the tops of the maple trees, causing the branches to brush against the white aluminum siding of the two-story farmhouse.

She picked a heaping six-quart basket of yellow waxed beans and then perched on the end of the picnic table where, with her legs swinging free, she stemmed and ended the beans into a dishpan. All the while, she contemplated the open house that evening in Vineland.

For a moment, her hands ceased their work and she sat motionless. I hate confrontation, she thought, her stomach suddenly churning. Why am I doing this? She broke a blighted bean in half and tossed it onto the grass. Her inner voice refused to keep silent: Because you have questions to ask, and that's what these open houses are for—or so the flyer said.

*   *   *

That evening, shortly before eight, Clifford, Marilyn, and Judy, the older of their two daughters, drove into the parking lot in Vineland. Picketers circled and waved their signs.

"The crowd's twice as big as it was on Thursday night at Wellandport," Marilyn observed.

As the three of them started across the parking lot, Murray Creed, OWMC's communications officer, stepped out onto a small porch. He set up his tripod and camera and began to take pictures. Suddenly, a male voice from among the picketers shouted in anger, "I don't want him taking any more pictures."

"Neither do I!" a woman yelled. "He just wants a record of everyone who's here tonight!"

Other voices rose in complaint. Frank Giovannone strode across the parking lot and over to Murray Creed. "No more pictures," he told him. "That's it."

Creed smiled blandly and appeared to take little notice. The picketers moved in a body behind Frank. Looking uneasy, Murray Creed said something to Frank, then held up his hand in appeasement to the picketers, motioning for them to stop. With his lips drawn into the semblance of a smile, he packed up his camera, collapsed the tripod, and disappeared inside.

Marilyn followed Clifford and Judy through the door. After the glaring sunlight outdoors, the conference room seemed dark. As she waited for her eyes to adjust, she spotted the OWMC's consultant on incineration. She threaded her way through visitors and displays as she headed for his booth. On her right, she noticed a huge map showing the proposed lines for hydro-electric power, sewage, discharge water, and other utilities.

I must be sure to have a good look at that display, she told herself. I want to check whether they've decided to dump into Lake Erie, Lake Ontario, or the Welland River.

OWMC—The "Wasted" Years: The Early Days

Marilyn waited her turn to ask questions about incineration. Mr. Griffiths eyed her coldly and did his best to turn his back to her. When the lady ahead of her left, Marilyn attempted to get his attention. Mr. Griffiths singled out the man behind her and took his question. The minutes ticked by as he talked to everyone but Marilyn.

Exasperated, she said, "Excuse me—"

He moved to the end of the table. Marilyn followed him. "Could I ask?—"

"I'm busy right now," he told her. He smiled at the couple beside her and launched into his presentation.

Marilyn could feel herself getting angry. She tried again. "Excuse me. I just wanted to ask—"

A voice at her elbow said, "Perhaps I could help."

Marilyn looked into the face of another of the OWMC's consultants. "I wanted to ask Mr. Griffiths a question. But he refuses to speak to me." She could hear her usually soft voice raised above those around her.

Another consultant joined the first, and Marilyn looked anxiously for Clifford. As she backed away, the consultants casually closed the gap. "Ask me your question," the first consultant said, smiling pleasantly, his face only inches from Marilyn's.

"I—I want to know about fugitive emissions."

"Well, I—" The consultant paused. Marilyn watched him mentally assess his ability to answer. "What exactly do you want to know?" he asked.

Flustered, Marilyn blurted out, "How can you burn fugitive emissions before the incinerator is up and running?" She realized that without the background quotes from Hansard, the question made no sense to the two consultants.

As the men exchanged amused glances, a third OWMC consultant sauntered over. "Can I join the party?"

Joking among themselves, the consultants stood with their hands in their pockets. They teetered up and down on their toes, and shifted their weight until they gradually diminished the distance between themselves and Marilyn. She again found them standing closer to her than she liked. She dropped back a step and found herself boxed into a corner, her back against the wall.

Instinctively, she searched the room for Clifford. He appeared at her side just as several irate residents joined the group. "It's nine-thirty," he told her. "You ready to go?"

"Not leaving us to the mercy of these snake-oil salesmen, are you, Clifford?" asked John Hitchen, one of Clifford and Marilyn's neighbours.

Richard Szudy, the consultant who had first approached Marilyn, said with a pleasant smile, "I resent that remark."

"Why?" Clifford snapped. "It's true."

"It's *not* true. We—"

Clifford interrupted him. "Are you—or are you not—here to sell us something we don't want?"

"You people—don't—understand—" sputtered the consultant.

"Just answer the question!" demanded Clifford, his face red and his eyes blazing.

John stepped up. "We understand, all right. You think you can come in here and walk all over us." He reached out in a parting gesture of disdain and flicked the consultant's tie loose from his vest. "That's what I think of you and your dump!" He elbowed a path through the consultants for himself and Gail, and slammed the door behind him as they left.

"C'mon," Clifford said to Marilyn, taking her arm. He motioned to Judy that they were leaving, and they pushed through the door into the fresh evening air.

"It's so much cooler out here," Judy said. Marilyn gazed up into the clear night sky, filled with stars, as Judy climbed into the back seat of the car.

Minutes later, as Clifford nosed the car into the line of traffic on Victoria Avenue, Marilyn exclaimed, "I *still* didn't get an answer to my questions. Or get to see the displays!"

She fell silent as she realized what had happened. She and Clifford, along with Gail, John, and several others, had been skillfully isolated. She had not seen the displays. She had not asked any embarrassing questions. Nor had she disrupted the carefully orchestrated presentations.

OWMC—The "Wasted" Years: The Early Days

She pulled the OWMC flyer from the pile of reports on her lap and glared at the large black letters in the glow from the dashboard. "OWMC wants to hear your views." Her voice had developed a hard edge as she read the words aloud. "No, they don't! This whole open-house thing is a sham and a lie."

She leaned back in the seat while the miles slid past. As Clifford turned onto Highway 20, she declared, "Cathy Vaughan was right. She told me that the OWMC's open houses were nothing more than a case of, 'If you want to have an opinion, OWMC will give you one.'"

The following day, a lengthy editorial in the *Pelham Herald* confirmed Marilyn's assessment of the open houses. "OWMC communications officer Murray Creed admitted the chances of convincing the OWMC to significantly alter its direction isn't that great," wrote editor Norman Nelson. "'We've done our homework,' Creed said, 'and the information we're getting is confirming that this is a rational choice among those open to us.'"

The editorial went on to speak about OWMC's mandate. "Whatever that is," Marilyn muttered to herself. Then the editor asked: "Is it really Dr. Chant's way or no way? Can't we even talk about alternatives?"

"No, you can't!" Marilyn told the name in the byline. "Because if you do, they'll push you into a corner." Marilyn was suddenly convulsed with laughter. "That sounds too ridiculous for words," she declared to the empty house, and promptly dissolved again into gales of laughter.

Over the next weeks, information came at Clifford and Marilyn from many sources. To her surprise, Marilyn discovered her mind had been functioning like a sponge. Even if she did not retain every detail, the concepts behind the ideas became clearer and more defined. Gradually, she became haunted by persistent voices from the past.

"Read the reports," one woman had said. "Write letters to the editor," another had advised. "Read," nagged one voice. "Write," nagged another voice. I wish I'd never heard either one of those women, Marilyn thought.

But the day came when Marilyn could no longer ignore the voices. She and Clifford had been hashing over the latest data to come into their hands. Marilyn, from her reading, had been able to shed significant light upon it for Clifford. In sheer frustration, she pressed her fingertips to her forehead and massaged her brow. "Clifford, how many of our friends and neighbours know the things we've learned?"

He stared at her, puzzled. "Probably none—except maybe those in citizen's groups. Why?"

"Because—" She stopped to choose her words with care. "Because I think we have to write a letter to the editor." She pushed her hair back from her face in a harried gesture. "It's not right for people like Fred, for example, to believe everything the OWMC tells them, just because no one tells them anything different."

"Don't you think you have enough to do?" Clifford questioned. His gaze shifted to the twenty-one quart jars of peaches cooling on the counter, and the bushel of peaches waiting to be canned.

"More than enough," she agreed. "But for whatever reason, I'm able to read these reports and go to meetings. I think—I think my years of Bible study have trained me to pick a fact from here, and one from there, and make the connection." Suddenly, overwhelmed by the enormity of the task she had just proposed, she wanted to back away before she committed herself too deeply. "What do you think?" she asked.

"You do whatever you want," he replied. She knew the tone. It meant she would be on her own in this.

"Clifford, we have to do this together—or I won't do it at all. We could be under a lot of fire if we get into the public eye. We both sign the letters, okay?"

His shrug was noncommittal. "You don't need me."

"Oh yes I do," Marilyn insisted. "We need to consider this very seriously. They might try to shut us up by digging up any dirt on us they can find." She gave him a steady look. "Has either of us ever been dishonest in any of our dealings? Have the income tax returns always been accurate?"

"As far as I know," he answered. "Don't you think you're getting a little carried away?"

"Maybe," she said, as she shook her head and shrugged. "I just don't want anyone to be able to dig up one true thing to shame us.

OWMC—The "Wasted" Years: The Early Days

They might use lies, and we can't do much about that, but I don't want to be dragged through the mud for some stupid thing in my past or yours."

He looked past her, his expression distant—the look of someone taking a mental inventory. "No," he replied at length. "I don't know of anything."

"Are we agreed, then? Do I write the letter?"

"Go ahead."

"Okay, one more thing—no, two. First, you must never sign any letter without reading it. If you don't agree with something, we discuss it."

He nodded in agreement. "Sounds reasonable."

"Second, we agree never knowingly to put anything that isn't dead honest in the letters—no twists, no deceptions. And we do no name-calling. We attack the issues, not the people."

"Okay." He settled back in his chair. "Now, can I catch forty winks before I go back to work?"

"Tonight?"

"Yes. The feeder chain broke on the manure spreader. I've got to fix it before chores tomorrow morning."

"Have fun," she told him with a grin. "I'm going upstairs to start this letter on emissions. The quantities look so minute—until you look at accumulations. Then we're talking *pounds* of heavy metals."

Marilyn worked every day for two weeks on the letter, which she eventually divided into two letters. In part one, she questioned the treatment of the leachate. She pointed out that, at the OWMC Open House held at Prudhommes', Murray Creed, under pressure, had finally been forced to admit that although the wastes to be solidified and buried would be "less toxic", they would nonetheless be "toxic". Next, she dug out her now dog-eared copy of Hansard, the Official Report of Debates of the Legislative Assembly of Ontario, sent to her by Mayor Colyn. From it, she quoted Dr. Chant.

> ...In the landfill, with the waste residues that have been turned into what looks like concrete, a slow process of leaching will go on as rain and other precipitation work

61

themselves through the [clay] cover and into the solidified mass. That leachate will be collected and tested...[and] treated as if it were a new waste...

Marilyn wrote, "Our question is: If the material in the solidified landfill could not be completely detoxified in the first place, how can the leachate, now a complex chemical soup, be detoxified at some later date?" This was another of the questions she had asked Mr. Griffiths at the Wellandport open house. To this question, too, she had received no answer.

She ended the letter with a second quote from Hansard by a Mr. McGuigan, one of the committee members, who told of the attempt by the Ministry of the Environment to site the OWMC facility in Harwich Township, near Cayuga. The Ministry's efforts had failed, Mr. McGuigan had told the parliamentary committee, and in looking back, he believed one of the things that contributed to the failure had been the determination of the people to fight:

> ...This really came down to a group of ladies...One lady quit her job—she had a very good job—and for two years she worked on this. They held cookie sales, auction sales, and had singsongs in church basements. I can hardly tell you about all the events they held. They raised some $25,000 back in 1981. In addition to raising that money, every person who attended one of those meetings and bought cookies or whatever, became part of their team. They just kept building and building that team so that it was overwhelming..."

Marilyn read the letter over one final time before sealing the envelope. She rested her head against the back of her old orange recliner. One woman urged us to read the reports, she thought. Another stressed writing letters to the editor. Now, she added to the litany, They just kept building and building that team so that it was overwhelming...

That's what the Task Force is doing, she told herself. It keeps building and building...Plans for a second rally, to be held in the St. Catharines' arena, were well under way.

OWMC—The "Wasted" Years: The Early Days

But where is God in all these plans? she wondered suddenly. She pressed her fingers to her forehead, in what was becoming a familiar gesture, and closed her eyes. "Oh Lord, forgive us, I pray, for trying to do everything in our own strength..."

The stifling heat and humidity of summer waned as August plodded to a close. Despite her heavy workload, Marilyn wrote several letters to the editor, pointing to some of the concerns residents had expressed regarding the questionnaire.

In response, Leslie Daniels, coordinator for OWMC's regional office in Smithville, told the newspapers, "The corporation tried to find out how people were feeling with a survey, but people refused to cooperate on instructions from one of the citizen groups."

Marilyn wrote more letters, refuting the accusation. In them, she reminded the readers that no explanation of the tabulation formula had been supplied. Without it, the data would be open to interpretation by the OWMC, regardless of the answers given. Because these concerns had not been addressed, some residents had refused to answer the questionnaire.

The letters prompted a telephone call from a young woman claiming she worked with surveys collecting similar information and, in general, all of these surveys followed a standard format. Marilyn's heart began to beat heavily and she could not get her breath. This was the moment she had dreaded, the time when she would be exposed as an ignorant fool.

The woman began to explain the thinking behind the questions—how they were set up, how they were administered. Gradually, Marilyn's fears melted away. Finally, the woman assured Marilyn, "Your concerns are valid. Keep up the good work."

"We were right!" Marilyn declared to Clifford when she hung up half an hour later. "We aren't being silly or stubborn, or any of the other names we've been called." She waltzed and pirouetted over to Clifford, and kissed him playfully on the tip of his nose. "I wonder if that young woman has any idea how much her phone call has cheered and encouraged me!"

- 7 -

In the middle of August, the Region of Niagara announced it had hired an environmental lawyer. Two weeks later, the Township of West Lincoln made a similar announcement. On September 4, Township Council held a public meeting in Wellandport Hall to introduce the lawyer and the newly hired Project Coordinator to the residents.

The night of the public meeting, Clifford and Marilyn found chairs about halfway back. Marilyn could feel the suppressed tension. Rumours that the council had sold out secretly to the corporation had stirred up doubt and ill-concealed animosity.

To the nearly 100 residents who attended the meeting, a spokesman for the council promised firm opposition to the Ontario Waste Management Corporation's proposal. The statement received polite applause.

When it came time for the question-and-answer period, Deborah Harrison demanded, "What gives the government the right to change the zoning from agricultural to industrial?"

The lawyer, Dennis Wood, reiterated what Marilyn already knew. "The Ontario Waste Management Corporation is a Crown corporation, and the province of Ontario is not bound by municipal laws."

The statement caused an uproar. One man shouted, "How can we fight a Crown corporation? Where are we supposed to get the money to pay lawyers and all that? Donald Chant just snaps his fingers and he's got all the money he needs."

Alderman John Schilstra, chairman of council's toxic waste committee, rose to his feet. He held up his hands, urging a return to order. "Ironically," he said, when the noise subsided, "funding for the Township's opposition must be approved by the OWMC, which grants the intervener funding."

Groans swept the hall. "It's hopeless," Marilyn heard a voice say behind her. From then on, it seemed that every question the residents asked, every reluctant answer elicited from the lawyer or the project manager, brought to light additional obstacles. Any small element of hope Marilyn might have harboured disappeared as she felt an attitude of defeat bear down upon the crowd, rendering it silent.

Mary Kovacs, taking her turn at the microphone set up for residents, put the sense of defeat into words. "Taking on the OWMC is like David taking on Goliath."

Following Mary, Lloyd Comfort, in his unhurried, hesitant style, agreed: "I guess you'd say we're—we're the underdogs. But you know, it—it might not be so bad to be the—the underdogs." A smile tugged at the corners of Lloyd's mouth as a twinkle lit up his eyes. "David was an underdog," he reminded the crowd, "and—and we all know who won *that* fight." Laughter relieved much of the tension. Marilyn suddenly felt lighter, her spirits lifted by Lloyd's homespun wisdom and humour.

Riding on the wave of optimism that Lloyd had kindled, Dave McCallum, the new project coordinator for the Township, promised, "We'll have specialists go over the Phase 4-A report. We'll be looking at what it says and what it might not say. For example, perhaps OWMC has not considered the alternatives."

Clifford leaned over to Marilyn and whispered, "It sounds good. But he looks too young for this job."

When lawyer Dennis Wood broached the issue of compensation—grants and other financial goodies designed to defuse public opposition and entice politicians to look kindly upon the proposed facility—the response was instant and definite. Rumblings, jeers, and angry shouts were replaced by applause when one resident told him flatly, "We're not going to give in on this project! If we discuss compensation, we're admitting defeat."

*   *   *

Soon after, the Task Force announced an eleven-week campaign, ending with a rally at the Garden City Arena in St. Catharines. Frank Giovannone, one of the organizers, admitted to the newspapers, "Too many people feel it's unfortunate that this area was chosen as the disposal site. But they've been told the toxic waste facility has to go somewhere."

Meanwhile, the *West Lincoln Review* responded to complaints that OWMC had not answered residents' questions or else it had answered them with technical jargon or patronizing "bafflegab." "People want straight answers to straight questions," the editor wrote. "With this in mind, the *West Lincoln Review* will offer page 5 of its October 1 issue to residents' unanswered questions…Dr. Chant will answer the questions briefly, succinctly and with as few technical terms as possible."

Gerald Lammers from the West Lincoln Task Force urged residents to take part: "Everyone's tax dollars are being spent on the issue," he said.

Day in and day out, Marilyn felt bombarded by OWMC activity. Corporation employees worked full time at promoting the toxic waste facility and could be everywhere at once, it seemed. The OWMC office, which had been opened in Smithville, attempted to discredit residents' complaints and letters to the editor. The Toronto office pumped out reports and newsletters faster than Marilyn could read them. Meanwhile, Dr. Chant and his associates made speeches to prestigious groups to which residents had no access.

Unlike the OWMC, except for letters to the editor, which might or might not be printed, residents also had little access to radio, newspapers, or television. In large part, the media considered them NIMBYs, proponents of the not-in-my-backyard mindset, and therefore not credible.

Marilyn felt overwhelmed by a sense of how unfair the whole process was. When she read the text of a speech made by Dr. Chant to a five-day conference of public health inspectors in Niagara Falls, she thought she would explode with sheer frustration. He had offered the conference two choices: do nothing—or build the toxic waste facility. He mentioned no other alternatives. He again attributed the cause of local opposition to fear of an unfamiliar technology. The wording all through the speech assured the audience, OWMC "will" do this, and OWMC "will" do that, as though the proposal was a done deal. He used the word "could" only when he referred to monitoring the facility.

Then he spoke about meetings he had held with small groups of West Lincoln property owners in their homes. Marilyn knew nothing of such meetings. "These are people who have an enormous stake in what we are proposing to do," he had told the public health inspectors. "They are successful farmers, with roots in their community that go back several generations..."

Marilyn frowned. She and Clifford had as big a stake in this as anybody, she thought. And Clifford's family went back three generations. Why hadn't Dr. Chant met with Clifford? Still frowning, she found the spot where she had left off reading.

"Successful farmers...that go back several generations. They are not interested in joining active protest movements against our proposal."

Marilyn saw the sentence as though it had been printed as a huge black headline. She threw the paper down, then picked it up and read the statement again. *"They are not interested in joining active protest movements against our proposal."*

"I can't *believe* this!" she exclaimed. "So, anyone who opposes the OWMC is to be left out of these meetings." She felt utterly powerless, as though she had been dropped like a newborn kitten into a huge cardboard box and left to struggle in vain to get out. We're trapped, she thought. Clifford and I and all the other residents have been boxed in by a mighty giant who makes all the rules.

By the time Clifford came in for supper, Marilyn felt physically and mentally whipped. "Everything is stacked against us," she told him in a tired voice after she read the paragraph aloud. "We dared to ask questions at the open houses. You saw how cleverly they isolated us. We've written letters to the editor. We're a threat because we don't buy everything they're selling, so we're to be left out."

Clifford shrugged in resignation and repeated the old adage, "You can't fight city hall."

Marilyn sighed. "I suppose not," she murmured. "But it's just not right."

* * *

On October first, the *Review* printed a letter received by the West Lincoln Task Force Against Toxic Waste, dated September 6, 1986. A resident in the town of Eden, twenty-two miles south of Buffalo, New York, had found one of the balloons containing an address, which had been launched at the People Power Rally in July.

"I'm very concerned," the resident wrote. "I found this balloon at the base of a beautiful black cherry tree...on my mom's property... Eden Valley has some of the most fertile soil in this area... I've noticed that the jet stream has been bringing us air from Canada almost all summer, which explains how the balloon arrived here. Please help me to be heard. God bless you for your commendable efforts."

* * *

Nearly a year had passed since the OWMC had selected LF-9C, the site nearest Clifford and Marilyn, as the preferred site. On the second Wednesday in September, the *West Lincoln Review* printed a long letter written by Dr. Chant. It took up nearly half a page— four to five times the amount of space allowed to any resident. In it, he responded to residents' letters to the editor. The writers claimed that the Crown corporation had spent less than one percent of its budget on the Four Rs, as they were called, and urged the OWMC to place greater emphasis upon reducing and eliminating toxic waste rather than building a huge facility.

In response, Dr. Chant wrote:

> Certainly we can do more to encourage the reuse and recycling of industrial wastes. Staff at OWMC are now working full-time to assist plant operators across the province in their efforts to reduce the amount of wastes that require treatment...

Marilyn sat back in her old orange chair, her eyes staring into space as she tried to recall something she had read about staff working in these areas. She rued the damage that spinal surgery fourteen years earlier had done to her memory. Although she considered the surgery itself a blessed success, she still had days—like today—when she felt as if someone had rifled the filing cabinet in her head and flung the files into the air like confetti.

She rested her head against the back of the chair and closed her eyes in an effort to concentrate. Dr. Chant, she thought, makes it sound as though an army of people is busily at work. Yet I'm sure that's not the case. If only I could remember...

"Hansard!" she exclaimed aloud. She jumped up from the chair and raced upstairs. She sorted through the stack of OWMC stuff piled on the sewing table, searching for the pages of Hansard sent to her months ago by the mayor.

As she pulled them from the pile, she upset a folder bulging with clippings. Impatiently, she stooped to scoop them up and stuff them back into the folder. Then she leafed through the document until she found what she was looking for. Dr. Chant had reported to the parliamentary committee on Thursday, June 6, 1986: "...Also, we have a staff of two engineers in the waste recovery, recycling, and reduction and abatement area."

"Two!" Marilyn exclaimed. "You lied in your letter to the paper!" Yet the analytical part of her brain refuted the accusation. Dr. Chant had not lied. But neither had he been as truthful with the residents as he had been with the parliamentary committee.

Ignoring the laundry and the outside work, which desperately needed to be done before the cold weather set in, she went to her old typewriter and pounded out a letter to the editor. In it, she lined up the quotes as a lawyer would line up the arguments in a case. First the statement from the *Review*: "Staff...working full time...across the province..." Then the quote from Hansard: "We have a staff of two..." She put the two in capitals and ended the letter with, "This is just one more example of the misleading kind of information being fed to those who deserve more honest and straightforward treatment."

Then she started the tedious task of addressing envelopes. Because the OWMC, through speeches, drop-in centres, kitchen meetings, and its newsletter called *Update*, had the resources to spread its influence far and wide, she had gradually convinced an increasing number of newspapers in the Niagara Peninsula to print the letters. Although editors usually frowned upon "round-robin" letters—letters sent to more than one paper—they heeded her pleas for a voice for the ordinary citizens caught up in this maelstrom called the OWMC.

Meanwhile, page five of the *West Lincoln Review* had become a popular forum for residents who wanted straightforward answers to honest questions. Staffing information came out as a result of a question sent in by Ken and Carol Haynes. OWMC, Marilyn learned, employed forty-two professional staff and engaged the services of twenty-seven consultants.

"Out of all those," she declared to Clifford at the supper table, "a pitiful *two* have been assigned to the Four Rs."

After her letter exposing what she considered linguistic trickery in Dr. Chant's speech, Marilyn expected the corporation to rectify the misleading statement concerning "staff...working full time." She expected the OWMC to align itself more closely with the truth in the future. So, a week later, when OWMC communications director Michael Scott wrote of "staff...working...throughout the province" to encourage more waste reduction, reuse, and recycling, Marilyn's fury almost consumed her.

Although her father had passed away several years before, he had laid the groundwork for her fury. Never in her entire life had he ever conceded a point she made, no matter how valid the point. Now, like her father, the corporation had made no concession to her arguments. She felt literally ill from the rage boiling within her.

A month later, almost to the day, Dr. Chant responded to a delegation that had presented a brief to Pelham council in another of his long letters to the editor. Concerning the issues of waste reduction, reuse, and recycling, Dr. Chant reiterated: "We now have staff working full time on this effort...throughout the province."

Marilyn wanted to do damage to something or someone. Shocked that she could be roused to such rage, she blamed the corporation for stirring passions in her that were contrary to her nature and contrary to her faith. As she struggled to control her fury and frustration, she felt a steely cold determination come over her. From now on, she vowed, every time OWMC gave the truth a half twist, she and Clifford would expose the half-truth to the world—carefully, accurately, and with meticulous documentation.

Marilyn wrote four letters to the editor in four weeks. She sent them to the *West Lincoln Review*, the *Pelham Herald*, the *Welland Tribune* and the *Guardian Express*. In one, she questioned, "Why—if the waste facility is so badly needed—will it take six to seven years to drum up enough business to run at the initial capacity of 150,000 tonnes?" To back up her charge, she quoted Dr. Chant's testimony from Hansard: "...Over any reasonable period of growth—we plan to hit the 150,000 tonne capacity in something like year six or seven..."

During this time, the West Lincoln Task Force Against Toxic Waste entered a float in the Niagara Grape and Wine Festival Parade. The float committee, headed up by John and Gail Hitchen, named it "Let's Keep Niagara Beautiful." The hardworking committee had raised contributions from local businesses. They borrowed green artificial turf from Lampman's Funeral Home, and evergreens from a local florist. Members built a miniature picket fence and rose arbour, and painted it a pristine white.

Decked with flowers and greenery, the float took first prize in the competition, which had over 160 entries in various categories. Marilyn, when she heard the news, danced up and down like a child, shrieking, "We won! We won!"

Though Marilyn had taken no part in the project, she felt as proud and fired up as if she had wielded a hammer or paintbrush herself. For the rest of the day, she felt her spirits soar. If a handful of residents could win first prize with their float in the prestigious Grape and Wine Festival Parade, maybe the residents could win against the OWMC.

– 8 –

Marilyn checked the calendar and felt a surge of expectation. Wednesday evening meant Bible study and prayer meeting in George and Lillian's home. For years, the Southwicks had acted like spiritual parents to Marilyn and Clifford, and their home had been, for Marilyn, the haven of rest she craved. Now, it had become a place to hide from the ever-encroaching shadow cast over their lives by the OWMC.

Just as Marilyn settled into her recliner to go over the Bible study for that night, the telephone rang. "Can you get that?" she called out to Clifford as he shrugged into his coat and opened the back door.

With a muttered complaint over being in a hurry, he picked up the receiver. After a minute or two, he held his hand over the mouthpiece. "It's Gerry Schouten. He says he knows it's short notice, but the OWMC is having one of those kitchen meetings at his house tonight for local hog producers. He says we're the closest hog producers to the site, and he and his wife don't think it's right we've been left out. So he's called to invite us."

Marilyn shook her head slowly. "I don't think we should go. If it was in a hall, we could say anything we wanted. But I wouldn't feel right about getting into a dispute in someone's home." Her mouth set in an angry line. "The OWMC knows that. They count on the community's innate good manners to act as a muzzle. That's why they hold these cozy little 'kitchen meetings,' as they call them."

Clifford relayed Marilyn's concerns, then thanked Gerry for the invitation. "We'll think about it. And thanks again." As he hung up the receiver, he turned to Marilyn and said, "Gerry says the

OWMC set up the meeting. He and his wife didn't ask for it. He says we should feel free to say anything we want to."

Marilyn thought for a moment, then heaved a reluctant sigh. "I'll call Lillian and tell her we can't come." Anger flared in her eyes. "I hate the way the OWMC invades every aspect of our lives." Until now, the Wednesday-night Bible study and prayer meeting had remained sacrosanct, and she could almost taste her resentment. "But it's only this once," she promised herself fiercely.

Marilyn postponed cleaning the house to spend the afternoon formulating questions and hunting up documentation. After a hasty supper, they drove to the Schouten home.

The communications officer from OWMC's Smithville office acted as if she didn't know their names, letting Clifford and Marilyn know from the start that they had crashed the meeting. Instead of feeling embarrassed or out of place, Marilyn merely filed the incident away in her mind as a ridiculous, petty deception, proof that nothing OWMC personnel said or did could be trusted to be entirely as it seemed.

After that, the meeting quickly deteriorated into a series of heated exchanges between the Graceys and Dr. Chant. Almost immediately, Marilyn raised the question of what the OWMC intended to do about the leachate and the contaminants it would contain.

"We'll treat it later, as a new waste," Dr. Chant told her, with an air of long-suffering patience.

Marilyn insisted, "But how can you treat the leachate later, after it's a complex chemical soup, if you can't treat it fully to start with?"

"We will treat it later!" Dr. Chant told her firmly.

"How?"

Dr. Chant suddenly pounded his fist on the table. His face reddened and his eyes flashed in anger. "I told you, we'll treat it later!"

"That's like telling me you can jump over the moon," Marilyn snapped. "I want to know *how*."

Dr. Chant reined in his temper and switched topics, only to clash almost immediately with Clifford over whether or not raw toxic waste would be shipped across the border from the United States to the proposed facility. Clifford refused to be pacified by the oft-repeated, pat answers he'd heard before. Dr. Chant's face flushed as he eyed Clifford with disdain. "You have sarcasm running out of both sides of your mouth!"

It's true, Marilyn thought, as she barely smothered a smile. Instead, she patted Clifford's leg under the table, a reminder to him not to lose his temper more than he already had.

The meeting lasted several hours. When Dr. Chant rose to leave, he remarked to the communications officer, "Well, this certainly was one of the more lively meetings."

As Clifford and Marilyn pulled from Schouten's driveway onto the highway, Marilyn said in a tired voice, "I don't think we gained an inch."

"Probably not," Clifford replied, "but at least they didn't get away with presenting a slick picture of the facility."

The next day at breakfast, Marilyn sat with her elbow on the table and her fist jammed into her cheek. "I think we'd have been better off to go to prayer meeting at George and Lillian's. At least I would have slept last night." She gave her head a small shake. "Amazing how I know all the right answers at four in the morning."

Less than twenty-four hours later, Marilyn received word that George Southwick had suffered a massive heart attack and died. Marilyn let out a wrenching cry and buried her face in her arms in a storm of tears. "This stinking OWMC stuff! It kept us from spending one last night with George," she sobbed. "We missed our last chance to see him alive." Her voice turned harsh with rage as tears poured down her cheeks. "I hate them! I hate them all!" Her eyes turned steely with resolve. "I'll never forgive the OWMC for this."

Despite her grief-filled tears, she baked a chocolate cake and iced it with her grandmother's old-fashioned brown-sugar icing. When it was finished, she and Clifford packed up a box of food, including some of their own homemade farmer's sausage, and delivered the food to Lillian's home. Later, unable to sleep, Marilyn sat for hours in her old orange recliner, wrapped in a blanket and plagued by regret. If only Gerry Schouten had not called. If only they had gone to George and Lillian's as they'd planned. If only...If only...

The following evening, Marilyn sat in front of the television. Though her eyes watched the screen, her mind roved restlessly over memories now bathed in sorrow. Suddenly, she jumped up from her chair. "I forgot to call Mom and let her know about George!" she exclaimed to Clifford as she dialed her mother's number. After fifteen rings, she hung up. Her mother found it difficult to get up from her armchair and often, if she was watching something of particular interest on television, she chose not to answer. Marilyn smiled at the irony. More than one long-distance phone call, if not totally ignored, had been cut off with a terse, "Don't bother me right now!" and the receiver had clunked down in Marilyn's ear.

Marilyn placed her last call at fifteen minutes past midnight. Then she gave up. Her mother would not answer the telephone late at night, convinced that no one but robbers checking to see if she was alone would call at that hour.

However, on Saturday morning, when her mother still did not answer the telephone, Marilyn suffered the first twinges of alarm. She called her younger sister in Niagara Falls. "Sharon, is Mom staying with you?" she asked without preamble.

"No," Sharon answered. "Why?"

"Because I tried to call her all last evening and then again this morning. I can't get any answer."

Worry coloured both of their voices. "Have you called Aunt Dora?" Sharon asked.

"Not yet. I'll do that right now." Marilyn hung up and called her aunt, who lived next door to her mother.

"I haven't seen her since earlier in the week," Aunt Dora told Marilyn. "I'll check on her and call you back."

Five minutes later, Aunt Dora telephoned Marilyn. "She doesn't answer when I call. Harry came along and he tried to force the door. But she's got that darned bolt on it. I've told her—and *told* her—that in an emergency, I wouldn't be able to get in."

"We'll be there in twenty minutes. Clifford will have to break the door down."

As Clifford collected a pry bar and whatever else he thought he might need, Marilyn alerted Sharon and then raced to the car. All the way to Welland, her heart thundered in her ears. What would they find when they got there, she wondered, dreading the answer. Tremors started in her stomach, then spread to her arms and legs. By the time she and Clifford reached her mother's house, the tremors had become so extreme that Marilyn could barely stand when she stepped from the car.

Clifford put his shoulder to the door and shoved. The bolt held, but the doorframe splintered like matchwood. Her mother lay on the floor, a deep gash in her scalp where she had struck it when she fell. She could not speak, but she was alive.

Too frantic to remember where her mother kept the phone book, Marilyn dashed next door. Aunt Dora dialed and Marilyn snatched the phone from her hand. "Hurry, please! We need an ambulance," she gasped, hardly able to speak as she gave the address. "I don't know what happened. I don't know how long she's been on the floor. But she's still alive. Please! Can't you hurry?"

She raced back to her mother's side. "It'll be okay, Mom. The ambulance is coming. They'll be here soon," she whispered, touching her mother's cheek. "They'll be here soon."

Marilyn tried to keep calm, but the trembling grew worse and she thought she would faint. The blood on the floor, the picture of her mother lying there for hours, possibly days, engraved itself in her brain. No medication. Nothing to eat or drink. Marilyn imagined her mother, helpless on the floor, listening to Marilyn's repeated attempts to call her, but unable to answer. She felt sick to her stomach. Where was that ambulance? What was taking it so long?

After what seemed like an eternity, she heard the wail of the siren and someone saying, "The ambulance is here."

Marilyn watched as the ambulance attendants checked her mother over, listening to her heart, and asking her questions although she could do no more than silently form the words with her lips. Finally, the attendants lifted her onto a stretcher and loaded her into the ambulance.

Marilyn tried to push past the attendant as he closed the doors. "I want to go with my mother!" she pleaded. "She'll be so afraid. Let me hold her hand."

"You can't ride with her," the attendant told her firmly. "Get someone to take you to the hospital."

Marilyn stood aside, tears streaming down her cheeks. "I want to be with her," she sobbed. "I want to be with her." Clifford took her arm and hurried her to the car. "Clifford, what if she dies before I get there?" George's death struck at her with renewed force and her weeping grew desperate. "Oh Clifford, not Mom, too!"

Clifford started the car and headed toward the hospital. "Marilyn, you have to calm down," he told her. "They won't let you see your mother if you keep crying like this."

Minutes later, they raced into the emergency receiving area. The nurse insisted they go to the waiting room. Sharon, white-faced and out of breath, hurried in just as the nurse stuck her head around the corner. "You can see your mother now."

By asking questions to which their mother could signify yes or no, Marilyn and Sharon determined that their mother had fallen late Thursday night. Marilyn shuddered to think of the hours she had lain there.

"But with her medication and the intravenous to combat the dehydration," Marilyn reasoned, "she should be back on her feet fairly soon." Sharon agreed, and Marilyn left her sister at her mother's side. Then she and Clifford hurried to her mother's house to nail up the door. From there, they raced home, changed their clothes, and drove to Fenwick for George's funeral.

The following morning, after church, they drove the thirty-two kilometres to the hospital. Although cognizant of everything that was said, Marilyn's mother could only whisper. Monday morning, Marilyn rushed through the week's laundry and again drove to the hospital. Contrary to her expectations and Sharon's, their mother had not regained her strength or the ability to speak aloud.

Worried and weary, Marilyn returned home to the farm to find Clifford in the foulest of moods. Because the OWMC took up so much of Marilyn and Clifford's time, the lawn had grown up like a hayfield. Clifford had spent the afternoon mowing wet grass and raking the soggy clippings into large mounds. As if that wasn't enough, Mary Lou Garr, the communications officer from the Smithville office, had showed up at the house.

"Along comes this car," Clifford told her, his face like a thundercloud. "Out steps Mary Lou Garr from the Smithville office. So I asked her straight out if we had been on the official guest list for the meeting at Gerry's the other night. 'Well, no,' she says. Then I told her I thought that was discrimination. 'Well,' she says, 'I suppose you could call it that.' So, I asked why we had purposely not been invited, considering we're the closest hog farmers to the site. Her answer? 'You're too loud and your wife is too emotional. And you stirred up too much of a commotion at the Open House at Wellandport Hall."

Marilyn's mouth dropped open and her eyes flashed fire. "Of all the nerve! I have that Open House on tape. We'll see who's 'too emotional.'"

"What about me being too loud?" Clifford inquired indignantly.

"Uh—well—" Recalling the loud voices and the angry exchange between Clifford and the consultant, Marilyn cast about for a diplomatic reply. Finding none, she spun on her heel and started toward the house. "I'd better start supper," she called over her shoulder.

In between frying pork chops and peeling potatoes, Marilyn began scribbling the first draft of a letter to the editor. She quoted Dr. Chant from the speech he had made to the five-day conference of public health inspectors in Niagara Falls just two weeks earlier. "I've been meeting with groups of West Lincoln farmers. They are successful farmers, with roots in their community that go back several generations." She emphasized the last sentence: "*They are not interested in joining active protest movements against our proposal.*"

"Let's make sure the public knows who gets invited and who doesn't," she reasoned aloud as she flipped the pork chops and waved the salt shaker over them. Next, she quoted another of Chant's statements from Hansard to the parliamentary committee where he had spoken about another of his cozy kitchen meetings. He had pointed to a group of West Lincoln poultry producers because they had not read the OWMC publications: "...Very intelligent people, very good people and deeply concerned people...Once again, I found that these people had read almost nothing about the Ontario Waste Management Corporation even though they were within a mile of the site."

Marilyn narrowed her eyes as she moved in for the kill, using Dr. Chant's own words ten pages later from the same copy of Hansard. "...There is nobody anywhere who could deal with that vast mass of technical information..."

"To read—or not to read," she murmured with satisfaction. "Dr. Chant, that is the question."

Marilyn then quoted from a recent letter to the editor written by Dr. Chant. In it, he accused another of his critics of not reading OWMC's many volumes of technical information. He then intimated that the resident should stick to the facts.

To finish off the letter, Marilyn related how she and Clifford had been treated because they *did* read the reports, *did* ask questions, and dared to challenge the data. She then listened to the tape of the Open House at Wellandport Hall and carefully listed some of the "facts."

> FACT: Figures for fugitive emissions aren't included in any figures released by the OWMC regarding emissions.

> FACT: The consultant dealing with Air Dispersion seemed to have little, if any, knowledge concerning the emissions to be dispersed.

As proof, she quoted his responses to her questions.

Meticulously, she listed fact after fact, each one a scathing denunciation of OWMC's lack of ability or desire to provide the public with anything beyond superficial information. Then she edited the letter—trimming, clarifying, and arranging her arguments to the best effect. But above all, she double-checked for absolute accuracy and documented every statement.

Finally, with her lips set in a straight line and a slight frown knitting her brows, she read her rough draft one more time.

Nowhere could she find anything that could be misconstrued as "emotional."

* * *

Two weeks later, the *Standard* carried an article on how the toxic waste fight had changed residents' lives. In it, individuals described the last few years as an endless blur of meetings, protest rallies, and wading through the corporation's lengthy technical reports. "I think it's unfair we have to take precious time away from our livelihoods to deal with this," one West Lincoln dairy farmer complained. A second resident, Edith Hallas, stated she had been attending meetings since 1983. Doris Migus, another long-standing opponent, had never participated in public lobbying until the OWMC came along. "But we don't own the earth," she said, "and we have a duty to preserve it for future generations."

The article concluded with an unwelcome reminder: "The hearings are expected to begin some time next year and take more than a year to complete." Marilyn felt a hollow spot form under her ribs. Every tick of the clock was bringing the nightmare closer.

As though triggered by the unwelcome reminder, ceaseless activity surrounding the OWMC's proposal swept over the community. The Regional Planning Committee toured the West Lincoln toxic waste site. Dave McCallum, the young coordinator for West Lincoln, traveled to Denmark for a world symposium on waste disposal. In the interim, residents' questions to the *West Lincoln Review*, seeking the promised straightforward answers from Dr. Chant, filled page five for four straight weeks.

Meanwhile, four or five times a week, Marilyn made the trip to the hospital to see her mother. With every visit, she felt increasing concern. Her mother was not responding as she and her sister, Sharon, had expected.

- 9 -

The face accompanying the interview on the front page of the *Evening Tribune* had become discouragingly familiar. "If Dr. Chant sneezes, his picture hits the papers," Marilyn grumbled to Clifford as he laced up his heavy workboots and started down the back steps. Clifford shrugged his shoulders, his attention fixed on the feed truck backing up to the steel bin at the north end of the hog barn.

Marilyn made herself a second cup of coffee and carried it into the living room. She sank into her favourite easy chair and put her feet up on the footrest. Since she and Clifford had begun sending letters to the editor, friends and strangers alike often alerted her to articles in the wide range of newspapers that served the Wellandport/Bismark area, or mailed clippings to her.

Today, an early morning phone call had sent her to the corner store to buy yesterday's *Welland Tribune*. The front page carried Dr. Chant's response to questions following his luncheon address to the St. Catharines Chamber of Commerce the previous day. Marilyn compared yesterday's answers in the *Tribune* to the statements he had made in a half-page letter printed ten days earlier in the *Pelham Herald*. They weren't consistent.

In the letter, Dr. Chant had assured residents they had no cause to worry about spills composed of concentrated wastes: "Most of the wastes in these trucks will have a high water content (as high as 95 percent)..." he wrote.

Yesterday, however, when the reporter from the *Tribune* suggested that liquid wastes, if spilled, would be hard to contain and clean up, Dr. Chant had replied, "Most of the wastes we would be carrying will be sludges, filter cakes—the kind of things that have a very low liquid content..."

As Marilyn circled the discrepancy between the two clippings, she told the face in the newspaper photo, "Dr. Chant, you can't have it both ways!" Cutting out the interview from the *Tribune*, she clipped it to the *Herald* letter. On the back of an old envelope, she jotted down the outline for a letter to the editor. Then she backtracked, reading both accounts a second time to make certain she had kept everything in context.

Beneath the carefully chosen words of the long letter to the *Herald*, she sensed Dr. Chant's impatience with the doubts expressed by residents. The delegation to Pelham council, he stated flatly, had ignored points the OWMC had emphasized repeatedly regarding air emissions. As well, he wrote, the idea that wastes would leak into the environment was wrong. The landfill would be equipped with a number of engineering features designed to isolate any contaminants from the environment.

Marilyn felt the old doubts creep in. Despite the discrepancy over the high/low water content, the letter had a convincing ring to it. Perhaps Dr. Chant was right. Perhaps the residents had become obsessed with chasing groundless fears.

Using hot-pink and fluorescent-orange markers, Marilyn highlighted the paragraph in the *Tribune* about the sludges:

> Most of the wastes we would be carrying will be sludges, filter cakes—the kind of things that have a very low liquid content and don't have nearly the potential hazard of a gasoline spill or a chlorine spill.

As she did when she studied the Epistles of the Apostle Paul, Marilyn stripped the statement to the bare bones. "Most of the wastes...don't..." Marilyn closed her eyes in reflection. That meant some of the wastes...do. She spoke the words slowly, carefully. "Some wastes *do* have the potential hazard of a gasoline or a chlorine spill."

She read the statement again, her attention galvanized by the word chlorine. Nothing coming to the plant would be as dangerous as chlorine, he had said. She furrowed her brow in thought as her memory struggled to grasp a vaporous recollection as elusive as shifting mists. Yet out of the mists she formed the vague impression of a report…a chart…on the left-hand page…

She ran upstairs and began to search through the OWMC's reports. She sat on the cold bedroom floor, one leg folded under her, leafing through volume after volume. Forty minutes later, her eyes locked onto a list of additives and reagents that would be trucked into the facility to be used in the Physical/Chemical treatment plant.

On the list: liquid chlorine—400 tons per year.

Marilyn took off her reading glasses and chewed the earpiece. Perhaps, she reasoned, in the world of industry, 400 tons would be considered a minuscule amount.

Half-afraid she would reveal herself for a fool, Marilyn made some discreet inquiries. One source informed her that the amount of chlorine to be trucked to the plant each year would equal close to six times the amount carried by the derailed tanker car that had caused the famous Mississauga disaster in November of 1979. That disaster had required mass evacuations and it had been days before residents could safely return home.

Marilyn found herself doubting the information on the train derailment, doubting her conclusions. Even yet, she could not bring herself to believe that the Chairman of a Crown corporation would purposely misrepresent the truth.

"There must be some circumstance I haven't grasped," she told herself, "some data I lack."

She read the interview in the *Tribune* again. "There are more dangerous loads on that highway…than *anything* we would be carrying *at all*."

Any doubts she had harboured vanished in the light of those words. She had not been wrong, or ignorant. Or unfair.

She marched upstairs to the front bedroom and sat down at the typewriter. With precision she lined up the quotes, first on the inconsistencies regarding the high/low levels of water content in the wastes, then on the conflicting statements regarding chlorine.

Then she stuffed the draft into her purse. With a glance at the clock, she ran a comb through her hair and set off to visit her mother.

As she drove to the hospital, she mulled over in her mind the letter to the editor. In the hospital parking lot, she scribbled several changes to the wording, then cut across the lawn with a spring in her step.

Yesterday, at long last, her mother had seemed greatly improved. The intravenous tubes had been removed three days earlier and, for the first time in weeks, her mother had made sense when she talked. Not once had she seen creatures on the walls or on Marilyn's shoulder.

Marilyn rode the elevator to the fifth floor. With a smile on her lips, she sailed through the door to her mother's room and stopped in her tracks. Her mother sat propped up in bed, scooping up food with her fingers and stuffing it into her mouth—applesauce, mashed potatoes, green beans. The creatures had returned and her mother turned nasty when Marilyn could not hand her the silver box that only her mother could see.

Marilyn left two hours later, her head throbbing and fear in her heart. For the first time, she faced the possibility that her mother might not recover.

* * *

As the weeks passed, letters to the editor, interviews, and articles related to the OWMC proposal continued to flood the newspapers. Marilyn read them all. Figures on stack emissions competed for space with complaints about the lack of funding for residents.

"The OWMC," wrote one irate taxpayer, "has received millions of tax dollars and had several years to prepare its case... Whereas the West Lincoln Task Force, formed in 1984, has absolutely no money with which to prepare a fair case."

Editorials urged the Niagara Regional Council to lobby the provincial government to force a change of heart regarding the OWMC. In response, the Region's OWMC steering committee held a meeting to give the public a chance to make submissions. "The trouble is," stated the *West Lincoln Review*, "the submissions are to be made tonight...and the public had only 48 hours notice."

Overwhelmed by the constant conflict, Marilyn struggled to separate the rhetoric from the facts. When Edith Hallas telephoned to discuss the upcoming rally to be held November 29 in the St. Catharines arena, Marilyn expressed her exasperation. "Wild accusations don't impress me," she told Edith bluntly. "I want facts. I may be losing faith in those who run the Crown corporation, but I don't know yet where I stand on the proposal."

Ten minutes later, Edith arrived at Marilyn's door. "Here's a copy of a paper I prepared a while back. Read that."

After Edith left, Marilyn started to scan the paper, intending to skim through it and no more. But she found the pages loaded with facts, each fact documented. Before she finished reading it, she had developed strong misgivings about the long-term effects of the proposed facility.

The next day was particularly warm and sunny for early November. With a lengthy list of errands ahead of her, Marilyn jumped into the car and drove to Fonthill. She parked in front of the hardware store and got out. As she hurried down the main street, she almost collided with the editor of the *Pelham Herald*, Norman Nelson.

After the usual pleasantries, Marilyn said, "I want to thank you for printing our letters to the editor. We'd have no voice if it weren't for people like you. We need all the help we can get."

A frown formed between his eyes as he replied, "Don't thank me. No media person is your friend."

"Oh, I—I wasn't trying to—to say—" she stammered.

"Listen to me." His voice turned harsh and he shook his finger under her nose. "Get this straight—no media person is your friend."

Marilyn felt as though she had been kicked.

"I only meant—"

"I know what you meant. I'm telling you to guard every word you say to the media. You and the other residents will save yourselves a lot of grief if you get this through your heads." He spoke in capitals: "No media person is your friend."

He left her standing in the street, the joy gone from her day. Later, as she drove home, she recalled the warning Edith Hallas had given her the first time they met. "Anyone who thinks he or she is smart enough to handle the media is a fool. No one handles the media."

Marilyn felt betrayed. Was Norman Nelson friend or foe? She could not decide. But his words of caution had enveloped her in a cloud of lingering menace.

※   ※   ※

The day following Remembrance Day, the newspapers overflowed with reports and editorials regarding the OWMC issue. Marilyn settled into her old orange chair with her feet up and a pile of papers in her lap.

The *Standard* announced that Kai Milliard from Pollution Probe was scheduled to speak at Brock University on alternative technologies for toxic waste disposal. Marilyn squinted her eyes as she searched her memory. Hadn't Dr. Chant been one of the founders of Pollution Probe? One for their side, I suppose, Marilyn thought glumly.

The *Pelham Herald* announced that the OWMC had held its first monthly drop-in session in Pelham at the municipal council chambers, with others scheduled for Vineland and Niagara-on-the-Lake. That's *two* for them, she thought, swept by a feeling of helplessness. With frightening ease, it seemed, the tentacles of the Crown corporation reached further afield with every passing day.

The *West Lincoln Review* printed a belated report of Dr. Chant's speech to the St. Catharines Chamber of Commerce. Once again he had made the concerns of the citizens appear ridiculous by telling the audience, "The controversy...primarily stems...from the fear that toxic waste treatment consists solely and simply of finding a dump, a hole in the ground, and backing trucks up to it to unload raw, untreated chemicals." Then he had pitched the line about "the numerous economic benefits" the proposed facility would bring to the area. "Sometimes I feel like a snake oil salesman when I talk about some of the economic benefits that would stem from the facility," he told those at the luncheon. But Marilyn had no doubt the dollar-shaped carrot dangled before the group of business people had held great appeal. A dispirited sigh escaped her lips. Three for their side, she thought.

The *Review*, however, questioned the validity of the figures:

"...450 man years during start up sounds impressive...but pales when one changes it to 450 people working 1 year or 45 people employed for 10."

"One for our side," Marilyn declared triumphantly, then stopped. When had her thinking suddenly shifted from neutral to a definite 'our side,' 'their side'?

Under the heading "Grass Roots Waste Battle is Growing," the *Review* also carried an interview with Doris Migus. She had recently given two speeches—one in Buffalo and one in Toronto. "The OWMC strategy has a major flaw," she had told the gathering in Buffalo. "The focus is on disposal rather than management, thus perpetuating the problem."

She had made the speech in Toronto to a citizens' hearing on Pollution in the Great Lakes. "I went as a member of the Citizens for Modern Waste Management," she told the *Review*. "This group supports the four Rs: recycling, reducing, reclamation, and reuse. I also represented the newly formed Ontario Toxic Waste Research Coalition. Professor John Jackson is our coordinator."

Marilyn had not heard of the Ontario Toxic Waste Research Coalition, but she remembered Mr. Jackson from the meeting in Silverdale church concerning the infamous questionnaire.

Marilyn dragged her attention back to the interview. Mrs. Migus had revealed to the audience in Toronto the extent of the stress she and others had come under since the OWMC had chosen the West Lincoln site. "Oodles of frustrations," she had told them. "Snapping at one another over the telephone, family relationships strained to the breaking point." She had told of the unbelievable workload, which translated into unbelievable expenditures of time, energy, and money. "Sometimes members have intense differences of opinion in approach to the solution. We have to make sure we don't diminish our credibility by aligning—or appearing to align—ourselves with partisan politics."

Marilyn pictured Doris Migus—her grey hair, her snapping blue eyes, and her wiry figure. It didn't seem right that a lady in her sixties should have to suffer such stress.

Marilyn had saved the editorial in the *Pelham Herald* until last.

Since her chance meeting with editor Norman Nelson on the street the previous week, she felt uneasy over where he stood on the issue. One thing she knew: she did not like the title of this week's editorial: "If We Get The Dump, Give Us The Goodies."

To Marilyn, it smacked of compromise. "Make it worth our while," she paraphrased aloud, "and we can be bought." Heaving a discouraged sigh, she read the editorial with care.

Nelson argued that if the people of Niagara were being asked to carry the burden of Ontario's toxic waste problem, then they should be rewarded with any good that came out of it. Whether or not the proposed facility should be allowed in the first place was quite another subject, but if it was approved, and in his mind, that remained a big if, Niagara should get the head office, with all the jobs it entailed:

> We know that (a head office in) Niagara is not as central as Toronto, but heck, if it's too far away, maybe sending forty-five trucks filled to the brim with toxic waste each day shouldn't even be considered.

Marilyn stared at the last paragraph, reading it over slowly, word by word. It held the same tongue-in-cheek sort of logic her cousin Peter used. She never knew for certain what Peter meant, and she could not say for certain what Norman Nelson meant. But unless I'm mistaken, she thought, this request for head office is, in reality, a backhanded slap at the proposed facility.

Unfortunately, most of the community took the editorial as a bid for the dump and bombarded the *Herald* with scathing letters denouncing the editor.

* * *

The People Power convention scheduled for Saturday, November 29 in the St. Catharines arena gained momentum as flyers and signs flooded the Region of Niagara. Notices appeared in the newspapers. Clifford made stakes for signs and posters along the roadsides. Marilyn, despite her dread of dealing with the public, reluctantly agreed to sell T-shirts at one of the booths.

Politicians took advantage of the publicity generated by the upcoming rally. Welland's other newspaper, the *Guardian Express*, declared in bold headlines: "No More Mr. Nice Guy as Dick Vows to Get Tough With OWMC Chairman." Marilyn felt a thrill of hope. Twelve municipalities made up the Regional Municipality of Niagara. As chairman of Regional Council, Wilbert Dick vowed to "go over Chant's head and get at the political people who are hiding behind Chant." At last! Marilyn thought: someone with enough clout to take on the OWMC.

About the same time, John Jackson made a presentation to the Lincoln municipal council. Lincoln County, located along the southern shore of Lake Ontario, was in the heart of the tender fruit lands and directly downwind from West Lincoln.

As Marilyn read the account in the *Standard*, she felt her spirits lift even higher. The OWMC no longer had the media coverage all to itself. She circled Mr. Jackson's main points in red, then highlighted them with fluorescent green.

"Don't be drawn into a debate with the OWMC over compensation," Mr. Jackson had cautioned Lincoln's local politicians. "If you protest only the narrow issue of the truck route up Victoria Avenue, you may get a better class road. But the toxic waste facility will still be there."

He told them that the Coalition intended to challenge the OWMC on two major fronts. One, the impacts of potential stack emissions and spills were far greater than anyone was being told, and two, alternative technologies had been dismissed too lightly.

Alderman Jill Hildreth expressed fear that opponents were doing an expensive 'dance of futility.' "Meanwhile," she added, "fruit growers worry that even if no accident occurs, there will still be the OWMC stigma and people will not want to buy Niagara fruit."

Mr. Jackson had ended his presentation to Lincoln Council on a positive note for opponents:

> Provincial politicians who showed unanimous support for the OWMC's West Lincoln site are now taking an interest in opposing arguments. This could result in a political vic-

tory...despite the corporation's unlimited technical and legal budget.

A week before the rally in the Garden City Arena, Marilyn attended a workshop on God's creation of the universe, held at Eden Christian College in Niagara-on-the-Lake. Afterwards, she drove straight to the hospital. Her mother had ceased to eat or drink, and had become extremely weak. Marilyn sat by the bedside, holding her mother's hand, with tears of grief rolling down her cheeks.

Her mother passed away in the early hours of the morning.

– **10** –

The two days following her mother's death passed in a blur. The night before the funeral, Marilyn sat in her old orange recliner in her housecoat, with her feet up on the footrest. As she stared into space, she recalled the words of one of the visitors to the funeral home earlier in the evening: "Dr. Chant's giving a speech at Brock University tonight. He never lets up, does he?"

Still staring into space, Marilyn felt tears of frustration mingle with her tears of grief. The OWMC, with its huge resources and its far-reaching influence, could go anywhere, do anything. Even while I stood beside my mother's coffin, she thought, Dr. Chant was out there, drumming up support for his project. She buried her face in her hands as she wept. Why can't this OWMC mess stop long enough for me to bury my mother?

The following night, with the funeral service over, Marilyn felt utterly drained. Besides all the demands that went along with a funeral, she, her sister Sharon, and her two brothers had felt the need to empty her mother's house over the past two days of as many valuables as possible. Now, she could hardly move anywhere in the downstairs of their farmhouse without climbing over boxes and bags heaped everywhere.

Marilyn picked up the funny green frog and the black elephant, and the criss-crossed wooden hanger with the dangling pegs. So much of this stuff would be worthless to anyone else, she thought. But to me, they mean home, and childhood, and precious memories.

After Clifford went upstairs to bed, she curled up with a cup of coffee and the *West Lincoln Review*. The two-line heading read,

"Regional Chairman Wilbert Dick Slams Chant Over Funding Issues." For the past year, the Region of Niagara had tried in vain to gain intervener status. Without it, the Region could receive no money from the province for technical studies or to put up a legal defence.

Chairman Dick had complained to the *Review* that Dr. Chant had taken a month and a half to respond to the Region's letters, and then said he was concerned over double-funding the Region and West Lincoln.

> The letters say, 'We will continue to assess your application,' blah, blah, blah, blah. He's using the old stall tactic on us...I don't think the Region is going to put up with Dr. Chant's pussyfooting any longer. We'll go directly to the premier. We will get the money out of him now, or we'll get it later.

The *Standard* reported that Regional council had determined not to be outmanoeuvred by the OWMC. By a unanimous vote, the councillors agreed to commit $355,000 to the first phase of the impending hearing for legal fees and technical studies. The steering committee estimated that costs to the Region would eventually amount to over one million dollars.

As for Dr. Chant's speech at Brock University the night before, Paul Forsyth of the *Tribune* reported that Dr. Chant had told the audience that research and development of new technologies was not at the top of his list. OWMC had looked at over a hundred technologies around the world. "The critical need," he said," is not research and development—the critical need is the will...to create these treatment facilities to come to grips with the problem."

* * *

The day before the protest rally, a representative for Greenpeace made a statement to the *Tribune* regarding both free trade and toxic waste: "We already have free trade. Unhappily, it's in pollution. 100,000 tonnes a year of wastes are shipped both ways across the border. It's been going on for at least 5 years, with Canada appearing to be the net importer of American wastes."

Clifford frowned when he read the article. "So, Canada ships in more waste than it ships out?"

Marilyn hunched her shoulders. "Sounds like it."

"And Chant wants me to believe the OWMC won't be treating U.S. wastes."

The following night, as Clifford, Marilyn, and their two daughters filed into the St. Catharines' Garden Arena for the second People Power Rally, friends and neighbours touched Marilyn's arm and murmured words of condolence. Pale and subdued, she sat in the bleachers with her family, grateful that the Task Force had found someone else to sell T-shirts.

One speaker after another—politicians, environmentalists, and labour leaders—climbed onto the platform erected by rally workers. Marilyn, inspired or bored by turns, found that, as always, rhetoric without substance held little appeal for her. By the time John Jackson stepped to the microphone, she felt she had listened to all the speech-making she could tolerate.

With his hands in his pockets and his shirt collar unbuttoned, he gazed out over the crowd. "The OWMC says, 'Contaminants will build up only during the five-month growing season. Then everything starts anew the following year.' The Coalition believes that soil contamination, and therefore the food chain contamination, will increase over the thirty-year operating life of the plant."

I agree, Marilyn thought, as the crowd roared its support.

"Another risk is leakage from the landfill. Every landfill ultimately leaks." Point by point, he drove home the message: the OWMC would cause as many or more problems than it would solve.

"An obvious part of the solution is reduction—not producing the waste in the first place—and recycling." He looked and sounded incredulous. "Why, after six years and thirty-seven million dollars, has the OWMC done virtually nothing to reduce waste? If the Crown corporation had acted immediately, the environment would have already improved."

After each statement, applause and shouts of approval rose from the bleachers.

"In addition, the Coalition believes as much waste as possible should be destroyed on-site. We recommend using mobile waste destruction and also three or four regional facilities.

"Finally, waste residues should be stored above ground, where they can be monitored."

Mr. Jackson had been one of the last to speak. Marilyn found that, as they drove home after the rally, his words lingered in her mind. Thinking aloud, she remarked, "Information falls from John Jackson's lips like leaves from a tree in autumn."

Over the next several days, the rally received mixed reviews from the media. Some of the workers had made the mistake of predicting attendance figures as high as 3,500 based on ticket sales. When only half that many showed up, opponents and media alike focused on the empty seats, not the full ones.

Marilyn, inflamed by the criticism, dashed off a letter to the editor. "The rally may not have been as well attended as we would have liked," she wrote, "but it drew far more people than the smattering that show up for OWMC functions."

That same day, Stan Pettit, Wainfleet's mayor and chairman of the Region's OWMC steering committee, announced that the Region of Niagara and the Township of West Lincoln had agreed to coordinate their campaigns to keep the OWMC out of Niagara.

While politicians filled the newspapers and the airwaves with the battle against the Ontario Waste Management Corporation, Marilyn and her sister filled box after box with their parents' household belongings. For Marilyn, often alone at the task while her sister was at work, the house held horrible memories. She could not erase from her mind the picture of her mother lying on the bathroom floor, barely conscious.

About this time, a second balloon released at the rally in the summer by Rochelle Haynes, 12, of St. Anns, was found tangled in a wild rose bush behind the Robin Hood Flour Mill in Port Colborne.

A week before Christmas, Norman Nelson, editor of the *Pelham Herald*, wrote an editorial on compromise. He argued that when one asked Dr. Chant to compromise, he could quite honestly and correctly state that he was only carrying out his mandate. On the other hand, if one asked the government to intervene, it could quite honestly and correctly state that the matter was in the hands of the OWMC. "So one fine morning, a year ago, Niagara residents woke up to find that they had won the province's toxic waste dump lottery—no giving back the prize."

Marilyn frowned in puzzlement. Nelson often used the word "mandate" in his editorials. She knew that "mandate" meant the authorization given to the OWMC to carry out the government's orders. She just wished she knew what those orders were.

* * *

The constant activity surrounding the OWMC proposal was beginning to wear Marilyn down. Even over the Christmas holidays, neither side took a noticeable break. Politicians expected a provincial election in September, and made statements to the press at every opportunity; yet, beneath the protestations of support for residents, Marilyn sensed a depressing attitude of passive assent toward the OWMC's proposal.

In the *West Lincoln Review*, Phil Andrewes, the local MPP for the official opposition, had harsh words for Donald Chant. "He has eyes only for the rotary kiln. He's frightening off the innovators who would develop mobile technologies for waste destruction."

He was no kinder to opponents to the proposal. He told them bluntly that the only way to stop the OWMC from locating in Niagara would be with facts and logic argued at the hearings. Saying that the toxic waste issue was a political issue didn't hold water—the provincial government could not be pressured into altering the process. "I tell you," he said, "and I tell Frank Giovannone of the West Lincoln Task Force, he's whistling in the dark."

In the same article, Regional councillor Gladys Huffman predicted that the OWMC would do a smooth sales job on West Lincoln Council, convincing them that the proposed toxic waste facility would attract other industry to their area. "The politicians have been making their case behind closed doors," she revealed. "I believe the OWMC will be made to do the best job possible…through the Region's opposition and study."

Marilyn slowly folded the newspaper and laid it aside. It's hopeless, she thought. The Region *expects* the OWMC to win. What with the assertions of backroom conniving and statements betraying the lack of political will to oust the Crown corporation, she decided that alderman Jill Hildreth had been right. Residents were indeed doing an expensive "dance of futility."

* * *

On Thursday, January 15, 1987, the OWMC held a press conference in Smithville at the Old Farm Inn. Marilyn waited just inside the banquet hall for Clifford while he hung up their coats. She glanced around the room. A table set up across the front of the hall held a bank of microphones. At one end of the table, a four-legged stand supported a large drawing of the proposed facility.

She and Clifford had barely settled into their seats before Dr. Chant and Michael Scott entered the room and sat at the table. Michael Scott leafed through a sheaf of papers and adjusted the microphone in front of him.

"It all looks so formidable," Marilyn whispered to Clifford. So unstoppable, she admitted silently to herself.

Michael Scott spoke into the microphone. "May I have your attention?" The room fell silent. "We have asked you here tonight to announce the release of the first part of the OWMC's draft environmental assessment report. Comprised of twenty-two volumes, the documents total 7,000 pages."

Above the stir and buzz of voices, Marilyn heard muffled exclamations of "Seven thousand pages!"

Then Mr. Scott introduced Dr. Chant. Marilyn watched the reporters at the table designated for the press. Some scribbled madly. Others looked bored as they sat back, jotting down the occasional word or phrase at arm's length.

Dr. Chant read from a prepared statement. He spoke of government agencies, reviews, and final submissions. Marilyn furrowed her brow in concentration. Having no experience with the procedures surrounding a project such as this, she understood the individual words. But, she thought in bewilderment, I don't know what he's talking about.

Dr. Chant slowly surveyed the occupants of the banquet hall. "I firmly believe that the E.A. process in this province has teeth," he told them in a voice that invited no argument.

Overwhelmed by the complexities surrounding the proposal, Marilyn slid down in her chair. She knew that E.A. meant Environmental Assessment. But what was the E.A. "process"? What was a two-phase hearing? And why did this set of reports warrant such fanfare?

Dr. Chant switched to a new aspect of the issue. Instantly alert, Marilyn straightened in her chair. "Many West Lincoln residents have urged the OWMC to finish its work as soon as possible to remove the cloud of uncertainty hanging over their community."

Marilyn's gaze slid obliquely to the faces nearest her, a sick feeling seeping into the pit of her stomach. How many is *many*? she wondered. And who are they?

People like the owner of the Silverdale gun club, she decided. Several months ago, he had written a long, strongly-worded letter to the *West Lincoln Review*. The *Review* had given it the title, "'Quit crying' and build the OWMC facility."

Dr. Chant's voice gradually wormed its way into her thoughts. "Industry recognizes the need," he told the audience. "They have told us the sooner the facility is available, the better." He looked out over the crowd with an attitude of disarming sincerity. "I honestly believe that this is the best site possible."

He lowered his eyes and continued to read from the statement. "The OWMC will release the rest of the draft E.A. later in the year."

Every word, every sentence, Marilyn thought, gives out the message that, except for required formalities, the facility is no longer a proposal but a certainty.

Then Dr. Chant raised the issue of funding. "The OWMC has provided more than $250,000 over the past two years to groups and individuals for their costs of reviewing the OWMC reports on site selection and engineering."

"Big deal!" muttered the man beside Marilyn. Startled, she

turned her head. He scowled at her over his glasses as he said vehemently, "The OWMC has spent $37 million, you know!"

Marilyn mustered a polite half smile and eased a little closer to Clifford. Staring straight ahead, she noticed an increasing restlessness come over the crowd. Chairs creaked, voices whispered, candy wrappers rustled. Dr. Chant continued to read his statement "Now that we've issued our first environmental assessment, I believe it would be improper for us as proponents at the upcoming hearings to be involved in the funding of those who decide to oppose—or support—our submission."

He outlined upcoming procedural matters, which again had no meaning for Marilyn. "He might as well be speaking a foreign language," she whispered to Clifford.

Clifford grunted, slid down in his chair, and folded his arms. In another minute, Marilyn thought, he'll be asleep.

Dr. Chant's gaze traveled slowly over the audience as he delivered his closing statement. "We hope that the public hearing process can begin before the end of this year."

The end of this year! The words sent a chill through Marilyn. In the seventeen months since the OWMC had selected the Bismark-Schram Road site, she had gradually fallen into a state of limbo—the feeling that things would go on indefinitely as they had. Now, she had been told that they would not.

— 11 —

The OWMC press conference gave municipal and regional politicians an ideal opportunity to blast away once again at the issue of funding. "...Without government funding," they complained to the newspapers, "we cannot hire consultants to read the 22-volume draft Environmental Assessment produced by OWMC."

As for the two-stage hearing, the opposing politicians gave it a unanimous thumbs-down. Stan Pettit, chairman of the Region's OWMC Steering Committee, declared angrily,

> I just don't know how...we would have the time to do the studies...and do them well. It's like the OWMC has kicked the barrel over [Niagara] Falls and we are already down in the rapids, just going along for the ride.

Niagara Regional Chairman Wilbert Dick snapped, "(Dr. Chant) is the proponent and he doesn't really care if other people have enough time."

Reporter Doug Draper revealed in the *Standard* that just a month earlier, Rod McLeod, deputy minister of the Ontario Ministry of the Environment, had written to OWMC Chairman Dr. Donald Chant, telling him that the Ministry would prefer the corporation to make a "single submission" of its proposal. The corporation had split the hearings anyway.

Dr. Chant explained that the OWMC had opted for a two-stage environmental assessment in the hope the public hearings could begin before the end of the year. The twenty-two technical reports covered site selection, alternative waste management options, and the province's need for a central disposal facility. If the corporation waited until it completed its reports on the impacts of the facility, the public hearings might not get started until late the following year.

Opponents charged that the split submission would place a serious burden on them. They would be embroiled in public hearings over the stage-one reports and reviewing stage-two reports at the same time.

Regional Chairman Wilbert Dick told Paul O'Brien of the *Guardian Express* that the OWMC's idea of splitting the hearings was "strictly a tactical move on Chant's part..."

Meanwhile, despite the murderous demands upon her time, Marilyn felt compelled to write a letter to the editor on the topic of acid rain. She had learned that acid rain did not assault the environment at an even rate, but came in spurts and peaks. It was during these periods of greater concentration that the most damage was done. "It is reasonable," she wrote, "to assume that 'toxic' rain will follow a similar pattern. Thus, the unique agricultural lands of Niagara will have to survive not just the onslaught of one or the other in heavy doses, but both—simultaneously and in combination."

Two weeks later, the editor of the *West Lincoln Review* reported on the air monitoring, which continued at the proposed site:

> A high-speed vacuum collects dust into a filter. It pulls in 40 cubic feet of air per minute. It is used because 'the air in this area is so clean,' it would take about three months to get a good sample if natural velocity was used.

"And clean," Marilyn muttered through her teeth, "is exactly the way we intend to keep it."

*   *   *

Monday morning, with the laundry basket filled to overflowing, Marilyn descended the narrow cement steps to the basement from the freezing back entrance of the hundred-year-old farmhouse. She ducked through the five-foot doorway into the old stone cellar. Cold whistled under the ill-fitting door as she dumped the first of five loads of laundry into the washing machine. Racing upstairs to the light and warmth of the living room, she settled back in her old orange recliner to prepare her Sunday school lesson. With her feet up, her notebook on her lap, and her Bible on the arm of the chair, she drained her coffee cup and flipped to the Scripture lesson in II Chronicles 20:12. "O our God…" she read, "we have no might against this great company that cometh against us; neither know we what to do:…"

Suddenly, tears gushed down her cheeks in an unstoppable flood. Jehoshaphat's cry of helplessness had caught her unawares, his impassioned plea a perfect parallel to the constant cry of her own heart. Sobs shook her body and teardrops marked the page of her Bible as, over and over, she whispered the words, "O our God…we have no might against this great company that cometh against us; neither know we what to do…" She wept for herself. She wept for the community. She wept for Jehoshaphat, whose cry had been just as desperate, his situation just as hopeless.

She knew that the words had not been written especially for her situation. Nor had they miraculously appeared on the pages of her Bible for her benefit and hers alone. Yet, in that moment, she felt as though they had.

She dabbed at her streaming eyes with a Kleenex and struggled to read the remainder of the verse. "…But our eyes are upon Thee."

"O God," she murmured, "You were Jehoshaphat's only source of hope. So, too, You are mine." Through her tears, she touched the verse gently with her fingers, as though she could absorb strength from the page through her fingertips.

Finally, her eyes almost swollen shut, she reached for her pencil crayons. She ringed each of the words, "But our eyes are upon Thee," with orange, then filled the circles with a golden shade of yellow. This verse would be her anchor, her source of hope.

In the days to follow, the page became soiled and dog-eared as the passage became her bulwark, her place of refuge and strength. She extended her reading to the verses that came ahead, the verses which came after. Seeking the bare bones of the chapter, she marked and labelled, awed as the path Jehoshaphat had been directed to follow jumped off the pages at her. Until now, unsure of her own direction, she had blindly hacked her way through the complexities of her days. Now, God willing, she would tread a path similar to the one laid out for Jehoshaphat.

"Not that I can know Your ultimate decision, Lord," she prayed. "Nor can I presume to know what You, in Your great wisdom, will do day by day. But like Jehoshaphat, I can urge the community to seek Your will." She stopped suddenly, her heart thundering in her ears. Her faith was her own, private and precious. She dreaded the thought of exposing herself and Clifford to ridicule. She pressed her hands together and touched her lips with her fingertips as doubts tumbled one over the other. Would she jeopardize the credibility of their letters to the editor? Would she offend other citizens' groups who had battled so long and hard to oppose the OWMC? Or, even worse, alienate their hard-won supporters? She closed her eyes as caution battled against her need to put God first. Finally, feeling as though she had stepped into a yawning elevator shaft, she made her choice. Like Jehoshaphat, she would urge the community to seek God's will.

Slowly, she opened her eyes. With her heart still thundering, she ripped a page from her Sunday school notebook. Then, handling each phrase with care, she began drafting her fifteenth letter to the editor.

She started by quoting the cry that had first wrung her heart. "O our God...we have no might against this great company that cometh against us; neither know we what to do..." Then she compared Jehoshaphat's situation to the presence of the OWMC in their community. Wanting no misunderstandings, she made clear the motive behind the letter.

"It's not our intention to use Scripture for any self-seeking purpose," she assured the reader. "Nor do we wish to give the impression that this letter is a means to 'get God on our side.' However, for those who seek to put God first in all things, we offer the following skeletal outline as a devotional that has been of great encouragement to us."

Next, she quoted Jehoshaphat's declaration of God's might: "O Lord...art not Thou God in heaven? and rulest Thou not over all...and in Thine hand is there not power and might, so that none is able to withstand Thee?"

To that, she added God's response to Jehoshaphat through His prophet, Jahaziel. "Be not afraid nor dismayed by reason of this great multitude; for the battle is not yours, but God's."

Then, point by point, she listed the plan of action:

Go ye down against them...Set [position] yourselves...Stand ye still...Believe in the LORD...Praise the LORD; for his mercy endureth forever.

She concluded by saying:

In the spirit of going against, positioning ourselves, and standing, we offer the following information:

- Incinerator slag will go directly to the landfill. No details, no figures on toxicity.
- The probability of transfer stations for untreated waste across the province is being kept quiet. Reason? [We believe] OWMC doesn't want to battle more than one municipality at a time.
- Although the Niagara Peninsula is the heart of Ontario's fruit belt, fruit is not part of OWMC's Risk Assessment.

To Marilyn's amazement, the *West Lincoln Review* printed the letter in its entirety, under the heading, "Deliver us, Lord, from OWMC."

*   *   *

Rarely a day passed when Marilyn did not turn to her "anchoring" verse to keep up her hope. Yet always, she left out one phrase: "wilt Thou not judge them?" No matter what the corporation did, she could not ask God's terrible judgment to fall upon those who ran it. She felt guilty, however, about editing the verse.

Then came the morning when she read I Samuel 3:13: "I will judge...for the iniquity which he knoweth..." Here, at last, was the solution to her dilemma. She could ask God to judge, not for the purpose of bringing His wrath upon the heads of OWMC personnel, but to prevail against intentional wrongdoing, untruths, and unfairness.

*　　　*　　　*

Several nights later, Marilyn had just finished loading the dishwasher when the telephone rang. She recognized Lynda Bradley's southern drawl, although she had encountered her only twice. "We're holding a meeting of local residents," Lynda told her. "It's on Thursday night in the basement of St. Luke's Lutheran Church in Smithville. We've read the letters you and your husband send to the newspaper, and we'd like you to come out and join us."

"I—I don't—think—"

"Oh, you *must* come!" Lynda urged. "We're going to insist that council take a stronger stand against the OWMC," she explained heatedly. "And we're going to demand that the mayor and anyone else on council who's in favour of the OWMC should resign. We're not going to stand for these backroom deals we keep hearing about."

Marilyn found Lynda's impassioned manner of speech alarming. All of her life, she had shrunk from forceful personalities expressing strong convictions. Anxious to end the conversation, Marilyn told Lynda, "We—don't want to join any groups."

"Why ever not?" Lynda asked. Her voice conveyed genuine astonishment.

"Well—" Marilyn began. How could she explain—without sounding holier-than-thou—that she feared group pressure might require her to compromise her beliefs or her values? She scrambled in her mind for an acceptable answer.

In the past, she and Clifford had discussed the idea of joining one of the groups. For a number of reasons, they had decided against it. Now, she resurrected one of the most genuine and highly applicable. "Well, for one thing, if we belong to a group," Marilyn told Lynda truthfully, "we'd be accountable to the other members. We don't want anyone telling us what we can or can't write in our letters to the editor." Marilyn paused, then voiced a deep-seated fear. "If I—we—make a huge blunder in one of our letters, we don't want anyone else to lose their credibility on our account."

"Oh, I don't for a minute believe that would happen!" declared Lynda.

"I'm not willing to take the chance," Marilyn told her firmly.

"Thursday night will be our first meeting," Lynda explained. "We don't have a membership, so there's nothing to join." A note of coaxing came into her voice. "Couldn't you at least come out and give us your opinion?"

Marilyn sighed. "I'll ask Clifford. But I'm not promising anything." Marilyn hung up the receiver to the wall phone in the kitchen.

"What was that all about?" Clifford asked from the living room.

Marilyn repeated the conversation, including the invitation to attend the meeting of the local residents.

"I think we should go," Clifford said, without a moment's hesitation. Flabbergasted, Marilyn stood with her mouth open. Clifford turned defensive. "I know I said I don't like meetings," he argued, "but I think the others are right. It's time someone spoke up against the Township's cozy little closed-door chats with the OWMC."

* * *

At the meeting of the local residents on Thursday evening, Marilyn counted eleven people gathered in the small, low-ceilinged church basement. Eight six-foot tables covered with worn oilcloth had been placed in the form of a hollow rectangle. She glanced over the faces in the room. Except for Lynda, she didn't know anyone. While Clifford stopped to exchange pleasantries with several of the local farmers, Marilyn slid into a seat in the corner. To her surprise, the meeting opened with prayer.

Like Marilyn, most had come to the meeting empty-handed. Even the organizers carried nothing more than a writing pad and a pen to jot down notes. Only one person, a woman, soft-spoken and self-effacing, came with her arms loaded. It did not take long for her to take control of the meeting.

Alasse Plains [2]—Alasse, with the accent on the last syllable—promptly changed the focus of the meeting from demanding the

---

[2] Not her real name.

mayor's resignation to demanding the resignation of the chairman of the Township's Toxic Waste Committee. The chairman, outspoken in his support for the waste project, made an easy target.

The group decided to make a public presentation to council on Monday night. They recruited Marilyn, because she owned a typewriter, to type up the draft and make photocopies. Clifford, to Marilyn's amazement, agreed to give the presentation.

The self-effacing Alasse receded into the background.

The next morning, working at the typewriter, Marilyn had second thoughts about the presentation. "Are you sure we want to do this?" she called downstairs to Clifford at lunchtime.

"Why?" Clifford questioned, mounting the stairs.

"Because this is a cruel, vicious document. I don't feel comfortable with it."

He leaned over her shoulder and stared at the page in the typewriter, then shrugged. "We said we'd do it. Today's Friday, and the presentation's on council's agenda for Monday night. That means you've only got tomorrow morning if you want to get it photocopied at the library." He turned and headed for the stairs. "Even if we knew who to call, it's too late to back out."

Reluctantly, Marilyn went back to picking out letters on the typewriter keys. This is exactly what I feared would happen, she thought, trying to throw off the heaviness that had settled over her.

– 12 –

Throughout the weekend, misgivings and guilt over the presentation Clifford was to make at council on Monday evening pressed down upon Marilyn. No matter how she tried, she could not justify what the residents were about to do. The document she had typed was cruel and mean-spirited, and she wished fervently that she and Clifford had not been swept along into taking part in it.

Monday night, as the group filed into the council chambers, Marilyn slid into a seat against the wall and sat with her head down, her fingers plucking at the pleats of her skirt. At the residents' meeting on Thursday night, John Dykstra had volunteered to arrange for them to go to council. As soon as he arrived, she handed him the photocopies of the presentation, which he immediately distributed to the mayor and the six members of council. The councilman who acted as Chairman of the Township's Toxic Waste Committee took one look at the statements attacking him and stormed from the council chambers into a back room, his face suffused with outrage. Marilyn felt sick with regret, and she longed to undo what had just been done.

At that moment, Alasse Plains opened the heavy door to the council chambers and slipped inside. Within seconds, the councillor under attack strode back into the room, his back rigid and his face set like stone. He hitched his chair up to the table, and the mayor convened the meeting.

Lynda Bradley signalled to Alasse and indicated the chair beside her. "I saved a place for you," Lynda told her in a loud whisper.

Alasse shook her head and perched on a chair in the corner, away from the group.

When it came time for Clifford to make the presentation, Marilyn stared at her hands clenched damply in her lap, her shoulders hunched together as she tried to shrink into an invisible ball. The chairman of the Toxic Waste Committee barely controlled his anger as he attempted to respond to the accusations. Marilyn's gaze shifted sideways to the other residents. They looked as ashamed and embarrassed as she felt.

Except for Alasse.

From the corner of her eye, Marilyn caught a glimpse of a secretive exchange between Alasse and the mayor. In that instant, Marilyn realized that she and the residents had been tricked: the scathing accusations, originally aimed at the mayor, had been cleverly diverted to the chairman of the Toxic Waste Committee.

Marilyn felt her expression harden with suppressed fury. You *used* us! she thought, her eyes narrowing dangerously. But it won't happen again!

The residents, their business finished, rose and quietly filed from the council chambers. Once outside, a blanket of uneasy silence descended upon them and, with eyes averted, they quickly dispersed and headed for their cars.

* * *

Ten days had passed since the council meeting, and remorse lay heavily upon Marilyn's spirit. As Clifford drove toward Smithville for the second meeting of the residents in the basement of St. Luke's church, Marilyn stared through the windshield into the night. The headlights silently picked out ragged patches of dirty snow along the roadside.

Ever since the fiasco at council, she had struggled with unrelenting shame made worse by bitter disillusionment. Not once in the past year and a half had it crossed her mind that anyone would claim to oppose the OWMC and then exploit the trust of the beleaguered residents to advance a personal agenda.

"If Alasse Plains shows up tonight," Marilyn told Clifford suddenly, "I'm leaving."

At five-to-eight, Clifford pulled into the parking lot. Marilyn, carrying a plate of homemade shortbread, reluctantly entered the rear door of the church. Partway down the stairs, she hesitated. By ducking her head, she could see the tables arranged in a hollow rectangle and the residents standing in groups of two or three. Although the scene looked much as it had the first time, the sounds of enthusiastic conversation and buoyant laughter were missing.

Amid subdued greetings, Clifford and Marilyn pulled out chairs and sat down. Paul Balint opened with prayer. Then, as Lynda read the minutes of the first ill-directed meeting, Marilyn stared at her hands, clenching and twisting in her lap. In the unnatural quiet that followed, she glanced around the room.

John Dykstra fidgeted in his seat, then laid one arm across the back of his wife's chair. "Before we get started," he said, studying the thumb of his free hand as he rubbed at a worn spot on the oilcloth covering the table, "I have to tell you—Mary and I are ashamed of what went on at council a week ago Monday night. As Christians, we don't want to be involved in anything like that again." He raised his head, his colour high, and surveyed them one by one. "I can't claim Christ as my Saviour, then turn around and bring shame on His name."

The frankness of his statement startled Marilyn. She could never have brought herself to speak so boldly, even to people she knew, much less strangers.

"I think we were set up!" Paul declared. "Did you see the look that—that passed between that woman and the mayor?" The room became noisy with angry assent.

"Our mistake," Marilyn said quietly, "was to attack a person rather than the issue."

"Be that as it may," John said, "but just so you all know, I intend to call the chairman of the Township's Toxic Waste Committee and apologize for our part in it."

"Please," said Marilyn, glancing at Clifford, "apologize for us, too."

"For all of us," said Paul. Murmurs of agreement circled the room.

"We didn't feel right about it when I was typing it up," Marilyn confessed, "but we didn't know what to do. Everything had been arranged and—and there was no time." Her gaze fell away. "I—I know that's no excuse."

"We're all to blame," John consoled her kindly. "But from now on," he said, his earnest gaze travelling slowly from face to face, "I think that if one of us doesn't feel right about something, we should say so. Agreed?"

"Agreed!"

Steve Dinga, a heavy-set man in his late fifties with sleepy-looking eyes, a shy smile, and a quiet voice, piped up: "Now, maybe, we can get on with the business we came for—fighting the OWMC."

With dispatch, the group decided to attend West Lincoln's council meetings and, whenever possible, to monitor the public appearances of both Dr. Chant and other OWMC personnel.

Lynda's husband, Bob, spoke up. "I think we should join the Ontario Toxic Waste Research Coalition." As the others discussed the advantages and disadvantages of uniting under one umbrella with farm organizations and other citizens' groups in the area, Marilyn felt herself draw back from the idea. She wanted no more involvement, especially with an environmental group, and definitely no additional meetings to attend.

Above the babble of voices, John suggested, "Before we decide, why don't we go to a couple of Coalition meetings?"

Bob whipped out his wallet and checked the date. "They meet the first Thursday of every month at Dr. Chant's old stomping ground, the Vineland Research Station. Next meeting's on March fifth."

"Okay," said John. "Let's meet in the parking lot outside the building and we'll all go in together."

Paul closed the meeting with prayer and gave thanks for the refreshments. For the first meeting, his wife, Marija, had brought home baking and served coffee. Tonight, each of the ladies had arrived bearing homemade cookies, squares, or tarts.

John popped the last bite of one of Marija's pastries into his mouth and licked his fingers. "You know," he said, swilling it down with coffee, "a couple of years ago, Mary and I had a bunch of cards printed for people to sign, protesting the OWMC's dump. The Task Force helped to collect the signatures, then returned the cards to me." A gleeful grin lit his face. "We've got twenty-five thousand signatures at home, just sitting in a box."

"Twenty-five *thousand*?" exclaimed Lynda. Her astonishment made her southern drawl more pronounced than ever.

"That's right," he declared, with a mischievous gleam in his eyes. "I think we should take them to Queen's Park and present them to Premier Peterson."

Marilyn's mouth fell open. The audacity of the man! she thought. Instinctively, she retreated from the prospect of confrontation. But from the excited voices around her, she realized that the others thought it a magnificent idea.

"If we're going to do this," John proclaimed, "we don't have any time to waste. It won't be long until I'm sowing clover seed."

An instant babble of voices erupted. Like Clifford, most farmers in the area planted and harvested a variety of field crops as well as engaging in some type of livestock operation. John and Mary Dykstra, she learned, milked a dairy herd morning and evening with the help of their four children. Paul and Marija Balint raised beef animals. Steve Dinga and his wife, with the help of their married son, had a large beef operation and raised pullets for commercial egg-laying establishments. Clifford, of course, produced fat hogs.

Lynda's husband, Bob, the only non-farmer in the group, worked shift work at the paper mill in Thorold.

The weariness of Marilyn's busy day caught up with her. No one, she thought, has the time or energy for this battle, much less a trip to Queen's Park. But by the time the group had demolished the goodies and finished the coffee, they had made a number of decisions. Purely as a courtesy, they would first present the 25,000 signatures to West Lincoln Council. At the same time, they would announce their intention to take the signatures to Premier David Peterson and Environment Minister James Bradley. Then they would invite council to go with them to Toronto to Queen's Park. "That way," said John, "we'll have official support for our trip."

Marilyn, characterized by careful and reasoned thinking, felt overwhelmed by the speed at which things were moving.

"I'll call the town clerk and make arrangements," John said.

"Who's going to make the presentation?" Clifford asked cautiously.

"Well—" John gave an embarrassed laugh. "They're my cards. I paid to have them printed. So, I guess I get to make them public."

"Good!" said Clifford. "I don't want the job again."

"If that's settled," said John, "I'll see you all on Monday night at council!"

* * *

The next day, the minute the St. Catharines *Standard* hit the newsstands, excitement raced like a raging forest fire through the community. PCBs from the abandoned D and D site had been found in wells monitoring Smithville's drinking water. Consequently, the main reservoir had been shut down.

In August of 1985, according to reporter Doug Draper's account, Pat Potter, formerly Pat Hayes, and her husband Chuck, along with several other members of a local environmental group called the Niagara Ecosystem Task Force, NET Force for short, had managed to dip a sample of gooey black ooze from the derelict fenced-in chemical waste storage facility owned by the D and D Group Ltd. They had sounded the alarm just one month before the OWMC had announced LF-9C as the preferred site. Marilyn had paid no attention to the discovery. She didn't live in Smithville. She knew nothing about PCBs or leaking transformers, or where to find the abandoned D and D site.

At the time, the Ministry of the Environment had hired Gartner Lee, a consulting firm, to assess the problem. After months of geological testing, Gartner Lee had advised the Ministry that it would take more than twenty years for the polychlorinated biphenyls, otherwise known as PCBs, to penetrate the clay underlying the site and enter the water table. Monitoring wells had been drilled as a safety precaution, and people relaxed.

Now, barely a year later, the monitoring wells showed the presence of PCBs.

OWMC—The "Wasted" Years: The Early Days

Bill Goodings, an engineer with Proctor and Redfern, another consulting firm involved in the issue, told the *Standard*,

> We didn't expect to find any PCBs in the monitoring wells. We were confident that the clay had the integrity when we did our soil tests. Now, the question we're asking is, how the blazes did the PCBs get down there?

This time, Marilyn paid avid attention to the discovery. For Gartner Lee, and Proctor and Redfern, were the same consulting firms that had affirmed the integrity of the clay for the proposed OWMC toxic waste landfill.

"It's the first good news we've had in ages!" Marilyn told Clifford with glee as she grabbed her purse and opened the back door. "No matter what the consultants say, West Lincoln clay is West Lincoln clay! Here's proof that it *doesn't* keep toxic waste out of the water table! Now they'll *have* to cancel the OWMC's dump!"

With errands to run, Marilyn ran down the back stairs. Her heart felt lighter than it had since the OWMC first announced its proposal in September of 1985. She hopped into the car and headed for Wellandport. Everywhere she went, the reaction was the same.

"If the same consultants were wrong in Smithville," Gary Dunning told Marilyn heatedly as they stood in front of the post office, "they're wrong about the OWMC's landfill."

His wife, Mae, agreed emphatically. "I think they should just call the whole thing off."

Outside the hardware store, Bev Johnson declared angrily, "Just this past November, Chant assured us the leachate wouldn't reach the ground water for at least 800 years. I didn't believe it then, and I *sure* don't believe it now!"

In one fell swoop, Marilyn realized, the community had swept aside OWMC's assurances of a safe landfill. Suddenly, the contamination of the local groundwater had become a personal problem, not just some hazy issue that a future generation might have to deal with 800 years from now.

Over the weekend, John Jackson challenged the credibility of Gartner Lee, and Proctor and Redfern, during a public debate with

OWMC's Michael Scott in Niagara-on-the-Lake. How could these consultants be right about the OWMC site, he questioned, when the calculations they made a year ago about the D and D site now appeared to be wrong?

Michael Scott argued, "At the Smithville site, there were raw, untreated PCBs in questionable storage containers..."

Murray Creed, acting as spokesman for the Crown corporation, insisted that there was no comparison between the current leaking-PCB scenario and the treatment facility proposal. He told Paul O'Brien, reporter for the *Guardian Express* in Welland, "We knew people might draw parallels between the two circumstances. But it isn't the same."

In people's minds, however, the fact remained: Proctor and Redfern, and Gartner Lee Ltd. had predicted low permeability of the clay, and they had been wrong.

* * *

On Monday night, with the OWMC's landfill consultants discredited in the eyes of the community, the jubilant citizens showed up at West Lincoln council en masse. Any day now, they assured one another, the government would kill the OWMC's proposal. But in the meantime, they would keep up their opposition.

The council members, noticeably reserved toward the residents, looked increasingly uneasy as the chairs in the council chamber filled to near capacity. When the time came, John heaved the blue recycling box containing the 25,000 signatures onto a small table. He read his two-page presentation, and at the end, he said, "Mr. Mayor and members of council, will you join us on our pilgrimage to Queen's Park?"

The Mayor, the Township Administrator, and the six councillors whispered back and forth behind their hands, then adjourned to a back room. Ten minutes later they returned, and the mayor informed the residents, "We'll give you our decision two weeks from tonight at the next council meeting."

Deflated, the residents filed into the parking lot. Lynda, her dark hair perfectly coiffed and her voice snappy despite the softening effect

OWMC—The "Wasted" Years: The Early Days

of her southern drawl, declared, "I don't think council's ready to part company with OWMC and all their million-dollar promises!"

*   *   *

Two days later, Marilyn searched Wednesday's *West Lincoln Review* for some mention of the 25,000 signatures presented to council on Monday night. Until the last few months, she had never been one to read the newspapers. Clifford, on the other hand, read the farm papers from front to back, and pored over any and all classified ads as though they held clues to a gold mine. But since the OWMC had come on the scene, Marilyn no longer waited on Wednesdays for Clifford to fetch the mail when he came for supper. The minute she saw the mailman's car, no matter how nasty the day, she shoved her arms into her coat and dashed to the box.

For on Wednesdays, both the *Pelham Herald* and the *West Lincoln Review* arrived.

Anxiously, she leafed through the papers. She found no mention of the signatures. The *Pelham Herald*, however, carried an article titled, "A Honeymoon with the Voters." On February 17, it said, David Peterson, premier of Ontario, had made a speech in St. Catharines at Brock University. Although no politician admitted as yet to being on the campaign trail, the media treated every political appearance as a bid for votes in the coming election.

Marilyn retired to her old orange chair. She felt the muscles in her neck and shoulders tighten as she shook out the newspaper and held it under the light. No politician could come to the Niagara Peninsula without referring at some point to the toxic waste issue.

Sure enough, the premier had told the audience, "I'm confident that the process, which put West Lincoln at the top of the list, has been fair." Marilyn's hands, still clutching the newspaper, dropped to her lap, crumpling the pages.

I should have known Dr. Chant would have the premier in his pocket, she thought. She closed her eyes and took several deep breaths as she fought down the sick, desperate feeling forming in the pit of her stomach. Slowly, she opened her eyes and let her gaze rest upon the pine boughs waving gently outside the living room window. But her mind focused on scenes from her childhood.

As a girl, she had watched the political manoeuvring that had gone on at home. She had listened to her father on the telephone late at night, garnering support for school board projects long before the meetings took place.

That's why I *never* wanted to be involved with politics, she thought. She swiped her hand angrily across the crumpled page of the newspaper to smooth it, then continued to read the detailed account.

"The investigation has been exhaustive by a scientist with impeccable credentials and no political axe to grind," Premier Peterson said of Dr. Chant. "The fact that many groups in Niagara oppose the dump is not reason enough to abandon the project, not even if a majority don't want it."

"No matter how many voters petition against it?" a student had asked.

The premier had retorted, "The minute we run a government because we get a lot of names and signatures or polls or people yapping is the day we've abdicated our responsibility…I sympathize with the people in the area," he continued. "But it has to go somewhere."

Marilyn stared at the sentence. *It has to go somewhere.* In helpless rage, she scrawled along the margin in red pen, "He makes it sound like a UFO that has run out of fuel and *has* to find *some*place to land."

A time for questions had been allocated at the end of the premier's speech. A member of the audience asked, "Why should a toxic waste dump be located in an agricultural community?"

The premier had parroted the often-repeated response, "Well, where would you have us put it?"

"That line, of course," editor Norman Nelson pointed out, "is stolen from Dr. Chant's arsenal…"

Marilyn felt crushed under the weight of the impossible odds. Our efforts mean nothing, she thought. We're just "people yapping." In a desire to run away, to hide from those who had all the power while she had none, she did as Jehoshaphat had done: she fell to her knees.

"Art Thou not God in heaven?" she inquired almost angrily, using Jehoshaphat's prayer. "Rulest not Thou over all the kingdoms... In Thine hand is there not power and might, so that none is able to withstand Thee?"

In the Scriptures, she thought, with a bitterness born of envy, the story had a happy ending. Jehoshaphat returned to Jerusalem, rejoicing over his enemies. "So the realm of Jehoshaphat was quiet," the Bible said, "for his God gave him rest round about."

"But what about us, Lord?" she whispered. "How will it all turn out for us?"

– 13 –

On Thursday night, the residents met for the third time in the basement of St. Luke's church. Heated speculation over council's delayed response to the upcoming trip to Queen's Park took precedence.

"The Township's talking compensation with the OWMC," declared Lynda in her lilting southern drawl, her dark brows drawn together angrily in her expressive face. "The OWMC has bottomless pockets and the Township wants all the kick-back it can get!"

Marilyn, like the others, gloomily shared Lynda's opinion. With an inward sigh of defeat, Marilyn sat with her chin propped up on her fist. The voices around her faded as her mind travelled its own path. Showing up in Toronto with a boxful of signatures won't change a thing, she thought. *"It has to go somewhere,"* Premier Peterson had stated decisively in his speech two nights earlier, which meant he had already made up his mind. Since then, Marilyn had lived with a constant overlay of anxiety. For, with that single statement, he had publicly thrown his support behind the OWMC's proposal.

Yet Marilyn forced herself to keep silent. It's too easy to spread discouragement, she reminded herself, and too hard to rebuild hope. With effort, she pulled her mind back to the present.

"You know," John said. "I think we should hire a bus to go to Queen's Park."

"A bus?" several voices exclaimed at once.

John held up his hands to quieten the clamour. "Look," he argued, "we'd all be together. No one would get lost in Toronto.

And we wouldn't have to worry about traffic or parking. A bus would take us right to the door."

"How much would a bus cost?" asked Paul.

John broke into his ready grin. "I've been thinking about that, too. If we invite enough people to go with us, it would cut the cost."

Paul looked doubtful. "Who would we ask?"

"Relatives, neighbours...politicians. We've already asked our own mayor and members of council. Why don't we invite the mayors and councils from other municipalities?"

Marilyn shook her head and pursed her lips in scepticism as she framed her doubts with care. "How can we do that? We don't even know if our own mayor and council will go with us."

"Can't hurt to ask," said John, with his irrepressible optimism. "So, I think the next step is to go to the Regional Council meeting next month and make our presentation there."

Overwhelmed, Marilyn watched silently as the residents made a series of rapid-fire decisions. "Mary and I will go to the Regional offices tomorrow to have the presentation put on the agenda," John said. He turned to Clifford. "Can you come with us?"

Clifford shrugged. "I guess..." He looked a little puzzled by the request.

"We need someone to go to the office of our local MPP, Phil Andrewes, to make arrangements to go to Queen's Park."

Marilyn said, "Can't someone just call?—"

John cut Marilyn off. "No! You get better results face to face." He turned to Paul. "You know Andrewes. Why don't you go?"

Paul took a little persuading before he agreed.

"See if your brother Dennis can go with you," John advised. "If not, give me a call. Maybe Steve could go. Or Clifford. I'll go if no one else—"

"John!" Marilyn slapped her hand impatiently on the table, her brows drawn into an expression of exasperation. "Why on earth do we need two or three people involved in every job? Don't we all have enough to do as it is?"

John turned to Marilyn. For once he looked deadly serious,

with no hint of his infectious grin. "I don't think any of us should go anywhere alone."

"Why, for Pete's sake? We're not children!"

"Because if there's ever a problem, we'll have at least one witness to prove what was said."

Mary spoke up, her normally soft voice intense. "And it's easier for two to remember everything that goes on."

Marilyn shrugged and shook her head doubtfully. "Okay, I guess. It just seems such a waste, when we're all spread so thin."

"One other thing," John said. "I think if we have a husband from one family and a wife from another, or vice-versa, a third person should go along. That way, there can be no hint of scandal."

Marilyn leaned her elbows on the table, instantly lost in thought. With her hands pressed together, she touched her fingers to her lips. "You're right, of course," she conceded. "We've all heard the rumors about—" She left the sentence unfinished.

In the silence that followed, Bob Bradley doodled on the notepad in front of him, then asked, "Who's going to inquire about a bus?"

Clifford shifted in his chair. "I can do that, I guess. I'll get a price from Farr's bus line, and maybe Greyhound."

John looked around the room, his grin back in place. "Have we forgotten anything?" He paused. "No? Then let's eat!"

* * *

The next day, Marilyn started work on her sixteenth letter to the editor. Head on, she attacked the talks between the Township and the OWMC on compensation. "You cannot win a war," she wrote, "by discussing in advance the terms of your possible defeat.'

* * *

For the third Monday night in four weeks, the residents filed into the West Lincoln council chambers. After two weeks of nerve-wracking waiting, the small group would find out tonight whether or not the Township would endorse their plan to present the 25,000 signatures to Premier David Peterson and Environment Minister James Bradley.

As the clerk stood to read from a prepared statement, Marilyn sat with her fists clenched tightly in her lap. "The committee has decided that council should accompany the group to Toronto in a way which establishes the credibility of the citizens—" Gasps of relief interrupted the statement as the residents broke into elated whispers and smiles.

"Ssshh!" Marija Balint cautioned in a hoarse whisper, her brows drawn together in a rare frown of fierceness. "You make them mad, maybe they change their mind."

With a glance at the mayor, the clerk continued. "We suggest that other politicians and groups should also be urged to go."

"We intend to!" John remarked triumphantly, his eyes aglow with victory.

John Schilstra, chairman of the Township's Toxic Waste Committee, turned in his chair and faced the residents. "However," he stated flatly, "it is my opinion that a trip to Queen's Park will do little to change the politicians' minds about the Ontario Waste Management Corporation's proposal."

Drat the man! Marilyn thought with annoyance. Didn't he know that discouraging words made great grist for the press? With suppressed fury, she watched the reporter from the *Review* scribbling down every inflammatory syllable.

Yet, despite his dark outlook and the angry rumblings of the residents, John Schilstra declared his intention to join the three other councillors who had agreed to accompany the residents on the trip: Ed Fulford, Joan Packham, and Lillian Zdichavsky.

He started to sit down, then straightened again. He looked squarely at the residents. "You should hold a press conference—same as OWMC does. Get all the publicity you can get."

Councillor Joan Packham spoke up. "I've been asked to tell you that Jim Green, the Township's Town Crier, would like to go on the bus trip to Queen's Park. He has offered to appear in his full regalia and to "cry" a fifty-word proclamation from the steps of parliament."

John nodded in acknowledgment. "Mr. Mayor and members of council, thank you for your kind attention."

With their business completed, the residents quietly left their chairs and crowded through the door. Standing outside in the cold night air, Marilyn said, "I didn't know that West Lincoln had a Town Crier. And what does 'full regalia' mean?"

"Oh, Mr. Green's outfit is very colourful," John informed her. "Actually, I think—" He got no further as peal upon peal of laughter erupted.

"Oh, no!" declared Paul. "When John starts thinking, watch out!"

John gave a good-natured laugh. "No, seriously," he insisted, "I think the press release and the Town Crier are good ideas. We need publicity, and the press loves anything that's different." His smile faded. "I just wish John Schilstra hadn't said what he did in front of the reporter from the *Review*." Then, as always, his unfailing optimism took over and his expression brightened. "Anyway, see you all Thursday night at the Coalition meeting."

* * *

Three nights later, Clifford and Marilyn drove across the ornamental stone bridge onto the manicured grounds of the Vineland Research Station. Earlier in the day, Marilyn had sent out letter number seventeen to the editors of seven newspapers in the Niagara Peninsula. "What goes up must come down," she had insisted, using quotes from several sources, including *Storm Warning*, an Environment Canada publication, to back her statement. Her mind still whirling with ideas and words, she fretted over the coming Coalition meeting. She had never wanted to get mixed up with any environmental groups, not with their reputation for aggressive and sometimes dangerous protest actions.

As they approached the parking lot in front of the second building, she saw members of the residents' group milling around, talking, gesturing, laughing. Always laughing, Marilyn remarked to herself. And behind the laughter, she realized with a surge of affection and admiration, lay incredibly brave spirits.

She and Clifford put on their own bright faces as they pulled into the parking lot. "Ironic, isn't it?" Marilyn proclaimed as she stepped from the car. "The Coalition meets where Donald Chant once reigned as the research lab director."

At twenty-five minutes past seven, John said, "Let's go." He pulled open the heavy door of the red brick building and led the way up the stairs to a large classroom. Varnished tables, three feet by eight, had been dragged into the shape of a huge rectangle. John Jackson sat at one end, with a wall of blackboards behind him.

Marilyn's gaze ran swiftly over the faces around the table. She had met Edith Hallas and Doris Migus, but no one else. Although she had encountered no stauncher opponents of the OWMC than the wiry, sixty-five-year-old Mrs. Migus or the lively Edith Hallas, both had struck her as forceful and determined, reinforcing her wariness toward environmental groups. She met their smiles and words of welcome with polite reserve. She had no intention of getting tangled up in anything she would regret later. Not again. Not after that shameful mess involving Alasse Plains.

With her back to the row of windows, she sat facing the door, silent and watchful as the meeting began. She had tossed and turned most of the night before. Flashes of news clips, showing environmentalists committing outrageous acts of defiance, had circled endlessly in her mind, driving sleep away until the first streaks of daylight crept across the night sky.

A burst of rippling laughter travelled around the table. Untouched by it, Marilyn watched the camaraderie among the Coalition members, observing all the while John Jackson's quiet, almost retiring manner as he steered the meeting from issue to issue. As in the past, information spilled from his lips in easy-to-understand language. Nevertheless, Marilyn cautioned herself to wait and see.

Determination, seasoned liberally with optimism, enveloped the group as conversation flowed back and forth among the members. Doris Migus turned in several donations she had collected on behalf of the Coalition. Edith Hallas reported on a presentation she had made to a group of schoolchildren. "I simply tell them of the alternative technologies already in use for dealing with toxic waste," she said, with enthusiasm. "Then the kids go home and tell their parents, 'There is a better way.'" Using the slogan coined by the Task Force, she emphasized the word *is*. "And just as important—" She let her gaze circle the room. "—we now have a teacher in that school who is no longer certain that OWMC's proposed facility is the only solution."

Yet, despite the outward evidence of high spirits among the members of the Coalition, Marilyn recognized the underlying strain in their eyes and in their manner, the bracing for an ordeal, which they knew to be both inevitable and overwhelming. For without exception, every decision, every proposed action, hinged upon the effect it might have on the upcoming Environmental Assessment Hearings.

More laughter broke out when John Dykstra formally introduced himself. The Coalition, it seemed, boasted a plethora of members named John—John Migus, John Bacher, John Burton, and John Jackson.

John Dykstra explained about the 25,000 signatures, the proposed trip to Queen's Park, and the group's intention to lobby Regional Council for support.

"Ruth and I'll be there," declared John Burton, with a nod toward his wife. "It's time someone breached the Region's hallowed halls." Long and lanky, he sat slumped in his chair, his feet sprawled in front of him as he tapped his pencil on the table. Marilyn had noticed he often expressed a good-natured cynicism toward government and anything connected with it.

"I'll make it if I can," Edith promised. "That's two weeks from tonight, right?"

Doris Migus caught Ruth Burton's attention. "Okay if I ride with you and John?"

As the members reorganized their lives to include a trip to Regional council, John Jackson gave John Dykstra his phone number. "If you need any help between now and then, just give me a call."

*   *   *

A week later, Marilyn stared at the calendar, with its scrawl of times and places. Including the meeting which had taken place on Tuesday afternoon at the home of Peter Antonio—"The Township's version of an OWMC kitchen meeting," Marilyn had called it—and the meeting planned for tonight in the basement of St. Luke's Church, she tallied up four meetings in eleven days. She gave a weary sigh. Not bad, she thought, for people who don't like meetings.

That evening, the meeting of the residents had scarcely begun before John Dykstra said, "I've been thinking—" He gave a gleeful chuckle as he waited for the groans and good-natured complaints to die down. "Seriously," he said, his grin as broad as ever, "I think we need to visit each council in the Niagara Region to gain their support."

Marilyn felt dread spread through her like a slow-creeping menace. The spectre of long, tiring days, followed by hurried suppers and upsetting confrontations with twelve unfriendly municipal councils, made her shrink from the idea. Slowly she removed her reading glasses and laid them on her notebook. Resting her elbows on the table, she covered her face with her hands and massaged her eyes. I can't do it, she thought. I just can't do it.

In the background, she heard the voices weighing the benefits against the enormous commitment of time and energy. As with everything the group decided, they took no vote. Either they reached a unanimous agreement or they scrapped the idea. In the end, as the objections gradually dwindled away, John volunteered to make the arrangements.

But during the discussion, Marilyn noted that one thing had become certain. With or without the backing of the politicians, the residents were headed for Queen's Park.

- 14 -

One week later, on March 19, 1987, John Dykstra made the presentation to Regional council on behalf of the residents. The four councillors from West Lincoln who had pledged to go on the bus trip showed up, smiling and glad-handing as they made a public display of support. Altogether, including members of the Coalition who had come out as promised, the delegation swelled its number to a respectable thirty-five.

* * *

Like the Apostle Paul, Marilyn often felt "troubled on every side." For as he had so graphically described, "without were fightings, within were fears." So, when Wednesday's editorial in the *West Lincoln Review*, titled, "Questionnaire Questionable—Who said, Get on with it?" poked fun at one of Dr. Chant's more troubling pronouncements, she let out a shout of jubilation.

Six weeks earlier, during the OWMC press conference in mid January, Dr. Chant had claimed that many West Lincoln residents had urged the OWMC to finish its work as soon as possible. Marilyn had taken the statement to mean, get the facility up and running. At the time, with a sick feeling in the pit of her stomach, she had glanced at the faces nearest her. How many is *many*? she had wondered, and *who are they*?

Apparently Judy McEwen, editor of the *West Lincoln Review*, had wondered the same thing:

In pushing for early Environmental Assessment Hearings

OWMC—The "Wasted" Years: The Early Days

and construction of OWMC's proposed toxic waste plant, Donald Chant has said the rush is partly because residents have urged him to get on with it.

Well, three people anyway.

According to the infamous OWMC questionnaire, only 3 out of 512 people interviewed expressed that sentiment.

Maybe three people can influence Dr. Chant. Or maybe there were lots of others out of the survey range—people from Wawa or Northerners along the Wabigoon River.

Or maybe there were Toronto people at cocktail parties...

Marilyn, feeling happier than she had in days, gave the editorial in the newspaper a huge, smacking kiss of approval. Suddenly, the day seemed brighter, her workload insignificant. She jumped up from her chair and whipped out a package of hamburger. To celebrate, she would make one of her and Clifford's favourite meals: her mother's meat pie, baked potatoes with homemade sour-cream-and-onion sauce, and honey-butter carrots. And for dessert, elderberry pie.

* * *

On Friday night, Clifford and Marilyn's younger daughter, Kathy, struggled up the back steps and into the kitchen, lugging a huge duffel bag crammed with clothes and books. "Looks like you have everything you own in there!" Marilyn declared. She stopped stirring the gravy for the roast beef long enough to give Kathy a kiss and a welcoming bear hug.

"Just about," Kathy answered with a grin. "Sorry I'm late," she apologized, "but you know buses..." She sniffed appreciatively. "Dad'll be right in. He's gone to close some windows in the barn."

Marilyn ladled the deep-brown gravy into two small bowls, and shoved one into the oven to keep it hot. "How's work?" she asked as she drained the potatoes.

Kathy had graduated from McMaster University with two chemistry degrees the year before, but had stayed on to work in the Nuclear Lab, and now in the Geology Department. "Well, this week—"

The telephone on the wall shrilled in her ear. Marilyn handed the potato masher to Kathy and picked up the phone. Marilyn listened, marked a string of dates on the calendar, and sighed as she hung up. In a harried gesture, she brushed her hair back from her forehead as Clifford came through the back door. "John Dykstra just called. We go to Beamsville council meeting on April sixth and—"

"Hey, April sixth!" Kathy exclaimed, with little-girl elation. "Don't you know what day that is?" Without waiting for an answer, she declared, "It's my birthday, don't-cha-know!"

"Yes, I know," Marilyn said. Then, for Clifford's benefit, she rhymed off the rest of the series of dates. "Like I said, April sixth is Beamsville's council meeting—don't forget that on the seventh, you go to see the accountant about the income tax—the eighth is the Township's public meeting at the Old Farm Inn, the ninth is the Coalition—"

"Maybe I shouldn't bother coming home next weekend," Kathy snapped.

"Of course you should!" Marilyn exclaimed. "We've kept the weekend free." By this time, Marilyn had placed the hot bowls and platters on the table. "Let's eat," she said, as she pulled out her chair and sat down heavily. With a glance at Clifford, the signal for him to give thanks, she bowed her head.

"Our Father, bless this food we take and bless us, too, for Jesus' sake. Amen," he said.

"Now then," said Marilyn, as they raised their heads and began passing the roast beef and gravy. Her gaze rested expectantly on Kathy. "What would you like for your birthday?"

An ominous silence fell over the kitchen. Suddenly, Marilyn could hear the ticking of the clock on the wall. She watched with a sinking heart as Kathy's face darkened into a scowl. "I'd like," she said at last, "*just once*, to have a whole day at home without hearing about the OWMC!"

*   *   *

Early in March, OWMC's plan for a two-staged hearing was vetoed. Speaking for the Crown corporation, Murray Creed, OWMC

OWMC—The "Wasted" Years: The Early Days

media relations officer, warned that as a result, the intervener funding might be stalled until the certificate of completion was available.

Marilyn promptly set to work on her nineteenth letter to the editor. The topic: Intervener funding.

> These are monies [that are supposed to be] made available to groups...to study the OWMC proposal.
>
> At first glance, it sounds like a goodly amount—in excess of a quarter of a million dollars...However, that's only two hundred and fifty thousand—less than one percent of the money spent thus far by OWMC. And the money has strings attached.
>
> 1) It can only be used to review studies undertaken by OWMC.
>
> 2) It cannot be used to conduct independent parallel studies.
>
> 3) Or for active opposition.
>
> 4) The raw data (consultants' actual reports to OWMC) are not available unless OWMC chooses to release them.
>
> 5) They can be subpoenaed once the Environmental hearings begin—but that could be too late to study them in depth.
>
> 6) OWMC's attempt to have a two-stage hearing...has been refused. In an interview, Murray Creed, of OWMC, warns that the funding "may also be stalled"...
>
> No wonder the Citizens for Modern Waste Management, a member group of the Ontario Toxic Waste Research Coalition, are appealing to the public for donations to help them prepare for the hearings.

* * *

Despite a whole host of solemn vows to herself the weekend of Kathy's birthday, Marilyn failed miserably in her efforts to keep the OWMC from encroaching upon their time together. Newspaper clippings and OWMC documents littered every room in the house. As

Kathy chattered on about her work at the university and her friends, Marilyn's mind slipped away to the escalating plans for the bus trip.

"How's this look?" Marilyn asked without thinking. She held up the rough sketch of a tattered scroll of parchment. "Hear ye! Hear ye!" the scroll proclaimed. Around the edges, a fleet of trucks streaked along the highway, past farmland, toward a tall smokestack and a series of engineered landfill cells. "I'm drawing up the flyers for the trip to Queen's Park," she explained. "We've decided to emphasize the theme of the Town Crier..."

Marilyn's voice trailed away as a look of recrimination leaped to her daughter's eyes. "You couldn't do it, could you Mom—not even for my birthday."

* * *

Monday morning, Marilyn contemplated the crowded dates on the calendar. The weeks ahead loomed like an endless nightmare. Right at a time when the preparations and work involved with spring planting had moved into high gear, she and Clifford had been locked into a killing schedule of "lobbying."

"I never knew what lobbying meant," Marilyn remarked to Clifford over breakfast. "I always thought it involved some mysterious political procedure that only a favoured few knew anything about." She screwed up her face in distaste. "Really, it's just a high-class word for wheedling and begging for support."

Clifford only grunted, and Marilyn recognized the signs. His mind had already shifted to the prices of fertilizer and seed corn. "Don't forget," she reminded him, as he hurried down the back steps, "we have to be in Beamsville before seven-thirty tonight for the council meeting."

* * *

Two nights later, on Wednesday, April 8, Clifford and Marilyn attended a meeting organized by the Township at the Old Farm Inn. West Lincoln Council and the Toxic Waste Committee wanted to introduce the Township's team of consultants to the public. Marilyn looked around in dismay at the pitifully small number of people who had bothered to show up. She nodded to several members of the Coalition and waved to John and Mary, and Paul and Marija, seated along the outside wall. Then she did a hasty head count. "There might be seventy," she whispered to Clifford, "but I doubt it."

Just then, twelve men filed into the banquet hall and took their places at a long table set up on a raised platform. Marilyn recognized a dark-haired man with a substantial moustache as Dennis Wood, the lawyer hired to represent the Township in the OWMC matter. Like all the men at the table, he wore a dark business suit. Dave McCallum, the Township's youthful Toxic Waste Coordinator, rose and introduced the consultants. At the end, before he sat down, he told the audience, "This is about half of our team."

One by one, the consultants gave a brief report on the status of things in their particular area of expertise. The toxicology consultant, Bob Willis, stated, "Either the toxic waste treatment facility won't be here—or it'll be here and be safe."

Marilyn felt a subtle eroding of her spirits. She had expected the Township to at least make a show of a fight-to-the-death battle. If nothing else, she thought, it would have improved council's bargaining position. Instead, they had publicly declared a mealy-mouthed willingness to straddle-the-fence and ultimately surrender.

"I see two flaws," the consultant told them. "The OWMC doesn't know what kind of wastes will be transported through its gates. Nor does it have concrete figures on how much waste will be processed."

Marilyn frowned as she leaned close to Clifford. "Then how does the OWMC know it would take too much space to store the wastes *above* ground?" she whispered.

John Jackson and the Coalition favoured above-ground storage. That way, the members argued, any leachate oozing from the mass could be dealt with before it became a problem. The OWMC flatly declared it to be an unworkable plan. In the last two years, no further mention had been made of using the solidified mass for roadbed or patio stones.

Questioned about putting the facility on good farmland, Doug Hoffman, the agriculture consultant, said, "It is hard to find poor land in Southern Ontario. The only alternative would be to find a site up north in the Precambrian Shield, for example. In that case, transportation costs would be prohibitive."

Another resident asked, "Have any members of the consulting panel ever fought such a battle before, and if so, did they win?"

Dennis Wood answered for the team: "The OWMC's proposal is a unique project." Which, Marilyn thought, is legal gobbledygook for 'no experience and no wins.'

Discouraged by everything she had heard, Marilyn screwed up enough courage to make her way to the microphone. Suddenly, she found her eyes swimming with tears. With her heart pounding against her ribs, she swallowed and took a deep breath. "How?—" She stopped, her fingers laced tightly as she struggled to clear her throat. "How would you advise us to fight this issue?"

Dennis Wood, unsmiling as always, fielded the question. "The residents' best bet," he told them, "is to create a political fuss with the Ontario government."

*   *   *

The following night, the citizens' group officially joined the Ontario Toxic Waste Research Coalition.

− 15 −

Three weeks after the "Who said, Get on with it?" editorial in the *Review*, Dr. Chant wrote a letter to the editor, which the *Review* called, "Questionnaire Questionable." In it, he insisted that at numerous informational meetings with residents in West Lincoln and the Niagara Region, many citizens had repeatedly told the Crown corporation that they would like to see the cloud of uncertainty lifted and the issue resolved one way or another.

Certainly, as our questionnaire revealed, most...would prefer that the facility was not built. But they are also distressed by the length of time the OWMC's work has taken and are eager to know how all of this will turn out.

Once again, Marilyn was faced with the same distressing questions: How many residents are "many"? she wondered. And which friends, neighbours, acquaintances are they?

*   *   *

On Good Friday, Judy arrived home from Toronto for the Easter weekend. Marilyn watched Judy's eyes widen with excitement and her face become animated when Clifford mentioned that the pastor of their little church, Richard Snyder, had an old MGA to sell. From then on, every time Clifford stepped through the door, their older daughter wheedled and begged him to run down to Jordan Station to take a look at it.

As Marilyn watched Judy finally wear Clifford down, she opened her mouth to remark in jest, "Where I come from, that's called lobbying." She bit back the words barely in time.

Judy had sided vehemently with Kathy regarding the mere mention of the OWMC in her presence. "I don't want to hear about it," Judy had declared angrily. "I just want to come *home*! I don't want to think about...about..." Her eyes had brimmed suddenly with tears and, turning on her heel, she had fled upstairs to her room. Since then, Marilyn had made a heroic effort to keep the OWMC out of their family life whenever either of the girls came for a visit. But it was never easy. It took but a single slip of the tongue to ruin an entire weekend.

"Come on, Mama," Judy urged. "Come with us to look at the MGA."

With a sigh, Marilyn set aside her Sunday school lesson and picked up her purse. Clifford and Judy discussed nothing but cars all the way there. "I would expect this from a son!" Marilyn declared with mock exasperation. "Not a daughter!"

"But Mom, it's always been my dream—to drive down the highway in an MGA, with the top down and the wind blowing through my hair."

Half an hour later, Marilyn took one look at the MGA and shook her head. "It's junk!"

"Mom!" her daughter protested, "That car's a classic." Later, all the way home, Judy pestered Clifford relentlessly. "Should we buy it? What do you think, Dad?" Then a minute later, "Well, Dad, what do you think?"

As the two debated the merits of the car, Marilyn's mind drifted to the two council meetings coming up on Tuesday night—the first in Fonthill, to meet with Pelham council; then, a race against time to the council chambers in Wainfleet. The struggle to gain support for their cause and for the bust trip to Toronto had begun.

* * *

The presentation to Pelham council went much as expected. Council members asked critical questions, and then deferred their decision. However, when the citizen's group rose to slip out quietly, several members of the public left at the same time, including Norman Nelson, the editor of the *Pelham Herald*. He quizzed John Dykstra on several points, then requested a copy of the presentation.

OWMC—The "Wasted" Years: The Early Days

Two others, a man with a swarthy complexion and a tiny grey-haired woman, waited outside in the parking lot. "I'm Marie Austin. This is my husband Don. We like what you said," the woman told John, her blue eyes snapping with enthusiasm. "We'd like to join your group."

"We can't talk now," John told her. "We're on our way to Wainfleet's council meeting." He raised his eyebrows questioningly as he took a survey of the group. "Anyone object if they come along?"

"The more, the merrier!" Lynda declared, with a tinkling laugh, while the others nodded their consent with varying degrees of enthusiasm.

*   *   *

On Wednesday, to Judy's joy and delight, Clifford trailered the little MGA home to the farm. Marilyn walked around the trailer, shaking her head as she surveyed their treasure. "It still looks like a basket case to me."

From then on, as Clifford and Judy debated endlessly over the telephone how they would go about restoring the car, Marilyn fussed to herself. Where would Clifford find the time or the money? Every spare moment and every spare dollar vanished down the black hole Marilyn referred to as the OWMC mess.

*   *   *

Over the next several days, the couple from the parking lot accompanied the group to the council meetings at both Niagara-on-the-Lake and Niagara Falls. Marie Austin, energetic and petite, her dark hair turning to grey, had a sparkling smile and a spirit of endless enthusiasm. Her husband, Don, with his deep voice, laughed as readily as his wife, his eyes flashing in roguish merriment and his teeth white against an incredibly dark tan.

No trip to any of the council meetings within the Region of Niagara ended without an impromptu meeting of the residents afterward, either on the broad concrete walkway outside the municipal building or over coffee in a nearby restaurant. John scrounged up tickets for the bus trip that had been designed and printed in the early months of the battle by the Task Force, but never used. He handed them out in groups of ten.

"I think we should keep a record of who bought them," Marie Austin stated firmly.

Before anyone commented, her husband gave a hearty, braying laugh. "When she's got that look on her face, there's no use arguing."

As news of the bus trip spread, more and more people showed interest in supporting the small group of residents. To raise awareness and support, Marilyn stepped up her letter-writing campaign. "We have been assured that emissions from the OWMC's proposed facility would be well within the established standards," she wrote. "However, these standards are based in part on what industry can reasonably afford to do concerning pollution control, not necessarily on what is a safe limit."

A week later, she typed and made photocopies of one of her more indignant commentaries. Long had the OWMC tried to convince the people of West Lincoln that the proposed waste facility would be good for business. "Whose business?" she demanded to know. "Certainly not the farm business!" Then she quoted a statement Dr. Chant had made the previous November. "Certainly there will be new opportunities for those with an entrepreneurial spirit."

"Contrary to popular belief," she wrote, "'opportunities' do not lie around loose like dust balls under the bed. For one person to take advantage, another has to be disadvantaged. In this case, the *instant* entrepreneurs would gain at the expense of the long-time entrepreneurs of the area—the farmers."

After consulting *Webster's Unabridged Dictionary*, she noted that "entrepreneur" meant the organizer of an economic venture, especially one who owned, managed, and assumed the risks of a business. "However," she said, "in this day and age, the word seems to have taken on other connotations—that of a mover and a shaker, an innovator, one who is assumed to be more enterprising than the rest of us. But take note," she warned, "these are people who are out to make a quick buck by skimming off the cream, with no long-term commitment to the area and no regard for the consequences."

She finished up with a quote from the Federal Minister of the

OWMC—The "Wasted" Years: The Early Days

Environment: "Virtually every change has not come about by political clout, but by citizens making their wishes known on an issue and putting pressure on politicians."

For good measure, she added a quote from Dr. Daley of Pollution Probe, the high-profile environmental organization regarded as Dr. Chant's baby. "Governments respond to public opinion rather than initiating public opinion."

"So join us," Marilyn urged, "as we demonstrate public opinion at Queen's Park. Make your opposition known by your presence."

* * *

On the last Tuesday in April, the group met in the basement of St. Luke's Church. Clifford reported that, because the citizens didn't have a date for their trip yet, Farr's bus line didn't know whether any buses would be available, since most of the buses had been booked for kids going on end-of-the-year school trips.

"Oh, boy!" John exclaimed worriedly. "I never thought about that."

Next, they went over the roster of speakers. The tally had grown from three to twelve. "We need someone to introduce the speakers," John reminded them.

Marilyn slid lower in her chair and tapped her pencil abstractedly on the table. Stronger and stronger came the urge to volunteer. Leaning forward, she put her elbows on the table and her face in her hands.

"Someone has to do it," John told the group, his tone insistent.

Marilyn heard the sound of voices around her, but not the words, as she fought against the insane compulsion welling up. Then, from deep within, came the words: "I'll do it."

Aghast at what she had done, she then discovered that, as an official spokesman for the group, she must also take part in the upcoming press conference.

* * *

Marilyn watched eagerly the next day for the mailman. As soon as the *West Lincoln Review* and the *Pelham Herald* arrived, she scoured them for the latest news on the OWMC issue. The front page of the *Review* carried a quote by the new president of Smithville's Chamber of Commerce. "The OWMC's gotta go someplace," the former accountant had told the reporter, "and it'll be excellent for business."

"You slob!" Marilyn snapped. "Your business is nine miles away. Ours is right here!" Her voice had taken on a hard edge of fury. These instant entrepreneurs were all alike. They were just a bunch of money-hungry barracudas, waiting to devour the ill-fated residents!

As she glared at the accompanying photo, her gaze strayed to the article beside it, "Help for stressed politicians."

Despite her own predicament, Marilyn had long felt a measure of compassion for West Lincoln's politicians. Yes, they had begged to be elected, but had they bargained for the monumental problems they now faced? At present, the Township was embroiled in three major environmental projects—the PCB contamination of Smithville's water supply; the OWMC; and the four-municipality garbage dump, still in the planning stage. Any one of them, she knew, would heavily tax the local politicians' patience and stamina.

"As a result," the article reported, "the Township's council members are showing serious signs of stress. To aid them in dealing with their problems, a consultant has been called in." In an interview with the press, Mr. Smith, the consultant, revealed, "I have not heard politicians say that the OWMC won't be successful. With that in mind, I will help them plan how to manage living with [the facility]..."

Marilyn couldn't believe the swift, sudden pain of betrayal that lodged in her chest. "Live with it," she repeated in a low, incredulous whisper.

"And plan," Mr. Smith continued, "to build an industrial base around it...for long-term industrial development."

Marilyn swallowed, fighting for control. She pictured a nightmare of industry springing up around the facility, its ugly tentacles reaching farther and farther into the peaceful, pristine countryside.

"And also," Mr. Smith added, "the council has to learn how to keep from reacting to public pressure."

Marilyn slowly lowered the newspaper to her lap, then pressed her hands to her face. Mr. Smith had been hired to coach the mayor and the aldermen how to ignore the pleas of the residents. Council's betrayal felt like a deathblow to her embattled spirit. She stared off into space, the events of the past months running through her mind. She and the others had been right. Council could not be trusted.

Finally, leaden with defeat, she leafed half-heartedly through the *Pelham Herald*, searching for Norman Nelson's report on the group's visit to Pelham council the previous week. Suddenly, her breath caught in her chest and she could scarcely believe what she saw: almost an entire page had been devoted to the residents' fight against the OWMC in an editorial titled, "A formidable battle against Dr. Chant." At the bottom of the page, Nelson had printed the entire text of John Dykstra's presentation!

Jubilant over the extent of the coverage, Marilyn retired to her old orange recliner to devour every word. But before she had finished the first paragraph, she felt a cold, creeping dread invade the pit of her stomach. Nelson wrote, "I've seen so many letters to the editors come in from concerned and well-meaning people—letters that have been shot like targets out of the sky by the OWMC's president, Dr. Donald Chant."

Marilyn recognized instantly the connotation implied by the term, "well-meaning." It meant people who were naïve, uninformed—and wrong. It means people like me, she thought.

Her hands turned ice cold and shook so much she could hardly hold the page steady enough to read. Nelson used the rest of the page to rip to shreds past presentations and letters to the editor. Then, he started on John's recent appeal to Pelham council for support.

Nelson voiced criticism after heartless criticism, reinforced by statements made in rebuttal at various times over the past year and a half by Dr. Chant or other OWMC personnel. In summary, he quoted a statement by Murray Creed, OWMC's Information Officer, made a year earlier in reference to concerns raised by residents in letters to the editor and presentations to council: "It is this kind of analysis, not based on fact, that doesn't do anyone any good."

Marilyn felt sick. It was a ruthless, unexpected attack, as vicious as they come. As if in a daze, she pushed herself up from the chair and stood gazing around her, like a stricken animal seeking a place to hide. Then, tears streaming down her cheeks, she wandered from room to room, too devastated to turn her hand to a single task.

"How can any of us ever show our faces again?" she asked herself. She felt the courage she had mustered for the press conference and for the speeches from the steps of Queen's Park drain away.

She pictured the staff in the OWMC offices. From the highest paid to the lowest, they would be whistling at their desks and gloating to one another. The OWMC had scored a huge win this time.

As if drawn to the instrument of her destruction, Marilyn picked up the paper again and stared at the article. Nelson had tried to excuse the cruelty of his words by claiming to play the devil's advocate. "You're his advocate, all right!" she told the photo next to the byline, her voice roughened by hurt. Yet, in the midst of her anguish, deep down inside, part of her acknowledged that much of what he said was true. The residents *did* sometimes make unfounded, emotional statements easily cut to ribbons by Dr. Chant.

Marilyn brewed a pot of tea and took an aspirin for her pounding head. She read the final paragraph one more time, then tipped back in her aged recliner, and closed her eyes. In those final sentences, the editor of the *Herald* had written in unvarnished terms what Marilyn already knew in her heart: Premier Peterson would not be swayed by the 25,000 signatures.

Nelson had warned,

> Therefore...the only other choice is to play the game and hope to win. But to do that—and this is addressed to the politicians and all the volunteer citizens who plan to make the trip to Queen's Park to state the Region's case—you have to keep on your toes.

Too tired to struggle any more, Marilyn felt her mind drift onto its own course. She pictured Norman Nelson as a stern schoolmaster, striking his desk with a long wooden pointer and demanding excellence from a classroom stuffed with politicians, residents, and Marilyn herself.

As the scathing denunciations continued to seethe and churn in her mind, the brutal words distilled into a harsh, clear message: Don't make ridiculous, unsubstantiated statements. Don't miss an opportunity to present your concerns by talking like fools.

Within the hour, the telephone had rung five times. Anger and despair coloured the voices of friends, neighbours, and members of the group.

Ironically, Marilyn found herself wading through her own misery in search of elements of hope to offer to the others. Referring to the damaging remarks made by the Township's stress consultant, Marilyn suggested to John Dykstra, "Maybe Mr. Smith has done us a favour." She gave a small, bitter laugh. "He hasn't done much to alleviate our stress, but at least he's told us exactly what's going on behind the scenes."

"He hasn't done much for council's stress, either!" John told her. "I complained to the mayor's office and—" he broke off with a gleeful chuckle, "seems I wasn't the only one."

Don and Marie Austin were particularly irate. They had been the recipients of a direct hit in the article in the *Herald*. "Like it or not," Marilyn told Marie, "he's warning us to 'get it right'. No stupid statements. No wild claims we can't prove. And," she paused for emphasis, "his article is a timely reminder that, as he once told me to my face, no media person is our friend. From now on, we have to be doubly careful of every word we utter."

Despite her brave words, however, Marilyn wavered between the frantic desire to absent herself entirely from the trip and the determination to make certain she made no blunders. Every time she sat down at the typewriter, she felt the sting of Norman Nelson's words. Whether it was a letter to the editor or the preliminary draft of remarks for the press conference, she checked and double-checked every bit of data. If she had the remotest qualm about the validity of a statement, she chopped it out.

"Better to say nothing than to say the wrong thing!" became her clarion call to the group as they prepared for both the press conference and the trip. Particularly troubling to her of late had been the tendency for people to grasp at any straw. "Put the facility up north," they urged. "Or send it to Sarnia." Any place, Marilyn recognized, but West Lincoln. To ensure that such a sentiment did not emerge to make grist for the media, she took on the task of drawing up a handout for participants of the bus trip.

She started with a quotation from Isaiah: "Come now, and let us reason together..." She urged everyone to a spirit of reason and dignity. Even more, she urged them to the utmost care and discretion, should they find themselves interviewed by the media. "One ill-considered statement could undo the arduous work of months," she wrote, adding the suggestion that they direct the reporters to one of the organizers of the trip.

She referred to the position the group had taken in their presentation to the various councils across the peninsula. "Their support was given, based upon this position. By your signatures and your presence here today, you have indicated your endorsement of them." Then, to make sure every participant was familiar with that position, she listed the points:

> OWMC'S waste treatment facility and engineered landfill should not be built on West Lincoln farmlands or anywhere in Ontario. OWMC should implement an aggressive programme of waste reduction, recycling, reuse, and recovery. Wastes that cannot be adequately processed at the present time should be treated to the maximum extent, solidified, and stored in above-ground warehouses until suitable technology becomes available.

She added a final admonition. "Over the past months, we have striven hard to gain credibility with our elected officials. On this basis we have also gained their support. It is our fervent prayer that nothing be done or said to undermine the trust they have placed in us as responsible citizens."

Marilyn stared at the final paragraph. She wished she dared to make it stronger. In her heart, a plea went heavenward. "Oh, Lord, please don't let anyone cause any trouble."

– 16 –

John Dykstra's telephone call to the mayor's office protesting the stress consultant's remarks printed in the *West Lincoln Review* had been but one in a storm of angry calls. The Township council, anxious to quash the furor, hastily arranged a closed-door meeting in chambers on Monday at noon with a delegation from the citizens' group. Included was Mr. Smith, the stress consultant, and Judy McEwen, editor of the *West Lincoln Review*, whose front-page article had sparked the trouble in the first place.

When the meeting was over, Judy McEwen stopped Marilyn on the sidewalk outside the municipal building and asked her bluntly, "Did you get what you came for?"

Marilyn's gaze slid away uneasily. Judy McEwen knew perfectly well what had taken place at the meeting. She had been present the whole time, not in the role of a reporter, but as one of the parties being blamed for the conflict and had witnessed the heated recriminations and the raised voices. Now, her constraints lifted, she was seeking an official statement, something she could print in Wednesday's paper.

Marilyn spoke guardedly. "The *Review*'s story about Mr. Smith's work here made council seem…suspect." *Suspect* was the most diplomatic word she could come up with on short notice. Council had made conciliatory statements during the meeting and, although Marilyn remained skeptical, she wasn't about to have her doubts hit the front page of the newspaper. "I figured the residents were being shot down…and the politicians were saying in private, 'We'll go ahead planning the facility and the attendant industry.' But as you heard, council assured us that this was not the case."

The cynicism in Judy McEwen's eyes was compounded by the half-buried, reproachful smile that chided Marilyn for being less than candid. Then, with an impatient shake of her head, Ms. McEwen consulted her notebook. "John Dykstra told the mayor that 'Mr. Smith's comments were damaging to the fragile trust the residents had built up in their leaders.' Do you agree with his opinion?"

Marilyn struggled to keep her expression inscrutable. She wanted to snap, We don't trust them at all! Instead, she carefully tempered her answer. "We were concerned," she admitted reluctantly. "For one thing, we were afraid that the support we have gained from other municipalities for our trip to Queen's Park would be jeopardized by Mr. Smith's remarks."

"Now you don't think it will be?" Mrs. McEwen asked, an eyebrow arched mockingly.

"I hope not," Marilyn replied evenly.

* * *

That night, the residents arrived at six-thirty at the St. Catharines' council chambers to make their presentation. So far the group had received the official endorsement of West Lincoln, Lincoln, Pelham, Wainfleet, Niagara-on-the-Lake, and Niagara Falls.

The warmth of the early May sunshine outside St. Catharines' council chambers did nothing to dispel the chilly reception the residents received inside. Marilyn watched John walk to the microphone. He stood a head shorter than Clifford and Marilyn, but had an inborn feistiness that stood him in good stead at times like this. Nevertheless, Marilyn wondered how he could find the courage to stand before these stone-faced councillors when, just five days earlier, Norman Nelson's editorial in the *Pelham Herald* had trashed the group's presentation. *I couldn't do it*, she thought with absolute conviction.

As soon as John completed his presentation, the residents filed outside and split into two groups. John and one group set off for Grimsby to repeat the presentation. Clifford, Marilyn, and the others hurried to West Lincoln's council meeting. They arrived in time to hear the council vote to help the group arrange a press conference.

Daylight had begun to fade into dusk as Marilyn and the others stepped out into the balmy spring evening. Above the darkening horizon of modest homes and village shops, crimson streaked the western sky. The reporter from the *Review* caught up with Marilyn. "What date has been set for the bus trip?"

Marilyn shrugged and shook her head. "We don't know yet. The letters we sent to Premier Peterson and Environment Minister James Bradley have not yet been answered. Tentatively," she said, "we've planned the trip for Environment Week—the first week in June."

"And the press conference?"

"That date has not been set, either," Marilyn told the reporter, her voice edged with frustration. "Nor has the location. We can't make those arrangements until we hear from the Premier."

The councils of St. Catharines and Grimsby both declined to support the group, and Marilyn privately blamed Norman Nelson's editorial. However, a spokesman for Grimsby council told the reporter from the *Review*, "It's up to the individual councillors whether or not to join the bus trip."

The following night, as the residents continued their whirlwind effort to lobby the Region's twelve municipalities, they made the forty-minute trip to attend the council meeting in Thorold at seven, then sped down Highway 406 to catch the one in Welland at eight.

Two days later, John Jackson met with the residents at Clifford and Marilyn's home to help them prepare for the upcoming press conference. "Make sure each and every member of your group gives the same answers to the same questions," he cautioned. "Don't say anything you don't want to see in print."

"In other words," Marilyn remarked cynically, "no media person is our friend." Norman Nelson's warning had become a byword with her, an ironclad rule to live by.

Lynda drew her dark eyebrows into a deep frown. "I don't see why we have to ask the mayor to speak at Queen's Park, or at the press conference. It's *our* trip and *our* press conference!"

John Dykstra gave a mirthless laugh. "That may be. But it wouldn't look good if we snub our own mayor."

Marilyn dug in her folder for a small clipping. "I cut this out of the paper the other day. 'As long as you achieve your goal, it doesn't matter who gets the glory.'" She raised her head and looked around the room. "So, in my opinion, as long as the mayor speaks, publicly at least, against the OWMC, that's all that matters."

That evening at eight o'clock, Clifford, Marilyn, and the others arrived at the Vineland Research Station for the monthly Coalition meeting. "I think they're in good shape," John Jackson assured the membership with a confident smile. Just then Harry Pelissero appeared in the doorway, nodding and offering apologies for arriving late. It was the first time the political hopeful had shown up for a Coalition meeting and Marilyn felt an atmosphere of reserve fall over the gathering.

* * *

Despite the long delay and the difficulty in getting everything arranged on short notice, the bus trip was finally set for June 3, 1987. The buses would leave the West Lincoln Arena at eleven-thirty and arrive at Queen's Park between one and one-thirty. The delegation was scheduled to meet with the premier and the environment minister from four o'clock until four-thirty.

The Township offered the residents the use of the room above the fire hall, adjacent to the municipal building, for the press conference. The citizens settled on a date: Wednesday, May 20, at ten in the morning.

* * *

At six-twenty the following Monday night, the residents met at Chamber's Corners. Fort Erie would be the final stop in their marathon across the peninsula as they lobbied for support. Of the twelve municipalities in the Region of Niagara, Port Colborne was the only one they had been unable to fit into the hectic schedule.

After the presentation to Fort Erie council, the group stood on the sidewalk in front of the municipal building in the evening sunshine. Everyone, it seemed, needed more tickets.

OWMC—The "Wasted" Years: The Early Days

"Looks like we'll need a third bus," Clifford said. "I'd better call Farr's."

John tilted his head in speculation and squinted one eye. "We'd better have one more meeting before the press conference." He whipped out his wallet and checked his calendar. "How about Friday of this week?" He gave a short, worried laugh. "We've got a lot to do before the bus trip."

"We've got a lot to do before the press conference!" Marilyn reminded him.

Marie burst into a merry laugh. "We've just got a lot to do!"

* * *

To the meeting on Friday night, Marilyn brought what she hoped would be the final draft of the statements she and John intended to make at the press conference. "As we read them aloud to you," she instructed the others, "look for even the smallest error. If you have the slightest doubt about something, speak up. *Now's the time to fix it, not after the press conference.*"

She had no need to remind them that the media would cut them no slack.

On May 13, the Liberal party for Lincoln Riding chose Harry Pelissero as their candidate for the upcoming provincial election. In the past, Marilyn had ignored politics. Now, she read the article on the Liberal candidate with an orange marker poised in one hand. The interview took up most of page one in the *West Lincoln Review* and spilled onto a good-sized portion of page two. In it, Pelissero spoke of the proposed OWMC site, just over a mile from his home.

"Such a disposal site is necessary," he had told the *Review*. Marilyn pushed her hair back from her forehead as the familiar sense of helplessness and defeat stole over her. As feared, he had mouthed the opinion of David Peterson, Premier of Ontario and leader of the reigning Liberal party. Nor, Pelissero stated, would he oppose OWMC's proposed toxic waste facility, "provided all concerns were met." Then, as though he had been unfairly wronged, he pointed out that he had been criticized in the past for not being more vocal. In his defence, he said, "I don't want to risk being labelled as a NIMBY."

As for the citizens' protest trip to Queen's Park on June 3, Mr. Pelissero said, "I'll have to see what I'm doing that day."

Marilyn muttered a string of disparaging epithets. She would need to do some serious thinking before election day arrived. She wanted to vote Liberal to keep Jim Bradley in office as Minister of the Environment. But she despised the cavalier, "It's got to go somewhere" attitude of Peterson and Pelissero.

- 17 -

Marilyn had written her first round-robin letter nine months earlier. Since then, she had enlisted friends and relatives to clip articles for her from newspapers across the peninsula and, by quoting from those articles and naming the publications, she had courted the good will of the editors. As a result, she had added the names of two more newspapers to her mailing list, making a total of nine.

Despite her crushing workload, Marilyn sat up late into the night, thumping away on her typewriter as she stepped up her letter-writing campaign. In May of 1987, Dr. Chant had voiced an old argument in the *Pelham Herald*: "...The vast majority of Ontario's wastes are generated in the highly industrialized belt [around Lake Ontario] called the Golden Horseshoe."

In rebuttal, Marilyn cited two OWMC documents, complete with page numbers. "Of the approximately 70% of the wastes generated in the Golden Horseshoe, only about 7% actually comes from the Region of Niagara—4% from Welland and 3% from St. Catharines."

Her next letter dealt with concerns over truck spills. OWMC's now-infamous questionnaire showed that of the 240 residents polled, an overwhelming ninety-six percent believed there was the risk of a truck spill. Nor, she cautioned, should those concerns be centred only on toxic waste. Additives and reagents, including 400 tonnes of liquid chlorine, were slated to be trucked onto the proposed site for use in the physical/chemical treatment plant.

*   *   *

On Friday, May 27, 1987, on the eve of Environment Week, the *Standard* featured a waste reduction success story. The 3M Company had cut its overall waste production by more than fifty percent, thus saving the industry $400 million. Marilyn felt a rush of relief. From the start, cutting waste at the source had been part of the Coalition's overall waste-management strategy. Yet, scoffers had seen the idea as impractical. Now, here was proof that it not only worked, but could be highly beneficial to the industry itself.

Dr. Chant, however, did not sound overly impressed by 3M's announcement. "Once the OWMC facility is up and operating," he told reporter Doug Draper, "the cost to industries for proper treatment and disposal of their wastes will increase substantially. When faced with these higher costs, companies naturally will try to do everything they can to reduce their waste output."

Critics like John Jackson, however, insisted that the facility would have the opposite effect. Industries would simply build the cost of using the OWMC's facility into the price of their product, and keep right on generating wastes.

*   *   *

On Friday evening, five days before the press conference, the residents held a last-minute meeting in the basement of St. Luke's Lutheran Church. John Jackson arrived by bus from Kitchener to help them make sure they had everything ready.

"I don't think I can do this," Marilyn told him.

"You'll be fine," John Jackson assured her. His glance transferred to John Dykstra, and then swept the others in the church basement. "You all will. Just take your time. Don't let anyone rush you into giving a hasty answer."

Judy McEwen, editor of the *West Lincoln Review*, dropped by to take pictures. John Dykstra heaved the recycling box containing the 25,000 signatures onto one of the tables, then stood behind it. Beside him, John Jackson and Marilyn appeared to discuss the box's contents.

After Judy McEwen left, John Dykstra reported that Doris Migus had telephoned with some strategy pointers for the trip to Queen's Park. "You know Doris," he said with a grin. "*'Don't,'* she said, 'under *any* circumstance, let a politician be the last speaker at the meeting with Premier Peterson or outside on the steps.' And she thinks John Jackson should be the last speaker both inside and out, because he's good at spotting points that have been left out and he can fix up anything someone might say that would leave a wrong impression."

Red-faced, John Jackson gave an embarrassed, self-deprecating smile. But he didn't disagree with Doris' advice.

* * *

Marilyn spent the five days leading up to the press conference polishing, refining, and making final copies of her notes. Wednesday, May 20, 1987, dawned bright and sunny. Upstairs in the bedroom, she held aloft for a final inspection the plum-coloured, long-sleeved, high-necked blouse she had chosen to wear, and hoped the temperature would not rise drastically in the next few hours. Clifford rushed in from the barn and took a quick shower while Marilyn cooked breakfast. Feeling as though her stomach had shoved up into her throat, she managed to swallow only a few mouthfuls.

They arrived at the fire hall shortly after nine. Already the pavement swarmed with reporters, photographers, and television cameras. To Marilyn, none of it seemed real. Ordinary people, like the Graceys and the Dykstras and the other residents, did not command the attention of the media—they stayed at home and watched strangers on TV get embroiled in controversial issues.

In the meeting room above the fire hall, Marilyn took her place at the table, facing the area where the press would sit. She rehearsed in her mind what she would say, and how she would say it, as she tried to steady her nerves. John Dykstra grinned nervously and leaned over to whisper, "We'll be okay. John Jackson said so."

Marilyn laughed and it relieved some of her tension. At the suddenness of the laughter, Mayor Colyn, seated on John's left, gave them a questioning frown, then turned back to the pages of his speech. Reporters and news photographers began poking their heads through the door, scanning the large room in search of the most advantageous positions. Crews from television and radio stations set up microphones on the table in front of them.

"Are they turned on?" Marilyn whispered to John. He shrugged. She viewed them with suspicion. As she recalled, many a politician had been caught unawares by a microphone believed to be turned off. She decided to assume that all microphones were "live," just as her father, a hunter, had cautioned his children to always assume that all guns were loaded.

Right on ten, John welcomed the press and all those who were present. Marilyn felt almost detached from the proceedings as John read from his revised presentation. He reported that the group now had official support from eight out of the twelve municipalities—West Lincoln, Lincoln, Pelham, Niagara-on-the-Lake, Niagara Falls, Thorold, Fort Erie, and Wainfleet. He finished by saying, "Emissions from the proposed facility will contaminate surrounding agricultural lands. Everyone in the province who consumes Niagara-grown produce should join in opposing the facility."

Next, Mayor Colyn added his concerns: "A glance at a newspaper almost any day of the week provides an example illustrating that landfills are not safe. Incineration also is not safe."

Marilyn released a slow, silent sigh. For the first time that morning, she truly began to relax. The mayor had been the wild card, the one unknown quantity in the group's meticulous plans. With no way of knowing what he might say, Marilyn had spent more than one restless night. As his speech began to wind down, he charged, "I question whether existing environmental legislation is sufficient to protect area residents in the event that the facility is approved."

Marilyn felt her lips compress in disapproval, then swiftly adopt a benign expression for the benefit of the media. But inwardly, she was steaming. He had spoiled a perfectly good speech by slipping in the possibility that the OWMC would gain approval.

When Marilyn's turn came, she dealt mainly with details about the bus trip—times, places, the list of speakers. In an attempt to undo the mayor's indiscretion, she emphasized the residents' unequivocal opposition to the OWMC's proposal.

John added a final word. "When this proposed dump leaks into the aquifer, where will the rural people get their water? Once it's lost, it's lost forever. You can *never* buy that back!" His voice became adamant. "That's why we're here—and that's why we're going to make our voices heard."

During the question-and-answer part of the conference, Paul Forsyth of the *Welland Tribune* asked Marilyn, "Can you tell us what will be discussed on June third?"

"Not at this time," Marilyn told him. "But we hope the meeting will allow us to voice our concerns and bring the issue to the attention of the rest of the province. We hope to provide Premier Peterson and Mr. Bradley with information they might not have considered to this point."

"Do you think you can convince Premier Peterson to intervene against the Crown corporation's plans?"

Marilyn hesitated. It was a direct question, one she would rather not answer. But she couldn't lie. "No," she admitted, "not really."

Later, outside in the hot sunshine, reporters crowded around Marilyn and John. "What's the name of your group?" the reporter from the *Niagara Farmers' Monthly* asked.

John gave an uncomfortable smile. "Actually," he said, "we don't have a name. We're just a group of concerned citizens."

"How many are going with your group?"

"Well," said John, "We're going on a day when people are at work, so..." He glanced at Marilyn. They had agreed to keep all predictions modest to avoid ridicule later. "We're hoping that at least a hundred people will show up."

A question from Norman Nelson about Dr. Chant's illustrious qualifications sparked Marilyn's temper. "Dr. Chant is wearing blinders," she snapped. "He is saying that if you don't do it his way, then you have to live with nothing being done at all." Catching herself, she toned down her voice. "Dr. Chant has spent a lot of time and taxpayers' money on this proposal," she admitted, in an attempt to be reasonable. "But quite suddenly, while he was working hard, a lot of new things cropped up."

"Like what?" Nelson asked.

"Well..." She suddenly felt short of breath, as though she had run a great distance. I'm no good at this, she thought, frantically ransacking her memory for an answer. "As—um—one of the seven member groups belonging to the Ontario Toxic Waste Research Coalition, we—think that industry should—" She could feel perspiration beading across her forehead. With her fingertips, she tried to pat it away without smearing her makeup. "—Industry should—try to reduce their waste and do more recycling—like the 3M company is doing." As the sun blazed overhead, a light breeze riffled the leaves. Marilyn wished her blouse had short sleeves and a cooler neckline. "Um—as many wastes as possible—should be dealt with on-site...The PCBs here in Smithville are a good example. Bring in mobile destruction facilities—"

Nelson interrupted. "Don't they have incinerator emissions?"

"Yes—but—" Under pressure, she tended to forget everything she ever knew, and she blamed herself for not having a firmer grasp of the group's position. She instinctively sensed a flaw in his comparison. Biting her lip as she desperately sought for the flaw, she raised her eyes to gaze above the crowd and watched the sun play over the fluttering leaves of the trees. "But—the emissions would be short term—a year or two at most—not thirty, forty, fifty years all in one place."

He gave a small nod and shoved his notebook into his hip pocket, along with a copy of the Coalition's position paper. Marilyn watched him head for his car, suddenly aware of the worried beating of her heart. *What will he write this time?* she wondered, trying to recall precisely what she had said.

* * *

The next day, Marilyn felt both relieved and elated. The press coverage in the daily newspapers had been all that the group had hoped for—extensive, accurate, and fair. A week later, Norman Nelson outlined in detail the Coalition's official position paper, calling it a responsible, effective, and far-sighted waste management strategy.

In addition, the group discovered that they now had an official name. For, without exception, every newspaper had taken John's explanation that they were just a group of concerned citizens and capitalized the words, thus christening them the Concerned Citizens.

* * *

The press conference stirred up additional support from an unexpected quarter. Several days later, the group learned that members of the Canadian Auto Workers, Local 199, hoped to have two busloads leave St. Catharines to join the bus convoy to Queen's park.

At the next meeting, for the first time, the residents found themselves divided into two camps—those who hailed the announcement of the auto workers' support as a tremendous boon and those, like Marilyn, who had wanted to keep the trip free from the slightest hint of partisan politics or causes.

She could not say exactly why she felt uneasy. It was the same sort of uncertainty she had felt about the mayor's speech. The CAW was an unknown quantity to her, and she feared some sort of altercation might occur that could destroy the residents' hard-won credibility.

Clifford, John, Mary, Paul, and Marija held the same reservations Marilyn did. Some members, unable to discern the fine line between an independent, no-strings-attached liaison and the feared perception of affiliation, remained undecided. Bob and Lynda saw nothing but benefit in the doubling of their numbers with the addition of the promised two busloads of union workers.

After considerable discussion, John came up with a solution that had the potential to ease the situation somewhat: "Suppose we make a banner, identifying our group. That way, we can publicize who we are and we won't be overshadowed by the CAW." He grinned as he looked at Lynda and good-naturedly reminded her of her own words. "After all, it is *our* bus trip."

Mary offered to donate a length of heavy, white, canvas-like cotton that she had on hand.

Marilyn said, "I can make the patterns for the letters and lay them out on the cloth, but someone else will have to make the actual banner. I just *can't* spare the time."

In their usual rapid-fire manner, the residents decided to keep the message simple. They would put the name, Concerned Citizens, in huge block letters, using green felt on white—green for the environment, white for pristine. Underneath, flowing across the banner, smaller, softer-shaped letters would display the slogan coined by the Task Force, "There is a better way!"

On the way home from the meeting, Clifford and Marilyn detoured to the Dykstra farm to pick up the fabric. That night, Marilyn roughed out the letters on sheets of typing paper, traced them with black magic marker, and cut them out. Then she placed them on the white material that would become the banner, and pinned them into position. After shifting, spacing, and viewing them with an eye to aesthetics, she finally felt satisfied. Shortly after four in the morning, she fell into bed and slept soundly for the next four hours.

In the morning, Clifford took the fabric, with the paper letters pinned in place, back to Mary. Home again, he stuck his head in the door. "Mary says she and the other ladies will get started today."

* * *

On Friday, just five days before the bus trip to Queen's Park, Marilyn hung up the telephone and stared at like it was a viper. "Dear Lord," she whispered, her hands pressed to her mouth as she verged on hysteria, "what am I going to do?" With trembling fingers, she reached for the phone, drew back, then reached again. Shakily she dialed John and Mary's number. "Please be home," she begged. "Please...*please!*"

Mary answered on the fourth ring. Marilyn knew she sounded frantic and near to hysterics, but she couldn't help it. She heard Mary say in a worried tone, "Quick, John, pick up the other phone. It's Marilyn."

Seconds later, she heard John's voice. "Marilyn, what's wrong?"

Her voice trembled with emotion as she named a person and a powerful organization. "Without consulting me, they've set up a debate between Dr. Chant and me on the radio—for Monday." Her voice became shrill with panic. "Why are they doing this? I'm no debater! And why Monday, only two days before we go to Toronto?" She swallowed and tried to calm herself. "I'll end up sounding like a fool and ruin the trip to Queen's Park! Ooohhh," she wailed, "what am I going to do?"

"Who set this thing up?" John asked.

Marilyn named a name, and John gave a short, caustic laugh. "Then I know who's behind it." He named a public official. Marilyn untangled the coils of the long cord and sank into the easy chair around the corner from the telephone. "Are you sure?"

"I'm sure."

"What should I do?" she asked again.

"Don't do anything. I'll call Paul. You get Clifford. I think as many of the group as possible should pay our friend a little visit."

Within the hour, John had set up a meeting. John and Mary, Paul and Marija, and Clifford and Marilyn formed a cavalcade to meet with the official. Also present was the person who had contacted Marilyn. She threw Marilyn the kind of look she might have given to a foolish child. "You're passing up a great opportunity to publicize your trip to Toronto."

"Some publicity!" Marilyn snapped. "Chant would chew me up and spit me out in pieces!"

"You're very knowledgeable. What makes you think?—"

Marilyn cut her off. "Because I'm not a debater. I couldn't argue that the sun dawns in the east every morning—and win!"

The public official turned to John. "I thought we discussed this when you and I met alone in my office—"

John said flatly, "I've never met with you alone in your office."

The official ran his finger around the inside of his collar and loosened his tie. "Oh, that's right," he said. "It was the day you and Mrs. Gracey came—"

"I've never gone anywhere alone with Mrs. Gracey."

The official looked doubtful. "Are you sure?—"

"I'm positive," John informed him. "It's group policy."

"And *I*," Marilyn stated, emphasizing almost every word, "have *never*, at *any* time, been in your office—*ever*!"

The official hitched ahead in his chair. He pulled a handkerchief from his pocket and mopped his face. "Hot day, isn't it?"

"I hadn't noticed," John told him with a straight face.

Marilyn pushed her tongue against the inside of her cheek and turned the other way. If ever I've seen a worm squirm, she thought, it's now.

"Anyway," John told him, "you can call off the debate with Dr. Chant."

"But—but it's all arranged!"

"You can just *un*-arrange it." John looked him squarely in the eye. "And from now on, if anything needs setting up that involves a member of our group, *we'll* set it up."

He turned on his heel and walked away.

Clifford started to follow, then stopped. Shaking his finger at the official and his cohort, he spoke in a dangerous tone. "I don't know what you're up to. But *don't* involve my wife again—either of you—in any of your schemes!"

His face red with anger, he took Marilyn by the hand and strode to the car.

* * *

On Tuesday morning, the day before the bus trip, the telephone shrilled just as Marilyn headed to her orange recliner with her second cup of coffee. She had been up early, going over everything again and again. "Oh Lord," she had prayed, "don't let me overlook anything important."

"Mrs. Gracey?" queried the voice on the other end of the line. Since the episode on Friday, Marilyn froze at the sound of an unknown voice. "I'm Paul Forsyth from the *Welland Tribune*. We met at the press conference. Could I ask you a few questions?"

Marilyn hesitated. "Depends on the questions, I suppose."

"How do you feel about four of the municipalities—St. Catharines, Grimsby, Port Colborne, and Welland—withholding their support from your trip to Queen's Park?"

"Actually, it's only three. Welland has recently given their support." Marilyn swallowed nervously. Naïve and candid by nature, she found it difficult to guard every word and gauge its impact before she spoke. "But to answer your question, in no issue are you going to get 100-percent support. But we feel that we've received an excellent response."

"Has your group tried to persuade the nine municipalities who've given their support to send representatives tomorrow?"

"We really don't want to put people on the spot," Marilyn told him. "Elected officials can't be pinned down to spending a lot of hours on a bus trip. We understood, when we approached them, that they had obligations that would take precedence." She could tell he was writing furiously. As a former public schoolteacher, she reminded herself to wait a moment to let him catch up. "Even if not one of them showed up," she told him with sincerity, "the fact that they've given us official support means we're ahead of the game."

"What will you be discussing with the Premier and Environ?—"

"I'm sorry," Marilyn interrupted, "I can't get into that today." Then she gave a soft, musical chuckle. "You asked me that same question at the press conference."

Paul Forsyth laughed in return. "So I did. Well, you can't blame me for trying." After several more innocuous questions, he hung up.

Marilyn picked up her now-cold cup of coffee and dumped it down the sink. After making a fresh cup, she headed for the recliner. As she settled herself, she tucked one foot under her and spread the pages of her introductions on her lap. Picking up the first page, hand-printed in large, easy-to-read letters, and carefully composing herself as she would tomorrow, she surveyed the living room and spoke with a clear, carrying voice.

"Ladies and gentlemen, may I have your attention please…"

– 18 –

The next morning, as Clifford and Marilyn stood in the parking lot at the West Lincoln Arena, the sun poured down like molten gold out of an almost cloudless blue sky. Three buses stood ready to take the crowd to Queen's Park. Marilyn felt as if she walked in a trance. Although the group had planned and worked and stayed up half the night every night for the past two weeks, the trip felt unreal.

"Mrs. Gracey," a voice called out. "Can you stand over here, next to the town crier?" Judy McEwen, editor of the *West Lincoln Review*, stood ready to take down quotes while photographers snapped pictures for the papers.

As Marilyn edged through the crowd, she caught her first glimpse of Mr. Green. Not overly tall, and somewhat on the portly side, he had a full moustache and bushy white mutton chops, streaked with grey. In his leaf-green, cut-away coat, with its wide, white lapels and brass buttons, he looked as though he had just stepped out of another century. A scarlet shoulder-sash, encrusted with medals and pins, spilled across his gold brocade vest and trailed down onto the left side of his white breeches. He wore a black tricorn, piped in white, and carried a large brass handbell, a short knob-headed cane, and a parchment-like scroll.

As he straightened the lace at his throat and wrists, he answered questions in a hearty voice, rich with the cadences of Banbury in Oxfordshire, England. "My uniform's patterned after those worn around 1788 by an ensign, which was a cadet officer, in the Newark Fencibles." Marilyn, intrigued all of her life by accents and speech patterns, loved the way he twisted his tongue around *Fencibles*, so that it came out "Fen-si-bulls."

"Why the Newark Fencibles?" a reporter asked.

"Because Newark is the old name for Niagara-on-the-Lake, and because, after considerable consultation with West Lincoln Council, we determined that the Township's colours were white, gold, green, and red. Since the colonial regiments wore green, it seemed a suitable choice."

"Sorry to interrupt," John Dykstra announced in a carrying voice, "but it's time we got started." Marilyn could feel the excitement of the crowd build as people began to board. The Concerned Citizens spread themselves throughout the three buses, partly to keep tabs on things, and partly to give the participants a few instructions, once everyone had settled down.

The convoy of buses left the outskirts of Smithville and took the Grimsby Road toward the Queen Elizabeth Highway. Gradually, the laughter and the exuberant chatter died down to a conversational hum. Marilyn rose from her seat and stood in the aisle. "Could I have everyone's attention, please?"

She opened her folder and read off the agenda—the expected time of arrival, when the speeches would take place, and when the delegation would meet with the Premier and the Minister of the Environment. In particular, she stressed the expected departure time for the buses. "*Please* be there! We have no means of knowing if someone is missing."

She picked up a pile of photocopied pages. "We are very anxious that no one accidentally undoes any good we might do. The Concerned Citizens have prepared a summary of the points presented to the various municipalities and groups across the peninsula." She held the pages aloft for all to see. "Their support was given based upon this position. By your signatures and your presence here today, you have indicated your endorsement of them."

She lowered the pages and began to read in her clearest voice, "'Come now, and let us reason together'...Isaiah 1:18. It is in this spirit of reason and dignity..."

She finished reading the handout and made her way down the swaying aisle as she passed out the pages. Then she returned to her seat and gazed out of the window. We couldn't have asked for a better day, she thought. It would be hot, yet there was a light breeze.

A rare kind of peace prevailed within Marilyn, despite the dazed disbelief that she would actually speak from the steps of the provincial parliament. Partly, it stemmed from the knowledge that the Concerned Citizens had done everything within their power to make the day a success, and partly it was the result of much prayer and the conviction that she was doing what she had been chosen to do.

The buses passed through Grimsby and turned onto the Q.E. Minutes later, as they crossed the Burlington Skyway, Marilyn opened her lunch bag and took out an egg salad sandwich. Although she had no appetite, she forced herself to take a few bites, knowing that if she didn't eat, she would feel faint long before the day was over.

The traffic on the Queen Elizabeth Highway grew heavier as the buses neared Toronto. The Lake Ontario waterfront came into view, punctuated by new apartment complexes and restored structures. Soon after, the bus driver announced, "There it is."

Marilyn craned her neck to see the huge, circular flower bed with its formal arrangement spelling Ontario. The buses circled the flower bed and took the sweeping curve that brought the residents almost to the steps of the legislature.

As the bus pulled away and the next pulled up, she stood gazing for the first time in her life at the darkened sandstone of the ornately massive structure. Huge cement pillars, arranged in blocks of six, three across and three behind, supported the inner and outer spans of three cavernous arched porticos. A variety of flags barely stirred in the slight breeze. To Marilyn's surprise, a portable barricade made of iron railing blocked entry to the stairs. Behind the barricade, guards suddenly stood at the alert as the protesters streamed onto the broad walkway.

Marie Austin and Mary Skrypnyk unfurled the Concerned Citizens' banner, and Marilyn whipped out her camera. Around her, flashbulbs went off as news photographers and protesters alike captured the sign on film.

Marilyn followed the other speakers through a small gate and climbed the wide expanse of stone steps that provided entrance into the legislature through the three porticos. The steps had been terraced, as one might terrace a garden, with every third or fourth step several times the width of the others. Three-quarters of the way up, where the stairs broadened into a landing eight or ten feet wide, Phil Andrewes, West Lincoln's Member of Provincial Parliament, instructed a workman to set up the lectern and microphone.

Marilyn and the other speakers assembled on the steps. Inside the cavernous porticos at the top of the stairs stood a bank of varnished doors. From her vantage point on the steps, Marilyn looked out at those gathered outside the barricade. Ray Konkle, the mayor of Lincoln, Eric Bergenstein, the mayor of Pelham, and Alderman John Schilstra of West Lincoln leaned side by side on the iron railing. A short distance away, Ria Laskarin, a member of the Coalition, stood talking with Pelham alderman Diane Hubbard.

Marilyn said to John Jackson, "I didn't expect to see barricades."

John gave an incredulous laugh. "I don't know what they're expecting."

Just then, a fourth bus, partly filled with members of the auto workers' union, stopped in front. Two men carrying a large cardboard carton got off and looked around for a prominent spot to place the box lunches they had brought for the protesters. Marilyn's attention shifted to Murray Creed, OWMC's Information Officer, as he set up his tripod to take pictures. Then he sauntered over to the large carton the auto workers' union had provided and helped himself to a box lunch.

After a brief delay, Marilyn stepped to the microphone. Because it had been placed behind the lectern, she had to lean forward at an awkward angle in order to be heard. "Ladies and gentlemen, may I have your attention please." In that instant, Marilyn realized that the incredibly impossible was happening. She, a nobody, and those with her, nobodies to the rest of the world, stood on the steps of the Ontario legislature and dared to make their concerns known amid blinding flashbulbs, hand-held microphones, and whirring television cameras.

She thanked the various municipalities and groups for supporting them. "All of this adds up to one thing," she said, her voice strong and sure. "We have the support of you, the people."

"Now we come to a very special event. The office of town crier is a part of Canada's earliest history, an institution brought to the colonies of the new world by settlers from England, France, Spain, and other European countries. It was the crier's solemn charge to make known to the people matters of importance. In accordance with that time-honoured tradition, we have with us today the official Town Crier for West Lincoln, Mr. Jim Green."

As Mr. Green positioned himself on the steps to shout his proclamation, the media pressed forward. Surrounded by television cameras and popping flashbulbs, he unrolled his scroll and threw back his head. In his booming voice, he cried, "The Honourable David Peterson, Premier of Ontario; The Honourable James Bradley, Minister of the Environment; Members of the Ontario Legislature; Ladies and Gentlemen. *Whereas* the Ontario Waste Management Corporation has…"

Marilyn allowed Mr. Green's award-winning voice to fade slightly into the background of her thoughts. Seven years ago, within the walls of this very building, by an order-in-council, the Progressive Conservative Party, then holding the reins of power, had created the Crown corporation known as the Ontario Waste Management Corporation. At the time, the Liberal Party, presently in power, and the ever-hopeful New Democratic Party had both endorsed its creation. How fitting, she decided, that these steps should be the site of the people's discontent with that decision.

As Mr. Green reached the end of his proclamation, he declared with great feeling, "God bless Canada, God save Ontario—and God *help* West Lincoln!" With a gallant flourish, he swept his black tricorn from his head, and bowed amid cheers and applause.

Marilyn glanced over the array of speakers standing quietly on the broad stone steps. The Concerned Citizens, seeking to remain as non-partisan as possible, had invited speakers from a wide spectrum of interests, including representatives from all three political parties.

John Dykstra, with his fair, curly hair closely cropped, looked smart in his maroon-coloured sports coat as he walked to the microphone. Once again, he read the presentation he had first read to Regional Council and then to eleven out of the twelve municipalities of Niagara.

From the corner of her eye, Marilyn watched Mayor Colyn pat the pockets of his suit jacket in frowning concentration. Then, with an almost imperceptible nod, his expression relaxed and he reached into his inside breast pocket. Marilyn stepped to the lectern to introduce Phil Gillies, the Environment Critic for the Progressive Conservative Party, and then stepped back. As Mr. Gillies began to address the crowd, Marilyn found herself distracted by the mayor. He was worming his way through the huddle of speakers, handing out lapel pins bearing the crest of West Lincoln.

Phil Gillies quickly won rousing applause from the gathering. "We're building, or contemplating building, a facility in West Lincoln almost indecently attractive to the U.S. border. I want to say one thing to Dr. Chant and Mr. Bradley: If you're talking about U.S. waste for West Lincoln, thank you very much, but we're not interested."

Then he responded to the claim by Dr. Chant that all enjoy the benefits of industry and, therefore, all must share in the responsibility of dealing with industrial wastes: "...Of course, we benefit from all the products that are manufactured in our province," Gillies countered, "But surely the onus for cleaning up the effluent from that production should rest at the source..."

Bob Bradley, who followed Phil Gillies, made an impassioned speech. "We are being asked, a few thousand people, to assume the pollution burden for the entire province...In doing so, you are putting a horrendous burden on those few...I've always wanted my own few acres in the country. Now I have a property that is worth nothing. I couldn't sell it or give it away. Is this the heritage I want to pass on to my boy?" His voice took on a grave note of sadness. "I hope not."

The New Democratic Party had also sent its environment critic. Ruth Grier stood almost a head shorter than Marilyn. She had a thick cap of white hair, lively blue eyes, and a pleasing Irish burr. "I question OWMC's research into the oft-quoted alternatives," she declared. "OWMC contributed only one percent to (reduction, recovery, and reuse) in their 1984-1985 budget. I don't call that commitment!"

She waited for the cheers and the applause to die down. "I'm here because we in the New Democratic Party share your frustrations..." She adjusted the microphone a trifle. "The government has...given no priority to reducing the amount of that waste, to encouraging industry to reuse that waste, or showing them how they can recycle it. Until they do, we've all got a problem." As the spectators shouted and cheered their approval, a barely perceptible breeze drifted under the portico, giving a moment of relief from the heat.

After Ruth Grier came Al Godin, president of the St. Catharines Labour Council. "The Niagara area is responsible for feeding nearly eleven million people," he stated. "Emissions from the facility would provide the opportunity to contaminate that food supply...Sick people can live in a healthy environment—but healthy people can't live in a sick environment!"

Mayor Allard Colyn told the onlookers, "The OWMC is wearing blinkers. Alternatives exist in many fields that simply lack full investigation and implementation at this time." He looked out over the crowd. "OWMC's...mandate...should be reviewed... To build a facility such as the one proposed is merely a new source that will contribute to the degradation of the environment."

As the mayor gathered up the pages of his speech, Marilyn looked anxiously for the Liberal Member of Provincial Parliament.

John Dykstra leaned toward her. "They're not sending anyone."

"What should I do?"

John gave her one of his infectious grins. "Just say they're too chicken to send someone."

Marilyn stifled a laugh as she leaned over the microphone. "The Liberal Party has declined to send a speaker."

Marilyn waited to introduce Lynda Bradley until the displeasure of the crowd had quieted. Lynda touched her hand to her dark, elegantly-coiffed hair, then stepped to the microphone. Her southern accent vibrant with feeling, she read a letter to the editor, written by Marija Balint some months earlier. Lynda's white print summer dress fluttered daintily in the vagrant breeze that drifted between the huge cement pillars of the portico.

Marilyn, growing increasingly warm, lifted the neck of her blouse away from her chest.

The next speaker, Phil Andrewes, the Progressive Conservative Member of Parliament for Lincoln riding, had helped to organize the rally at the Toronto end. His voice rang with confidence as he praised the delegation. "You've focused on the issue...You have come here to express your views, and you are doing it in an orderly, democratic way."

He then criticized the direction taken by the OWMC: "Part of their mandate was to look at alternatives, not proceed with a mindset." He elevated his voice to a rallying pitch. "I think by coming here today, you demonstrate clearly that the people of West Lincoln are not going to sit on their hands...They're going to be actively out in opposition to this proposal."

Marilyn introduced John Jackson by giving a lengthy résumé of his achievements. Red-faced with embarrassment, he shoved his hands into the pockets of his lightweight trousers as he assured the crowd, "We're making a difference—and...it's important for us to see that. It helps us to keep on fighting... Today, we listened to people from both opposition parties support our position. We developed it, and now others are...saying, yes, it does make sense... But," he warned, "that doesn't mean the OWMC has been swayed in its view that the facility is needed...Therefore, we must focus our energies on convincing other people in Ontario that we have developed a proper position to proceed with." His voice rose in determination and hope. "We *can* change the way in which hazardous wastes are dealt with in the province of Ontario."

John received cheers and whistles from the crowd with his usual shy, self-deprecating smile. Marilyn hastily patted the beads of moisture from her forehead and stepped to the microphone. "Last, but by no means least, we have with us a well-beloved lady from Vineland, Mrs. Margery Coffman. Honoured jointly with her husband, David, as Citizens of the Year for the Town of Lincoln, she is noted for her admonition to the women at any meeting to cut down on the toxic waste going down their kitchen drains."

Marilyn turned to greet the sweet-faced, eighty-one-year-old retired schoolteacher. Wisps of white hair had escaped from the coil at the nape of her neck and trailed over the collar of her navy suit coat.

Mrs. Coffman placed the pages of a lengthy speech on the lectern and tipped the microphone downwards. "I came to Canada seventy-four years ago this month. My first memory is of strawberries. I was seven years old. Mother went picking, and I went along…Next came cherries…then peaches, plums, apples, grapes—I'd never seen such fruit!…We had come to a veritable garden," she declared, "a tiny stretch of land south of Lake Ontario,…marvellously suited to the cultivation of choice tender fruits." She adjusted her rimless glasses as her smile of remembered delight faded. "In those days, we had…never heard of toxic waste…Unfortunately, today, we are very much aware of our endangered environment. I'd like to read to you a poem written by Gracia Janes from PALS, the Preservation of Agricultural Lands Society:

Let us walk with a light foot upon the land.
Caring that it be there forever…

Marilyn listened to the clear, gentle voice through a sudden haze of overwhelming fatigue. The minutes turned into ten, and then fifteen. The constant standing made the pain in Marilyn's back so excruciating, she could barely tolerate it. Under the sun's steady glare, the faint breeze all but died, and the heat became intense.

As Mrs. Coffman's recollections came to a close, she admonished, "There are reasonable people on both sides of this issue. Surely we can get together to find a reasonable solution." She concluded with a prayer that began with a poem she had written for the occasion.

Great, wide, beautiful, wonderful world,
With the wonderful water around you curled…

She ended her prayer with a plea for help. "This dear Niagara, its orchards, gardens, lakes…Father! Help us to keep it clean and wonderful."

OWMC—The "Wasted" Years: The Early Days

Mrs. Coffman stepped away from the lectern. Her heart-rending speech drew the warmest and longest response of the afternoon from the crowd.

Marilyn raised the microphone and leaned toward it. "For whom do we speak today?" she asked. "We speak for the elderly neighbour down the road who is concerned for the future of her grandchildren and for the family farm...We speak for the Greek neighbour, the Polish neighbour, limited by language. We speak for those who have fought before us and, wearied, can fight no more. We speak for those who have lost heart...

"We are told that our anxieties, our sleepless nights, are self-inflicted and needless—the result of hysteria brought on by being uninformed." Suddenly Marilyn could not keep anger and frustration from colouring her voice.

"I tell you this: my husband and I were better off when we were less well-informed. The more knowledgeable we become, the more we realize that our worst fears of a year ago were nothing, compared to the concerns we now have.

"Thank you all for coming. You are free to go into the visitor's gallery of the legislature."

Unexpectedly, a young man stepped forward. To Marilyn's relief, he gave directions to the public restrooms and to the cafeteria.

Marilyn hurried down the stairs and onto the broad walkway where Clifford waited with a cold drink for her. After a quick stop at the restroom, she hastened up the stone steps toward the legislature. Half-running, she rushed through the heavy doors and into the spacious red-carpeted foyer at the foot of the Grand Staircase. Suddenly, a security guard let out a shout.

"Hey!" he bellowed to a fellow officer. "Get that leprechaun out of here!" Marilyn froze in mid-step, her heart thumping wildly. The guard pointed past her to Mr. Green, seated in the cafeteria with a cold drink. "Nobody wears an unauthorized uniform inside the parliament buildings!"

Marilyn didn't hear what transpired next, but Mr. Green left almost immediately, clutching his drink, his bell, his scroll, and his cane. Marilyn pressed her hand to the front of her blouse, her heart still thumping as she started up the Grand Staircase. John Jackson came up beside her on the stairs. "I thought he was yelling at me for running," Marilyn confided, sounding out of breath. "What was that all about, anyway?"

John shrugged, grinning in puzzled amusement. "Some sort of regulation. I guess they don't want Mr. Green starting a revolution."

With a smothered laugh, Marilyn followed John into the visitor's gallery and sat in the front row with Clifford, John Dykstra, and Mary. As fatigue took over, she felt as though she were drifting in a fog. Gradually she became aware of persistent whispers, which grew louder as a message passed from mouth to mouth. "Premier Peterson won't be at the meeting!" she was told.

"Why not?" she gasped. She felt as though the bottom had fallen out of everything. The weeks of planning, the struggling—all for nothing.

"It's got something to do with the Meech Lake Accord," John Dykstra told her. For once, no hint of a smile teased at the contours of his face. "They say Peterson's sleeping after being up all night at a federal-provincial constitutional session."

Marilyn wanted to scream in frustration. What about all the nights I've been up? What about John, Mary, Paul, Marie, and all the others? Tears sprang to her eyes and she forced them back. She stared at her hands, clenched tightly in her lap. She had to be reasonable, she told herself. She had to take it in stride. She could feel the eyes of the other residents watching her.

"I suppose," she said, with a weary sigh, "that in the bigger picture, the constitutional debates rate higher than a toxic waste dump." But beneath the outward appearance of resignation, she could not quell the frustration or the disappointment.

Minutes later, John Jackson signalled to the delegation. It was time to meet with Environment Minister, the Honourable James Bradley.

Marilyn followed their guide into the wide hallway and along the high-ceilinged gallery. She paused to peer over the balustrade onto the immense red-carpeted hallway below, awed by the magnificence of the ornate varnished pillars and handrails. The delegation converged with the Minister and his aides as they strode toward a door at the far end of the gallery.

The Minister stood aside as Marilyn, Mrs. Coffman, and the ladies from the Concerned Citizens entered the long, narrow room. Marilyn, John, and Paul sat at the table across from Mayor Colyn and Lincoln's Mayor Konkle. The rest of the Concerned Citizens sat in hastily provided chairs along the wall.

Minister Bradley, at the head of the table, greeted the delegation in a brisk manner. He apologized for Premier Peterson's absence, explaining he had been in Ottawa, along with the other provincial premiers, to add his signature to the Meech Lake Accord. Then he settled down to listen. Paul Balint started by hoisting the blue recycle box containing the 25,000 signatures onto the table. Then, one last time, John Dykstra read his presentation. At the end, several aides discreetly moved the box to a table at the end of the room farthest from the door.

Marilyn had no presentation to make. Instead, she studied the faces around the table as the Minister's aides took notes and looked increasingly bored. Mr. Bradley appeared to listen, especially to white-haired Margery Coffman; yet, as the meeting drew to a close, he did not respond directly to any of the concerns the residents raised. Instead, he talked about funding and assured them of a fair hearing. Marilyn, who had never for a moment thought the hearing would be anything but fair, felt as though he had sidetracked the meeting with meaningless promises and had slid right past the more important issues.

He stood and shook hands with the two mayors, John Jackson, and several others. Marilyn realized he had said nothing about bringing the OWMC nightmare to an end. As she rose from her chair and started toward the door, she noticed the minister's aides lift the recycle box filled with the 25,000 signatures and slide it underneath the table.

Mr. Bradley stood with his hand on the doorknob, his aides and the residents collected around him, ready to leave. Suddenly, he flung open the door. Flashbulbs exploded in their faces as voices shouted questions. Marilyn was buffeted by the crush of bodies pressing into the doorway and those behind her, trying to get out.

Trapped in the hallway by the reporters and cameramen, she heard snatches of Bradley's affirmations concerning a fair hearing. He extolled his government's efforts to compel industries to clean up wastes at the source.

"We're also investigating the possibility of using tax incentives to encourage more waste reduction and recycling," he told reporters. Then he issued a major statement about ensuring generous intervener funding for opponents. Marilyn felt a streak of anger run through her. He's using the media for his own purposes, she thought. He's the one who raised the issue of money, not us.

"Are you going to terminate the OWMC?" a reporter shouted over the others, his microphone shoved in the Minister's face.

"It would be dangerous," Mr. Bradley stated, "for government to politically intervene in hearings scheduled to take place before an independent panel of the Environmental Assessment Board." His gaze swept over the trapped residents. "But I've encouraged these people to participate in the process...They are taking a responsible approach. They are not simply putting forward arguments that say, 'not in my backyard'. They are saying, 'look at alternatives to what the OWMC is proposing.'"

Suddenly, Marilyn was blinded by a floodlight. "Turn this way," a voice ordered. Confused, she found a microphone shoved close to her face. "How do you feel about the meeting, about Premier Peterson not showing up?"

Flustered, she felt tongue-tied, unable to think straight.

"Aren't you disappointed? Angry?" the voice demanded to know.

"I—I suppose, in the Premier's eyes, Meech Lake and the constitution is—is more important than meeting with us." She hesitated, weighing the consequences of a frank answer. "Yes," she admitted, "it was disappointing. But under the circumstances, I guess—it was all we could hope for."

As Environment Minister Bradley shouldered his way through the reporters, they followed him like a swarm of bees. Marilyn felt light-headed in the sudden absence of commotion. Clifford sought her out and took her arm as he guided her around the railing of the gallery and down the long, wide flight of carpeted stairs.

Outside, reporters cornered John Dykstra. "Mr. Dykstra, can you tell us what went on inside the meeting room with Mr. Bradley?"

"Well, for one thing, we told him the OWMC has been railroading us into believing the hearings are just a formality. We told him we're feeling intimidated by this. He indicated it would be an independent process." He squinted up at the hot sun, loosened his tie slightly, and patted his face with his handkerchief. "We told him that the OWMC wants to put in the facility and then look at the alternatives." John's face broke into one of his infectious grins of boyish delight. "We say, look at the alternatives—and *no* facility!"

"Mrs. Gracey," shouted a reporter over the other voices, "how do you think the meeting went?"

Marilyn glanced nervously at John. "I—I think we gained a tremendous amount of credibility. For one thing, both Mr. Andrewes and Mr. Bradley praised the manner in which the rally was handled. Which means," she continued, "they will be approachable another time."

Just then, the buses pulled up. The weary protesters boarded swiftly and began the long trip home. Marilyn sank into one of the plush seats, seeking relief from the misery in her back. But as they left the outskirts of Toronto, it became obvious that those who had attended the rally needed to know how things had gone at the meeting, especially since Premier Peterson had failed to show up.

Marilyn stood in the aisle. Relief that the trip was over and the knowledge that everything had proceeded smoothly and without a hitch made her voice ring with optimism as she described the meeting. "I think we accomplished what we set out to do!" she told the busload of residents. Cheers and laughter brought a light, festive mood to the atmosphere.

Marilyn returned to her seat and gazed through a haze of exhaustion at the scenery sliding past the window of the bus. We did it, she thought. We actually did it. She tipped her head back against the headrest. Everyone worked so hard, she thought. As she recalled the scene from the steps of parliament, she felt a smile touch her lips. All during the speeches, the recycle box with its precious cargo of signatures had resided next to Don Austin, who had taken over supporting one end of the banner. And no one could have guarded it more closely than Don Austin did.

Her smile vanished. The recycle box!

It had been shoved under a table, out of sight, in a room barred to the media. No press pictures had been taken of John and Paul presenting the signatures. No one from the media had even referred to it.

A slow, burning indignation gripped Marilyn. The residents, disadvantaged by their inexperience, had been outfoxed. Without the box packed with thousands of signatures to draw the media's attention, the Minister of the Environment had highlighted his own agenda. Not once had he mentioned the signatures or the issues raised by the residents. Our agenda, she thought, got shoved under the table—literally.

* * *

That night, Marilyn sat in the easy chair closest to the television, a cup of coffee at her elbow. She flipped from station to station, trying to catch as much coverage of their protest as possible. They had received wider exposure than they might otherwise have done, thanks to Mr. Green and his colourful appearance as town crier.

She paused as CHCH-TV showed Environment Minister James Bradley making his declaration of generous funding for the opponents. Then came a split-second clip of Marilyn, saying, "It was all we could hope for."

Marilyn sat bolt upright in the chair. "That's totally out of context!" she exploded. "I wasn't talking about money—I was talking about not getting to meet with the Premier." She jumped to her feet and stomped back and forth. "How dare they do that! They make it sound as if our only purpose in going to Toronto was to get money!"

Later, once she had calmed down, she remarked to Clifford, "At least it's over. Now, maybe we can take it easy for a bit, and catch up on all the work we've let slide."

* * *

One week later to the day, the funding hearings began.

– 19 –

The two-day funding hearing, scheduled to be held in the banquet room of the Old Farm Inn, had been switched at the last minute to the meeting room in the Legion Hall. Marilyn sat sweltering in the heat, trying to take notes. Trying, she thought, because half of the time, she hadn't a clue as to what everyone was talking about. Terms, names, and references to reports and guidelines passed over the lawyers' lips with ease. With similar ease, they passed right over Marilyn's head. Not since her teens had she felt so ignorant and out of place.

She flipped through the fourteen-page handout, complete with appendices A and B, which comprised the most recent order-in-council regarding the OWMC's facility. The order, itself, appeared to be one long, run-on sentence, punctuated by the legal hallmarks of 'whereas' and 'therefore.' Marilyn looked around the room at the other residents. I don't know about you guys, she thought, but I'm way beyond my depth.

Because she and Clifford lacked experience in the environmental assessment process, they had attached little significance to the hearing on intervener funding. Funding was theirs by right. What was there to hear? So, to them, it was just some obscure formality. Not like "The Hearings," marked off in their minds with capital letters and quotation marks. *Those* hearings would determine the course of the rest of their lives. Uninitiated and uninformed, Marilyn listened, and struggled to understand the incomprehensible.

Mary Munro, a member of the Environmental Assessment Board and one-time mayor of the city of Burlington, had been appointed to chair the meetings. Petite and dark-haired, she had a pleasant, no-nonsense manner. She sat at the head of a long rectangular table. Her advisor, a lawyer named Herman Turkstra, sat on her right.

Marilyn had not expected a whole flock of lawyers, coordinators, and Regional bigwigs to be present. They occupied both sides of the table and, as she scanned the faces, she recognized only Dennis Wood, legal counsel for the Township of West Lincoln, and John Jackson. On the end, facing Mary Munro and with their backs to the residents, sat the OWMC personnel.

As each of the lawyers stood to formally introduce himself to Mary Munro, Marilyn jotted down the name. Beside Dennis Wood sat Roger Cotton. Stan Stein and Dan Kirby represented the Region. John Martin appeared for the Ministry of the Environment. John Jackson sat beside Joe Castrilli, the lawyer for the Coalition.

Ms. Munro called upon Ian Blue, counsel for the OWMC, to make the opening presentation. Tall and trim, younger than his white hair would suggest, he stood to report on the status of the OWMC project and to make submissions. Not until he introduced his two associates did Marilyn recognize Michael Scott, OWMC's Director of Communications, and Richard Szudy, OWMC's Coordinator of Environmental Projects.

For Marilyn, the first startling revelation came when she learned that the OWMC would dole out the funding to its opponents. The second was the realization that the Crown corporation, in the person of its lawyer, Ian Blue, would do everything it could to limit that funding.

But the third and worst shock came with the discovery that funding would not automatically be granted to the citizens' groups, as theirs by right. Not only were they required to apply, but they needed also to be found "eligible" to apply. And to be eligible, the residents had to show that they had made a serious effort to raise the thousands and thousands of dollars needed to pay the fees of the lawyers and consultants.

Feeling as though the floor had fallen out from under her, Marilyn looked around at the other residents. They all had a bewildered, defeated look about them. Suddenly, she wanted to jump up and shriek in outrage at the injustice. She wanted to pound her fist on the table and demand that the enormity of their struggle be recognized.

She wanted to curl up in her old orange recliner and let her tears fall into her lap.

Any vestige of Marilyn's hard-held belief that the government would in the end be both fair and benevolent vanished in those minutes. She stared, hard-eyed, at Ian Blue. The haughty tilt to his head, his challenging manner of speech, the way he pivoted on his heel to face the residents each time he made some inflammatory statement—all bespoke of a man convinced he held the upper hand. As he wound up his lengthy presentation, it became cruelly clear to Marilyn that while the OWMC would grudgingly and stintingly fund both the Township and the Region, it would fight to deny one penny of funding to the Coalition.

The hearing broke for lunch. John Jackson hurriedly passed the word among Coalition members that Joe Castrilli, the Coalition's lawyer, wanted to meet over lunch with as many of them as possible.

In the restaurant, John and Mary, Lynda, Clifford, and Marilyn pushed several tables together as Edith Hallas pulled up extra chairs. Joe Castrilli, dark-haired and slight of build, obviously felt ill at ease. He sat on the edge of his chair, like a bird poised for flight, and came straight to the point. "My legal firm is not willing to absorb the costs of my working for legal aid rate. I would have to drop most of my other cases and work exclusively on this one. That's okay if we're talking about a few weeks, or even a few months. But this hearing is expected to take a year, maybe two." He glanced around the table. "Is there any way you people could raise some additional money?"

John Dykstra gave a frustrated half-laugh. "Yeah, well, I guess Mary and I could take out a second mortgage on the farm."

Marilyn recognized that John had been joking, to emphasize how impossible a request it was. Yet, she had been so incensed by the way everything from start to finish had been stacked against the residents, despite the "fair" hearing promised by Jim Bradley on the day of the bus trip, and the declarations of "generous" funding, that she began to think in literal terms.

When she spoke, her voice shook with the enormity of what she was about to propose. "If we have to, we'll take a second mortgage on our farm, too." Her eyes met Clifford's startled gaze. "If the OWMC comes in," she told him defiantly, "our farm won't be worth a plug nickel anyway."

Edith nodded her assent. "Some of us have borrowed money from the bank in the past for the Coalition. We can do it again."

Joe pushed back his chair. "That's settled, then." He rose from the table and left the restaurant. Suddenly, Marilyn had no appetite. What had she done? What had possessed her to put everything on the line? She bunched her hands in her lap to hide their trembling. Part of her blamed herself for risking the loss of everything they had worked for, but another part argued that it had already been lost the day the OWMC picked West Lincoln as its target area.

When the afternoon session convened, Dennis Wood, the lawyer for the Township, argued that West Lincoln was a small municipality and that the OWMC should fund the entire cost for consultants, not just sixty percent. The Region argued that it, too, should be fully funded for its consultants. Nor did the Region believe that the legal aid rate was appropriate for its lawyers. "Legal fees based on the legal aid rate do not reflect the cost of long-term environmental legal help," Mr. Stein told Ms. Munro. "Most lawyers in the field charge twice the legal aid fee."

The Region also turned down the suggestion that it should share legal counsel with the Township. "The Region does not have exactly the same interests as the Township," asserted Mr. Stein. "Regional Council may decide to co-operate, but does not wish to be tied to West Lincoln."

Dennis Wood agreed. "For example, West Lincoln might find itself at odds with a hydrogeologist hired by the Region."

Joe Castrilli told Ms. Munro, "The Coalition represents 25,000 individuals." He, too, argued that the sixty-percent funding for consultants should be upped to one hundred percent. "Also, the Coalition asks that the funding be made available in the next few months. Waiting thirty to sixty days before the Environmental Assessment Hearings begin will be too late."

He glanced up from his notes to appeal directly to Ms. Munro. "There are 7,000 pages in the E.A. report. Interveners cannot critique the report without funding," he said, "and funding cannot be granted until the critiquing is done. Madam Chairman, this is your classic Catch-22."

Then he attacked the limit placed upon the total amount of legal aid tariff that the lawyers would be allowed to claim: "Preparation time has not been taken into account. It takes roughly one day of preparation for every day of hearing." He shifted his glasses up on his nose and leafed through his notes. "The legal aid rate is woefully inadequate. It establishes two classes of counsel—lawyers for the OWMC, who get paid at whatever the market will bear, and all others at the legal aid rate."

The residents broke into a defiant round of applause, despite John Jackson's admonition earlier to the Coalition members that there should be no public demonstrations. Ms. Munro's pleasant demeanour turned stern with warning.

Marilyn whispered to Clifford, "I don't care! It's not fair, and she might as well know we don't like it." She felt herself break into an impudent grin as she glanced around the room at the other residents. "So what's she going to do—put us all in jail?"

The next submissions came from Ray Konkle, the mayor of Lincoln, and Stella Cuban of Pollution Probe. Considering Pollution Probe's roots with Dr. Chant, Marilyn found it surprising that the organization had, as yet, taken no formal position on the proposal. However, Ms. Cuban did voice the organization's support for the interveners on a number of the issues raised that day. The sixty-percent ceiling for consultants' work was not workable, the legal aid tariff was inadequate, and duplication of some studies might be necessary. She checked her notes briefly, then said, "Probe does not believe, regardless of the outcome of the hearing, that the interveners should be required to repay funding."

Marilyn felt her mouth drop open in astonishment. "What?" she exclaimed in a loud stage whisper, her brows deeply furrowed with outrage. "What does she mean—repay?"

A rising rumble of displeasure swept the hearing room. Marilyn, for one, made no effort to keep her voice down. John Jackson looked anxious as he shook his head and made discreet calming gestures to the residents, urging them to silence. Marilyn slapped her pen down onto her clipboard and slumped low in her chair. Gradually, the muttering turned to sullen silence.

Mr. Martin, for the Ministry, stated he was there as an observer and kept his submissions to a minimum. "However," he noted, "there will be no money before the final environmental assessment reports are released."

Joe Castrilli, in obvious frustration, demanded to know whether he and his clients should drop their work on the environmental assessment and devote themselves to fundraising. "Delays in funding will greatly compound the problems facing the Coalition."

At the end of the day, Mr. Martin informed those present that the Ministry's review of OWMC's draft E.A. would take about four to five months.

Marilyn stood and stretched her cramped limbs, feeling as though she had been run through a wringer. She glanced at Clifford. He looked no better. "Now," she muttered to herself, "I can go home and make supper, and Clifford can do his chores." She watched John and Mary hurry toward the door. "And *they* can go home and milk cows."

*   *   *

The next morning, residents greeted John Jackson with a storm of questions. "Is Pollution Probe on our side or not?" Lynda demanded.

Edith Hallas gave one of her swift, fleeting smiles. "Probe backed us yesterday—but won't take an official stand."

"Like it or not," John admitted, "Donald Chant has always supported in theory the recommendations raised by Stella Cuban yesterday."

Marilyn disagreed. "OWMC's lawyer doesn't support those suggestions." She was interrupted when, promptly at ten a.m., the

OWMC—The "Wasted" Years: The Early Days

Hearings Officer commanded, "All rise." Ms. Munro and Mr. Turkstra took their seats at the table, and the second day of the funding hearing began. To Marilyn, it was little more than a rehash of the issues from the day before, with nothing settled.

*   *   *

Marie Austin telephoned to alert Marilyn to an article in the *Standard* on the funding issue. Just before she hung up, Marie asked, "What's that clicking noise on your phone line?"

"I don't know," Marilyn answered. "It's been doing that for months. Anyway, thanks for calling."

After Marie hung up, Marilyn left the vacuum in the middle of the living room floor and zipped up to the corner store for the newspaper. Doug Draper, the reporter on environmental issues for the *Standard*, had quoted extensively from an interview with Joe Castrilli regarding the funding hearings.

"There won't be a lawyer in Ontario willing to represent opponents of the OWMC," Castrilli stated, "unless the province provides adequate intervener funding... The legal aid tariff is completely out of whack with reality."

Castrilli explained to Draper that the tariff for lawyers representing clients before quasi-judicial government boards and commissions was only $389 a day, and only allowed for a maximum of five hours of preparation. In the case of the OWMC proceedings, lawyers would have to devote as much as ten hours a day, which at $389 a day translated into thirty-nine dollars an hour. "No lawyer in the province is going to get into a hearing of this magnitude at that rate," Castrilli told Draper. "It would be like asking to go bankrupt."

Castrilli then switched his focus from the legal aid rate for lawyers to the plight of the interveners and the costs associated with hiring their own experts and preparing their case. In February, he said, Provincial Environment Minister James Bradley had stated that opponents would be eligible for only 60 percent of those costs. The Coalition expected to spend about $250,000 on experts and at 60 percent, residents would face a burden of $100,000...

Marilyn stared at the number in blank dismay. We'll be bankrupt, she thought, totally bankrupt. She felt close to despair. "We can't start over again," she murmured. "Not from scratch." Her mind traveled back to the early days, when the roof leaked like a sieve, when she hauled all their water with a pail, when she tried on her hands and knees to keep the worn plank floors clean. All that struggle for nothing, she thought, as the ache in her chest swelled against her ribs. "All the hardships," she whispered, "all those wasted years…"

– 20 –

Having gained the public eye with their rally at Queen's Park, the Concerned Citizens decided to keep up the momentum. They immediately started work on a simple float to enter in West Lincoln's Canada Day Parade, less than three weeks away. Marilyn begged off, opting instead to type personalized thank-you letters to anyone and everyone who had aided the Citizens with their historic trip.

Meanwhile, the OWMC's proposal continued to spawn controversy. The fire chief from the neighbouring municipality of Lincoln expressed concern over chemical spills that might occur en route through urban Vineland or Beamsville. Both towns had grown up at the foot of the steep escarpment that skirted Lake Ontario. Trucks carrying toxic waste would be forced to negotiate a sharp incline right in the heart of each of these communities.

"What worries me most," Chief Rouse told the *Standard*, "is that government listings tell me thirty-nine of the wastes to be transported through these towns would require evacuation, and thirty-eight are violently or explosively reactive."

That same week, angered by the dinky little rectangle consistently pictured for landfill in OWMC diagrams, Marilyn sent letter number twenty-five to editors across the peninsula, stating that the people of Niagara had not been given a true picture of the magnitude of the landfill site proposed by the OWMC:

> ...The West Lincoln site, LF-9C is the smallest of the eight candidate sites. If 335 acres are enough, why was the Elcho site 640 acres? Or PI-1N at Milton 935 acres? Because

according to OWMC's consultants, "LF-9C is free to expand to the west."

Citing the report and the page number, she then explained, "Free means there are no engineering constraints—no roads, no rail lines, no major utility easements. The Kszan farm, of course, doesn't count."

About the same time, the *Shopper's Guide*, published by the *Dunnville Chronicle*, printed a report on Alberta's toxic waste facility at Swan Hills. Composed of a high-temperature incinerator, a deep well for treated liquid wastes, and landfill cells, it was expected to be in operation by October. Just the name Swan Hills sent a shiver of dread through Marilyn. Extolled at every opportunity by the OWMC as an example of government initiative rapidly approaching fruition, the facility cast a long, ominous shadow over the residents of West Lincoln.

*   *   *

On the last Wednesday in June, three days before the Canada Day Parade, Norman Nelson of the *Pelham Herald* published another of his inflammatory editorials. Marilyn took one look at the headline, "Chant is doing a great job," and blew her stack. But then, once she had read the editorial, she recognized the same satirical twist that her cousin Peter gave to things. Again the editor raised the issue of the OWMC's mandate, except this time, he provided details.

Dr. Chant, he said, like a good soldier, was simply doing his job. The 1981 act was still the OWMC's guiding light, and the objectives of the corporation remained the same: to research, develop, establish, operate and maintain facilities for the transmission, reception, collection, examination, storage, treatment and disposal of wastes, including sewage; and to perform such other duties as may be assigned to it under this or any other act.

Nelson went on to explain:

> The key word is facilities. I interpret that to mean that Dr. Chant is supposed to construct a toxic waste dump. And that is what Dr. Chant is doing...The OWMC was created

to address the results of [society's] pillaging of the environment...Then, the government effectively washed its hands of the toxic waste issue...

Next, Nelson quoted part of the speech John Jackson had given on June 3 from the steps of Queen's Park. "We in the Niagara Peninsula have been forced to...become experts on waste management. We've become aware of better ways. And it's exciting that we are getting the message across."

Nelson said bluntly:

> Unfortunately, the message can get no consideration...Dr. Chant's got enough on his hands designing his toxic waste monument, without the responsibility of implementing a toxic waste strategy for the province at the same time.

Nelson quoted John Dykstra's remark after the rally when he said, "The OWMC wants to put in the facility and *then* look at the alternatives."

Wrote Nelson, "But again, those are Chant's orders."

Taking another quote from the speeches made on the steps of Queen's Park, Nelson wrote: "NDP environment critic Ruth Grier told us, 'We're giving the government suggestions...and they're saying, "Don't tell us, tell the Environmental Assessment Hearing".'

"But," Nelson reminded the reader, "it's the responsibility of government to institute a toxic waste policy for the province—not the Environmental Assessment Review Board..."

*   *   *

On June 24, the same day as Nelson's editorial, the *Standard* revealed that the OWMC had organized a trip, early in September, for Niagara politicians to tour toxic waste plants in West Germany and Sweden. "The Toxic Tourist Trade," the *West Lincoln Review* later called it.

Excluded from the trip, John Jackson told Doug Draper of the *Standard* that the Coalition had not received an invitation. Dr. Chant stated that the tour would have to be limited to about ten people, including Corporation personnel.

Said John:

> The Coalition has always felt it was important to go to Europe, not to visit facilities like the one at Biebesheim, but to go where they have alternatives like waste reduction and recycling. We've never been convinced the trip should be part of any OWMC tour where they could control what we see and what we do.

Stan Pettit, mayor of Wainfleet and chairman of the Region's steering committee, assured the *Standard* that the Region would plan carefully to determine what sort of information they wanted to gather. "We want to ensure," he said, "that there is an element of independence and we aren't just being led around by the nose by officials from the OWMC and the Ontario Environment Ministry."

The hot, glaring days of late June continued into July. Except for one brief rain that left everything steaming, the landscape danced with the heat of the sun. Two weeks after Marilyn complained in the newspapers about the dinky little rectangle pictured for the proposed landfill, the OWMC publication *Update* arrived in the mail. It carried the same misleading diagram, depicting only a small corner of the site devoted to landfill. Incensed, Marilyn dashed off another letter to the editor.

> The latest issue of the OWMC Update is less than up-to-date. The...dinky little rectangle labelled "landfill" in no way reflects the actual size or design of the proposed landfill. We think it is high time that something a little closer to the truth be presented to the public.

Jim Micak, Project Manager of Environmental Projects for OWMC, responded in a letter to the editor, stating that OWMC was not "free to expand to the west" at any time it liked, as Marilyn had written in a previous letter. Such an expansion would have to gain approval from the Environmental Assessment Board.

Marilyn fired off letter number twenty-seven, calling Micak's response another example of OWMC's half-twist. "As we clearly stated at the time, it is a direct quote by OWMC's own engineering consultants and refers to the lack of physical constraints to expansion purely from an engineering point of view."

In the weeks following the excitement generated by the trip to Queen's Park, the Concerned Citizens attracted two new members—Zygmunt Sojka, known as Ziggy, and Mike Kicul, both from Fenwick. Mike, an old schoolmate of Marilyn's from Welland South Public School, was the artist responsible for the newspaper cartoons Marilyn had been saving.

Marilyn had never had a great deal of stamina. Now, little by little, she could feel the unending struggle draining her strength. When she wasn't rushing to the Concerned Citizens' weekly meetings, she was attending meetings with the Coalition, a meeting with the Ontario Federation of Agriculture, another with the Township's toxic waste committee. During the day, she drove the baler in the field while Clifford loaded the bales directly onto the wagon. Then, back at the barn, she unloaded the bales onto the conveyer, which carried them up to Clifford, who piled them in the mow.

When she wasn't working in the field or going to meetings, she picked and froze beans from the garden, carried water from the pond to the cucumber and tomato plants, canned sweet cherries for fruit, and sour cherries for pie. At night, after Clifford had gone to bed, she studied the latest OWMC reports and drafted letters.

And hanging sadly over each day was the problem of her mother's vacant house. Marilyn and her sister, Sharon, couldn't believe how difficult it was proving to sell the family home.

The float entered by the Concerned Citizens in the Canada Day Parade had been simple and hastily assembled; however, encouraged by the enthusiastic response it received from the people lining Griffin Street in Smithville, the group made plans to build an elaborate float to enter in the Labour Day Parade on September 7 in St. Catharines. Ziggy offered to help paint the miniature picket fence, recycled from the prize-winning float entered in the Niagara Grape and Wine Festival Parade by the West Lincoln Task Force Against Toxic Waste the previous autumn. Mike, a superior artist, drew up a diagram of the float and agreed to paint the signs.

To try to raise money, as decreed at the funding hearings, and to keep the issue of the OWMC ever before the public, the group decided to sell hats. Mike designed the logo. Steve Dinga put up $1250. Paul and Marija, John and Mary, and Clifford made the two-hour drive to check out the company. Satisfied, they placed an order for 500 hats.

* * *

In mid-July, John Dykstra alerted Marilyn to an article in the *Hamilton Spectator*. "The OWMC's put out a new report," he told her. "It talks about a two-kilometre nuisance zone for dust, noise, and visual impacts." He exaggerated the word nuisance. "Property taxes are expected to go up—"

"Up!" Marilyn exclaimed in disbelief. "I thought—"

"—and property values will come down," he interrupted, "just like we've said all along." Marilyn could hear the anger and frustration in his voice. "They figure the extra truck traffic could be as high as 50,000 vehicles a ye—" John broke off in mid-sentence. "What's that clicking noise on your line?"

"I don't know. Marie asked me about that the other day. It's been going on for months, possibly a year or more."

He gave one of the mirthless half-laughs that often prefaced some outrageous remark. "Maybe your telephone's bugged."

"Maybe," Marilyn joked in return. Yet, one corner of her mind half-believed it. "Talk to you later," she said as she hung up the phone. She grabbed her purse and rushed to the corner store for a *Spectator*. What possible brand of logic could the OWMC have employed to relegate the impacts of its proposed mega-facility to the nuisance level?

– 21 –

The following Wednesday, both the *Pelham Herald* and the *West Lincoln Review* carried the story of OWMC's proposed Nuisance Zone on the front page.

Of the fifty-six properties within the two-kilometre Nuisance Impact Zone, the *Review* explained, only fourteen would be eligible for buy-out. For others, including those who lived outside the Impact Zone, the OWMC proposed to implement a hardship policy.

"But," the reporter for the *Review* wrote, "the onus will be on the resident to 'demonstrate a plausible cause and effect relationship' between the OWMC and the problems experienced. And claims of effects must be comparably experienced by others."

Marilyn tossed the *Review* onto the couch in disgust. We'd never live long enough, she thought, or have money enough to prove a claim against the OWMC.

The *Pelham Herald* noted that the report, titled "Background Material for the Ontario Waste Management Corporation Phase 4B Public Consultation Site Assessment Meetings," had been released with little media attention. OWMC Information Officer Murray Creed admitted in an interview, "No public announcement was made... We still wanted to meet with all the residents directly in the Nuisance Zone."

Inside, on page four, however, the *Pelham Herald* carried a contradictory account written by David McCallum, West Lincoln's Environmental Project Coordinator. The OWMC had held a special meeting in Toronto on Thursday, July 9, 1987, to release the document.

McCallum revealed that the meeting had been restricted by invitation to the OWMC, their consultants, representatives of various government ministries and agencies, and representatives of the Niagara Region and the Township of West Lincoln. Approximately half of the close to seventy in attendance had been OWMC staff and consultants. Other interested parties such as the Ontario Toxic Waste Research Coalition, who intended to be at the hearings, had not been invited.

Instantly, a wave of impotent rage washed over Marilyn. Then with equal swiftness, she burst into laughter. It's so childish, she thought, so small-minded. It made her think of the OWMC's kitchen meetings, only on a larger scale: Don't invite the Graceys to the hog meetings or Danny Dobrucki to the chicken meetings. Don't invite the Coalition to go on the European tour. And definitely don't invite the Coalition to the launching of a major report. Don't invite anyone who poses a threat.

Although the realization made the insult cut less deeply, she felt the laughter leave her eyes as her chin lifted dangerously. If Dave McCallum hadn't kicked over the traces and informed the media, she thought with a fresh surge of angry resentment, who knows when we would have learned about this latest snub to the Coalition.

* * *

Later that same day, the real estate agent called to say that, despite the multitude of obstacles with regard to property lines and such, she had finally sold Marilyn's mother's house. Although the empty house had posed a worry, the sale hit Marilyn hard. Her parents had lived there from the day they were married, and it was the only home she had ever known while growing up.

* * *

Early in July, the OWMC had produced another document, its draft Environmental Assessment "Mockup", as they called it, for use by ministries and agencies. It, too, had been released without public fanfare. Several weeks after its release, John and Mary brought Marilyn the copy designated for the Concerned Citizens. Marilyn added the nearly-inch-and-a-half-thick tome to the pile of reports and documents already stacked upstairs on Kathy's bedroom floor.

Downstairs, open on the arm of the chesterfield next to her orange chair, Marilyn kept the inch-thick controversial report, "Background Material for the Ontario Waste Management Corporation Phase 4B Public Consultation Site Assessment Meetings," called "Background Material" for short. It contained the data on the OWMC's proposed Nuisance Zone. Because none of the reports ever had an index, Marilyn had begun listing topics and page numbers on the front cover. Slowly, a system evolved—blue dollar signs referred to finances; information circled in red meant it was more important than data highlighted in yellow or green; orange indicated something super-important. And anything marked in squiggles of one or more colours and encased with outward dashes like a flashing light shouted, "Hey! I'm what you're looking for!"

* * *

With election fever heating up in the province and a date soon to be declared, the *Standard* carried an interview with Ron Hansen, the New Democratic Party's candidate for Lincoln Riding. A founding member of Citizens For Modern Waste Management, he stated that he would campaign "on the belief that an Ontario Waste Management Corporation toxic waste disposal plant in West Lincoln would not be needed if waste was handled at the source." Hansen commented that "OWMC chairman Donald Chant is doing his job according to the government's current mandate, but that mandate is too restrictive and needs to be changed."

On August 7, when the Concerned Citizens met in St. Luke's basement, John and Paul piled the cartons of newly arrived hats on one of the tables. "I'll keep a record of the sales, if you like," Marilyn said, as she plopped a white peaked cap with its green logo on her head and tipped it at a jaunty angle. "How much are we going to charge?"

"The woman at the factory," Paul told the group, "says we could probably sell the hats for five or six dollars."

Troubled, Marilyn shook her head. "I think we should keep the price as low as possible. We want them on people's heads, not sitting around in boxes, collecting dust."

"You mean like the T-shirts the Task Force has left over?" Lynda remarked.

"Exactly!" exclaimed John.

Marilyn said, "I worked at their table last year at the Beamsville Fair. Kids *wanted* to buy them, but they couldn't afford them. You know, the Task Force could probably have sold two for the price of one and still made a little profit. Just think what an impact they could have had if people all over the fairgrounds had been wearing anti-OWMC T-shirts!"

"Plus," Paul added, "the sooner we sell the hats, the sooner Steve will get his twelve-hundred and fifty dollars back."

"Look," said Clifford, "if we clear at least a dollar apiece, that's five hundred dollars. I say we sell them for three-fifty."

With the price agreed upon, the residents plunged ahead to the next order of business—the booth at Smithville Fair in four weeks' time. Clifford offered to build a large display board, and Marilyn volunteered to make up a display using photos from the bus trip and blow-ups of newspaper clippings of remarks made from the steps of parliament. They would pin their banner to the wall, decorate the booth with produce from their vegetable gardens, and sell hats. Then they drew up a timetable. Marilyn tried to weasel out of manning the booth, claiming she would have more than done her share, to no avail.

"I've booked our booth for Friday and Saturday," John told them. "We have to pay for Sunday, even though we won't be there—"

"I don't see why we're closing for Sunday," Lynda interrupted heatedly. "The OWMC booth won't close for Sunday!"

A strained silence fell over the group. "Because," Mary answered in her quiet way, "we agreed at the last meeting not to man the booth on Sunday."

When Ziggy and Mike sided with Lynda and Bob, and Don Austin declared he didn't care one way or the other, Marilyn could feel her heart begin to pound. She hated conflict and confrontation, but she couldn't keep silent. "We open our meetings with prayer. We ask God's guidance in everything we do," she said. "How can we expect His blessing if we don't honour Him in all things?"

Though clearly unhappy with the decision, Lynda let the

matter drop. But Marilyn had the feeling the issue was not dead.

"Okay," said John, "As I said, we'll be at the Fair September fourth and fifth. That's Friday and Saturday. Then, on Monday the seventh, we'll put our float in the Labour Day Parade in St. Catharines." He spoke to Ziggy and Mike. "Will the funeral home and the florist let you have the greenery for the float, like they did for the float you guys built for the Task Force?"

"I'll find out," said Ziggy. Mike agreed to paint signs promoting the Four Rs for the parade and for the fair. Lynda and Bob offered to set up a table at Beamsville Fair, the week after Smithville's Fair, to sell hats.

"Another thing," said Bob. "I think we should publicly endorse Ron Hansen, the NDP candidate. He is definitely on our side and he'll fight for us." From the very first meeting of the Concerned Citizens, Bob and Lynda had pushed for political alignment with the New Democratic Party. John, Marilyn, and some of the others, fearing alignment with one party would alienate supporters who adhered to other parties, fought to remain politically neutral.

"Politically, we're a mixed group, and that's to our advantage," Marilyn argued. "That way, our elected officials don't know where we stand." She shrugged to demonstrate their uncertainty. "For all they know, we might be some of their own supporters, so they'll be more accessible to us."

Bob disagreed. Political alignment would garner supporters. It would give the residents power. Lynda, Ziggy, and Mike staunchly defended his stand.

Finally, Don Austin declared, in his deep voice, "Let's leave politics to the politicians. Now, what about this dad-blasted parade?" The discussion shifted to practical matters as the men decided on a night to work on the float.

But all the while, Marilyn sensed a tension that had never been there before.

※ ※ ※

The sultry heat of August did not slow the impetus of the upcoming summer election. Harry Pelissero, in an interview for the

Pelham Herald, said, "Two of the major issues are free trade and the proposed Ontario Waste Management Corporation toxic waste dump...I don't think we want to play politics with the environment. I don't think any candidate would want to say that a vote for him would get the facility stopped." He remarked that all three parties—including Lincoln MPP Phil Andrewes—had voted to give the OWMC its current mandate. "Having said that," Mr. Pelissero added, "The OWMC is blindly going down a one-way street...They've given token lip service, and token dollars...(to look) at the alternatives." He said he favoured another look at the mandate. "Is it relevant to 1987?" he questioned. "It may not be."

On August 15, 1987, eight weeks after the *Standard* first announced that the OWMC had invited Niagara politicians to tour toxic waste plants in West Germany and Sweden, reporter Doug Draper reminded his readers that Dr. Chant had promised just such a trip to Europe in September 1985, shortly after the corporation announced its plan to build the toxic waste facility in West Lincoln. At the time, Dr. Chant said he wanted some area politicians and residents living near the proposed site to view some of the European facilities in operation. "However," Draper wrote, "not one resident living near the...site...has been invited to go."

When interviewed for the same article, Dave McCallum had revealed that earlier in the summer, the corporation had asked West Lincoln to choose some township representatives for the tour. Now, suddenly, the OWMC was saying that citizens should have been involved. Said McCallum, "The corporation is trying to make (the Township and the Region) look bad and itself look good...I have very little patience left for that kind of silly game."

John Jackson confirmed that the Coalition had not been approached by the OWMC regarding the tour. "The OWMC has not lived up the promise it made...to include residents...[And] there wasn't enough lead time to do the kind of advance research needed to make the trip worthwhile."

Four days later, the *Review* reported that OWMC officials were unhappy that no residents were joining the tour, scheduled to leave September 5 and return on September 20.

OWMC—The "Wasted" Years: The Early Days

Those working on the Labour Day float gathered every Friday night at John and Mary's. On Thursday, August 20, the Concerned Citizens held a meeting to tie up any loose ends regarding both the Fair and the Labour Day Parade. John stated he had purchased the special liability insurance needed to use his tractor and wagon in the parade. Marilyn admitted to being far from finished with the display. "I don't know why," she declared in exasperation, "but it is taking me much longer than I expected it would."

"Yeah, well," said John, with his familiar grin, "the float's not ready, either. Can everyone come tomorrow night?"

"Clifford can," Marilyn volunteered. "But I can't. Kathy's coming home again this weekend to work on the sheaf for her wheat display at the fair—and I need to be there."

On the last Thursday in August, the OWMC office in Smithville called: would Clifford and Marilyn like to join the European tour?

"I...I don't know," Marilyn stammered in confusion. A multitude of problems raced through her head. September was one of the worst possible months to be away, even for a day, let alone for more than two weeks. With the last of the sweet corn coming on, and everything in the garden ready to can or freeze, crops to harvest and the wheat crop to plant...

At such short notice, could Clifford make all the arrangements to leave the livestock? What kind of clothes should they take? Who would take over her Sunday school class? How much money would they need? The questions piled up in her head. Passports, shots, medications. And paramount among them all, how could she prepare herself for the trip? She would need to read, read, read, and formulate questions. It couldn't be done, not in ten days.

As she began to express some of the problems, the voice on the phone suggested that she come alone and leave her husband to take care of things at home.

"Oh no!" she exclaimed. "I don't go without Clifford!" Suddenly, she felt angry at being manipulated. The papers had had a field day with the fact that no residents had been invited. Now, at the last minute, the OWMC personnel were trying to cover themselves, and Marilyn was supposed to turn cartwheels on their behalf.

"I'll discuss it with Clifford and let you know," Marilyn said, her voice cool. But she already knew the answer. It was impossible for them to make so many arrangements in so few days.

When Marilyn called back several days later, the voice replied, "Well, you can't say you weren't invited."

* * *

To prepare the delegates taking part in the European tour and to gain input from the public, the Region called a last-minute meeting for Tuesday morning, August 26, 1987 at ten a.m. But with only two days' notice, Marilyn and the others decided they didn't have enough time for the kind of research needed to ask worthwhile questions.

* * *

The men worked on the float two more nights. Then, on Friday morning, shortly before the fair opened at noon, the group met at their booth to assemble their display. Marilyn snapped candid photos and again tried to wriggle out of manning the booth. "The public's going to grill us, and I don't do well under fire." She pressed her hands together in supplication. "Please! I *really* don't want to be here!"

"Nobody *wants* to be here," Lynda stated, a brittle edge to her voice.

Hurt, Marilyn turned away. Fighting a sudden welling of tears, she busied herself checking the display, which needed no checking. Steve Dinga, with a kindly smile, moved to Marilyn's side. "I'll be here this afternoon when you are. You sell the hats, and I'll talk to the people."

"Thank you," she whispered. She gathered up odds and ends they didn't need for the booth and followed Clifford to the car. She walked with heavy steps, her shoulders drooped under the shadow of Lynda's repressed anger.

* * *

Kathy won first prize at Smithville Fair for her sheaf of wheat.

– 22 –

Although the Labour Day Parade didn't start until ten a.m., the floats had to be at the Penn Centre shopping mall in St. Catharines, twenty-three kilometres away, by nine o'clock, ready to be lined up. It meant Clifford had to get up at four-thirty to do his chores. After a quick shower, he gobbled some breakfast and set off for John's just before dawn to pick up the tractor and float, and John's son Jeff. Marilyn left at eight, her stomach churning from excitement and lack of sleep.

Most of the group, sporting their Concerned Citizen hats, had already assembled in the mall parking lot by the time she arrived. As soon as Ziggy and Mike drove up with a van filled with pots of shrubs and greenery, the group hurried to put the final touches to the float. Marilyn took a few pictures. Then she lifted two boxes of hats out of the trunk and passed them up to Mike. "Here's hoping we sell some of these along the parade route."

"Hey," John called out, "didn't anybody notice? The Four Rs signs are hung along the wagon in the wrong order."

Suddenly, the tension that had been building over the past several weeks erupted. Bob Bradley's temper turned ugly, and his face reddened in anger. He turned on John and jabbed his finger against John's chest. "You want to boss everybody and everything!" he shouted. "Well, you're not my boss! And you're not Lynda's. And I'm telling you right now, Lynda and I are selling hats at Beamsville Fair next Sunday, whether you like it or not!"

John pushed Bob's hand aside. "The group agreed—"

"I didn't agree!" Bob bellowed, giving John a shove. "I

wasn't there for that meeting!" John threw up his fists, ready to defend himself.

Lynda declared heatedly, "I didn't agree, either!"

Marilyn shrank back against the float as Clifford and Don Austin rushed to separate the two men. "Come on, you guys," Clifford insisted, his tone gruff. "This is no place to settle this."

John, his temper as fiery as Bob's, glared in anger. Then Marilyn watched as his fists relaxed and the fight left him. "You're right," he said. "We're supposed to be fighting the OWMC, not each other."

Bob's face remained drawn into a dark scowl. Lynda turned and flounced with venomous fury toward the float. She climbed up onto the wagon, stepped over the miniature white picket fence, and plunked down on one of the lawn chairs positioned amid bales of golden wheat straw, lush green corn stalks, and fall harvest produce. In a harsh, determined voice, she announced, "We don't care what you say—any of you." She made a sharp, sweeping gesture with her arm that took in the whole group. "Bob and I are selling hats next Sunday, and that's that!"

By this time, the ruckus had begun to draw spectators. John let the ultimatum drop unchallenged. Just then, Jim Green, in his town crier's finery, drove up and parked. "Beautiful day for a parade," he declared in his jovial, award-winning voice. He took his position in the middle of the wagon, holding on to the ornamental lamppost erected purposely to steady him on his jolting ride. In front of the lamppost, Paul and Marija Balint's teenage daughter, also named Marija, perched on Marilyn's kitchen stool, dressed in the colourful national costume of Slovakia.

Cheering crowds lined the street as the parade wound its way out of the mall parking lot and began its journey through St. Catharines. Terse and white-lipped, Bob and Lynda barely said a word. At every halt in the parade, Mr. Green shouted his revised proclamation. Each time, as he neared the end of it, he warned;

> Therefore, let us, the Concerned Citizens of West Lincoln, remind you, the people of St. Catharines, that should an accident occur, the prevailing winds in Ontario blow from

us in West Lincoln to you in St. Catharines. You, the people of St. Catharines, should also be concerned, you should be very concerned.

He then proclaimed the place and date, and ended with his now famous, "*God bless Canada, God save Ontario, and God* help *West Lincoln.*"

All along the parade route, crowds cheered the Concerned Citizens' float. People noted the sign, "Hats $3.50," and reached out with money, often saying, "Keep the change." The weather turned beastly hot, and it turned out that no one had thought to bring water.

But, at the end of the day, the float won first prize in its category.

* * *

Three days later, Kathy won first prize at Beamsville Fair for both her sheaf of wheat and her wheat display, which included a second sheaf of wheat and a pot of growing wheat.

As a result of their high visibility, the Concerned Citizens attracted two new members, Albert and Kathy Vanderliek from Bismark.

But neither the victory of winning first prize in the Labour Day parade nor the two new members dispelled the trouble brewing among members of the Concerned Citizens. The quarrel grew and festered daily. True to their word, Bob and Lynda sold hats on Sunday at Beamsville Fair. Mike and Ziggy, obviously torn by the strife, eventually sided with them as other issues arose over money and politics.

Finally, the differences became so great that the group split up. Albert Vanderliek, in casting his decision, stated bluntly, "Kathy and I are staying with the spark plugs—the Concerned Citizens."

Marilyn, devastated by the rift, wept for days over it. Her only consolation came from a verse in Romans that kept running through her mind: "And we know that all things work together for good to them that love God, to them who are the called according to his purpose."

"I have to believe it," she told herself. "I have to believe something good will come out of all this heartache."

* * *

The golden days of early September kept Marilyn on the point of collapse. Chili sauce simmered on the back burner while icicle pickles sat in a crock in the basement, demanding her attention every day for fourteen days. She canned tomatoes, pears, and plums. She picked the sweet corn, blanched and froze it on the cob. Gradually, she cleared the dead and dying plants from the garden, and prepared to work it up for the winter.

On the night of September 10, 1987, Marilyn sat glued to the television, watching the election results with the same intensity her father had during her youth. After much puzzling, she had cast her vote for Ron Hansen, not because she necessarily wanted the NDP to win, but because she believed a Liberal minority government would best serve the residents' battle. That way, she had reasoned, Jim Bradley would remain as Minister of Environment, while the other parties would have the leverage they would lack if the Liberals won a majority.

As the results began appearing on the screen, her gathering despair settled in a heavy lump in her midsection. The Liberals, with Peterson as Premier and Peliserro as the local MPP, were headed for a landslide victory. There would be no stopping the OWMC now that Peterson had a huge majority government.

* * *

With the first hard frost, the maple trees along the front fence turned translucent yellow in the sunshine. Over on the hill to the east, the huge Canadian maple turned a brilliant scarlet. Then the leaves began to fall—by the half-ton truckload. Day after day, Marilyn tackled the cleanup. It was the one time in the year when she did not treasure the huge old trees in the yard.

Her letters to the editor dwindled to zero. She had no time to read documents, no time to polish letters. Nor did she have the heart. Night after night, the *Standard* published reports sent back by Doug Draper, the environmental reporter assigned to travel with the delegates on the OWMC's European tour. Even when he made one of his rare adverse observations, he generally managed to include some redeeming element, driving Marilyn's will to fight to an all-time low.

OWMC—The "Wasted" Years: The Early Days

On Sunday, September 20, 1987, the $50,000 OWMC-sponsored two-week trek to Europe ended, and the delegation returned home.

Two days later, on Tuesday morning at ten, the Concerned Citizens attended the meeting of the Region's Steering Committee at the Regional Offices across from Brock University. An hour and a half later, as they straggled from Room 4, Marilyn exclaimed, "Geez, I felt like *we* were the enemy—not the OWMC!"

With his usual infectious laugh, John declared, "I kept wiping my forehead, looking for blood!"

"So, what side is the Region on?" demanded Don Austin.

"Who knows?" Clifford replied, a glum note to his voice.

Not until some days later did Marilyn learn that Stan Pettit, chairman of the Region's OWMC steering committee, had chaired a press conference shortly after the Concerned Citizens left the Steering Committee meeting. Ensconced in her big orange chair, she read the account in the *Tribune*.

"I was favourably impressed with the European facilities," Pettit told reporters. "There's no doubt the technology chosen by the OWMC for its West Lincoln facility would work. Sure it'll work." Then he added, "It'll take several months of sorting through a mass of gathered information before the Region can decide if its position in opposition to the OWMC's proposal will change at all."

Marilyn felt the strength drain out of her. Despite the chilly reception the Concerned Citizens had received during the meeting, she had clung to the hope that the delegates would, in the end, reveal serious concerns about the European facilities, which would bolster opposition to the facility. Now, the Region had all but declared itself in favour of the OWMC's proposal.

She dropped her head into her hands. "Oh Lord," she whispered in despair, "we have *no might* against this great company that cometh against us; neither know we what to do…" In her mind, she lumped the Region and the Township in with the OWMC. Suddenly she found herself crying, helplessly, hopelessly, crying with streaming eyes and harsh, inconsolable sobs. Struggling to control her tears, she whispered the last words of the verse that had become her haven in times of despair: "But our eyes are upon Thee."

Never had she needed the reminder more. It seemed like all she did these days was cry. She pushed against the back of her old orange recliner until the footrest came up and she could lay prone. Closing her eyes, she brought to memory more of the words from II Chronicles. "O Lord...art not Thou God in heaven?...and in Thine hand is there not power and might, so that *none* is able to withstand Thee?'

As her weeping subsided, she gradually became aware of the sparrows outside the window behind her. Chirping cheerily, they feasted on the round, scarlet-coloured berries still clinging to the honeysuckle bush.

I wish I were a sparrow, she thought, with nothing to do but to chirp and eat. The word eat reminded her of the double batch of raisin scones she had made the day before from her mother's recipe. She always purposely burned the bottoms of the last batch for herself. Rousing herself from the chair, she slathered two scorched scones with butter, made a hot cup of coffee, and curled up in her chair.

*   *   *

A week after the tour returned home, the *Standard* published an article by Doug Draper in *Spectrum*. It covered a full page and a bit. Marilyn groaned when she saw the huge photo of workers picking string beans in a field adjacent to the Biebesheim plant, with the incinerator stacks looming just behind them. His description was that of a pleasant pastoral scene:

> On the outskirts [of Biebesheim], the vans carrying the Niagara delegation turned down a narrow asphalt road...passing countless tidy rows of cabbage, lettuce, spinach, and string beans...

As for the image of farm workers going about their business so calmly in the shadow of HIM's huge incinerator stack, he said, it had seemed tailor-made for the OWMC, which had spent the past few years assuring farmers downwind from its proposed site that emissions would not pose a threat to their health and livelihood.

The article quoted Norbert Zielke, an environmental reporter with a daily newspaper near Biebesheim. Zielke reported several instances of fires and one serious spill, which had required a $500,000 cleanup. Otherwise, he noted, the incinerators had been given a clean bill of health. "I think you can believe HIM plant manager, Klaus Conrady, when he is talking to you about the incinerator tests," Zielke had told Draper.

The more Marilyn read, the more discouraged she felt. Germany had a long tradition for burning toxic wastes, Draper wrote. BASF had operated kilns since 1960. It now had six kilns and planned to add a seventh. In total, West Germany had twenty-four kilns in operation and planned to build about a dozen more.

Column after column of newsprint seemed to laud the technology. Only toward the end of the long article did criticisms surface. Even then, the analysis seemed to wear a veneer of acceptance:

> While some of the toughest critics of toxic waste facilities in the country acknowledge the existing incinerators are a 'necessary evil,' at least in the short term, there seems to be a growing opposition to building new ones.

Draper reported that a coalition of environmental groups in Hessen were opposing plans to construct a third kiln at the Biebesheim facility. In addition, the Biebesheim town council had sent a petition to the state government opposing the third kiln, arguing that it was already bearing enough of the state's toxic waste burden.

Meanwhile, Draper said, in Rhineland Pfalz, the Green Party was raising strong resistance against plans to build a rotary kiln incinerator in the small village of Kaisersesch. Dr. Harold Dorr, a Green Party environment spokesman and member of the Rhineland Pfalz parliament, told the *Standard*, "We don't want to see an expansion of these incineration facilities so that industry does not have the incentive to reduce the waste or change their processes to produce waste that is not hazardous."

The article then pointed to government policies designed to encourage waste reduction. At HIM, the state's environment ministry tried to keep the cost of waste treatment high enough to encourage waste reduction, yet reasonable enough so that industries would not dump wastes down sewers.

In Marilyn's mind, waste reduction had always meant produce less waste. Now, she was confronted with the realization that the term had more than one meaning. Quoting a Hessen ministry official, Doug Draper revealed that most of the waste reduction carried out by industries amounted to little more than "dewatering" wastes at their plants. This practice, while sometimes drastically reducing the volume of waste an industry had to ship to a disposal facility, also usually resulted in waste which was far more concentrated with hazardous agents and, therefore, far more dangerous to deal with.

Draper then detailed West Germany's recent attempt to effect real waste reduction as a result of an upsurge of opposition to toxic waste facilities. According to the managers of some of the waste facilities, a decade ago there had been virtually no opposition to establishing facilities. Now it took years to get a new facility approved. In one case, the HIM company battled opponents for seven years to open a landfill site about forty kilometres southeast of Biebesheim.

The knowledge that not all Germans embraced the waste facilities in their midst gave Marilyn a glimmer of encouragement. That spark of encouragement was quickly extinguished, however, when Draper took a backhanded slam at those opposing the OWMC by linking opposition to the German facilities with a German version of the so-called Not-In-My-Backyard syndrome.

"The Germans have a different phrase for it," Draper wrote. "They simply refer to it as St. Floriau (the patron saint of fire). This basically translates into: 'If you must burn, don't burn my house. Burn somebody else's.'"

– 23 –

Two nights after Doug Draper's account in *Spectrum* of the politicians' tour of toxic waste facilities in Europe, Clifford and Marilyn drove to Smithville. Mayor Allard Colyn had called a press conference to express his views regarding the trip. A small, dilapidated-looking frame house adjacent to the municipal building had been acquired by the Township and turned into the Toxic Waste Project Office. As always, the Concerned Citizens gathered in the parking lot and pressed through the door in a group.

Inside, in the dreary light, Marilyn recognized Eleanor Tait from the *Hamilton Spectator*, Doug Draper from the *Standard*, and Judy McEwen from the *West Lincoln Review*. Smiling and affable, Mayor Colyn greeted the residents and shook hands as he offered them chairs.

He glanced at the clock and took his place at the head of the table. Dave McCallum passed out copies of the four-page press release, printed entirely in capital letters. To begin with, Mayor Colyn gave background information, and then he read, "All incinerators and chemical treatment facilities toured were located in industrial areas, which is mandated by government, on previously contaminated sites such as old ammunition factories..." He lifted his head and scanned the listeners to emphasize his next statement. "Biebesheim was the *only* facility with crops being grown up to the gates on one side with a chemical industry on the other side."

Marilyn watched Doug Draper tap his pen on his notepad, his expression restrained and tense. Draper's photograph in *Spectrum* of workers picking string beans in the shadow of the Biebesheim incinerators had had a huge impact on anyone who had seen it. And many who hadn't seen the photograph had heard about it. Now, the mayor had attacked the validity of the photo. Yet, Marilyn knew it was too little too late. The mayor's press release wouldn't begin to get the coverage Doug Draper's full-page write-up had received.

The mayor went on. "Cosmetically, I would say Biebesheim (HIM) was the cleanest facility, but...it was not in operation at the time of our tour."

"Not in operation!" Marie Austin exclaimed. "Then why bother going?"

Gracia Janes from PALS, the Preservation of Agricultural Lands Society, looked across at Marilyn and lifted her eyes heavenward in exasperation.

The mayor again deviated from his prepared speech. "Generally the facility is reduced to eighty-percent capacity. It's been shut down several times. They've had a number of fires. It broke down in January and was shut down until April for not meeting emission standards.

"At the SAKAB national hazardous waste treatment plant in Sweden," he continued, "compensation lawsuits from property owners are before the courts because of the fires there in which facility operators didn't know what chemicals were stored where." His expression turned severe. "It's not supposed to happen—but it did."

He turned back to his prepared speech. "Four times our consultants have tried *unsuccessfully* to tour the Biebesheim facility while in operation. I do question why it is shut down every time a tour takes place... As for the engineered landfill sites, they were all above ground and had leachate collection systems."

He concluded by inviting citizens from the Niagara Region, not just West Lincoln, to attend a public meeting to be held ten days later on October 8 in the Wellandport Hall.

The minute the mayor opened the floor for questions, Doug Draper jumped in. "Stan Pettit said he was favourably impressed and that he believed the technology could work in West Lincoln."

Mayor Colyn pushed his glasses up on his nose and again surveyed the residents and the media. "Things aren't so rosy. Every rotary kiln we saw had problems such as not meeting emission standards, improper feeding of wastes into the kiln, improper testing causing fire, and an inability to monitor dioxins released from the burning of wastes."

"Mr. Colyn," Draper charged, "you were one of the seventeen people to make the $50,000 OWMC-sponsored two-week trip to Europe. Why weren't you at the Region's press conference?"

"Because the positive attitude expressed was not the consensus of everyone," came the mayor's reply.

"How is it, then, that some of the delegates stated that they felt better about the technology since the tour? Are you telling us you're more concerned now than before?"

"Not more concerned," the mayor answered. "The trip strengthened my doubts. My opinion has not changed. But now I have more reason to defend my position." He folded his arms and leaned his elbows on the table. "European governments," he stressed, "limit hazardous waste treatment facilities to industrial land. They go where the waste is generated."

"So you're concerned about not gutting a viable farming area?" Gracia Janes interrupted with a good-natured aside. "In Europe, they seem to have less of a fascination with clay soil."

"Mr. Mayor," Eleanor Tait asked, "do you consider that the trip was a success?"

"Yes, I'm glad I went. I took half a day. I saw what I wanted to see, talked to the people I wanted to talk to—local farmers, doctors, dentists, and labour representatives." He broke into a satisfied grin. "But I got my hands slapped by OWMC representatives for diverting from the pre-arranged tour schedule."

* * *

The next morning, heartened by the mayor's report, Marilyn set to work on letter number twenty-eight. She divided it into three segments and sent it to the editors of twelve area newspapers. She began each segment by declaring: "The logic escapes us."

In part one, she told how Premier Peterson, while on the campaign trail, had visited the farm of Howard Staff. At 1,500 acres, it was the largest vineyard in Canada. Mr. Peterson had remarked,

"This is one of the special and unique areas of this country…" Yet, two weeks later, in Port Colborne, he had said of the proposed OWMC facility for West Lincoln: "It's got to go somewhere and [West Lincoln] is the best place we know of at the moment."

"Are we to understand," she asked, "that the best place for a toxic waste facility is just a hop, skip, and a jump upwind from one of the special and unique areas of this country?"

The second segment dealt with Dr. Chant's latest hobbyhorse. In the past, he and OWMC personnel had rode to death other pet statements, such as the one about staff working full time across the province to promote the Four Rs, when in fact the staff numbered only two. Or the one picked up by Premier Peterson, "It has to go somewhere." This time, Dr. Chant had zeroed in on an observation made in Germany by the state's environment ministry.

"Again the logic escapes us," Marilyn wrote. In Dr. Chant's most recent letter, he had stated that if industries were faced with higher treatment and disposal costs such as those that would be charged by the OWMC facility, they would spend more effort on reducing their wastes to avoid these costs. OWMC's facility, therefore, would provide a much-needed incentive for increased reduction. Marilyn wondered, "Are we being told that it is necessary to put up a mega-bucks toxic waste facility to show waste generators that they can't afford to use it and should hunt up alternatives such as the 4 Rs?"

Her final segment dealt with resistance in Biebesheim to plans to build a third rotary kiln at the HIM facility. "We are told that the facility proposed for West Lincoln, complete with rotary kiln, will be good for the environment, good for those with an entrepreneurial spirit, good for the local economy, good for jobs, etc. Therefore," she argued, "logic says that if one rotary kiln is good, two would be better and (a third) should be welcomed with open arms."

Yet, according to Doug Draper of the *Standard*, September 9, 1987, the town of Biebesheim was in the midst of opposing plans to build a third high-temperature rotary kiln incinerator at its nearby site. One Biebesheim farmer had told Draper that if the people of West Lincoln had the chance to stop the OWMC proposal, they should.

OWMC—The "Wasted" Years: The Early Days

Despite Dr. Chant's repeated assurances that OWMC would use only the world's best-proven technologies, she wrote, it only took the logic of a Biebesheim farmer to put the world's best-proven technology into perspective: "If you have a factory like HIM, then you need a very good fire brigade."

*　　　*　　　*

The early days of October 1987 brought a nip to the air as the sun dipped lower on the southern horizon. On Saturday afternoon, Marilyn scanned the calendar and almost gave up in despair. The month had scarcely begun, and already she had circled eleven dates.

The following Tuesday, Marie Austin called, her words tumbling together with excitement. "Tricil is planning to build a rotary kiln down near Sarnia! Isn't that great?" she exclaimed. "They won't need OWMC's facility if Tricil builds one!"

Marie raced through the first paragraphs, her voice animated by hope. Then her tone suddenly changed. It became flat and angry. "As usual, Draper called the OWMC." She took a deep breath and began to read.

Murray Creed had stated that it was too early to tell what effect, if any, Tricil's plans would have on those of the Crown corporation, which was required at the upcoming public hearings to demonstrate a need for the technologies it was proposing. He added that a great deal would depend on what percentage of the province's toxic wastes Tricil's kiln would be capable of handling.

Marilyn listened with only half an ear as Marie read the remainder of the article. Despite the initial surge of relief, she couldn't quite share Marie's elation. The residents around the existing Tricil facility had endured hearings, hearings, and more hearings; her heart went out to them, for without fail, they had lost each and every battle.

Marie promised to save the article for Marilyn. But this time, Marilyn wanted to read it for herself. She needed to pore over it, pick it to pieces, and decide where she stood on the question. She picked up her purse, dug out her car keys, and headed for the corner store.

*   *   *

On October 8, 1987, ten days after the mayor's press conference, Clifford and Marilyn arrived at the Wellandport Hall for the Township's public meeting. The place was packed. No longer did she and Clifford seek the obscurity of chairs halfway back, but joined the Concerned Citizens seated near the front.

At seven-thirty, Mayor Colyn rose from his chair behind the table at the front. "Good evening ladies and gentlemen. I am pleased to see such a good crowd. First, I'd like to introduce my companions." On one side sat John Schilstra, Chairman of the Township's Toxic Waste Committee. Beside him sat Dave McCallum, the Township's Environmental Project Coordinator. On the mayor's other side sat Dennis Wood, the unsmiling, meticulously attired lawyer for the Township, and Cam Watson, one of West Lincoln's consultants. Rob Mens, the Township Administrator, sat in the front row near a slide projector.

Mayor Colyn picked up a sheaf of notes and began to read from them. "I will start with the facility that OWMC is proposing to clone here in West Lincoln..." He gave a brief run-down on the Biebesheim facility, then came to the part where he and reporter Doug Draper had skipped out of the tour and met with a group of local farmers. From them, he had learned that the Biebesheim site was previously an old oil-burning incinerator facility, which generally spewed black emissions and looked wretchedly dirty.

"They all felt that this new facility is kept cleaner and the burn seems cleaner," he said. "But...they do not know the long-term health effects...of the facility...and their biggest concern was the application to install a third incinerator...They made it quite clear that two incinerators were enough..."

When he finished his report, Dave McCallum, who still projected the gauche, untried eagerness of the very young, gave an update on recent project activities. In closing, he noted, "As for the funding hearings with Mary Munro in June, the recommendations have not yet been passed." Administrator and Treasurer for the Township Rob Mens then presented his slides of the trip.

OWMC—The "Wasted" Years: The Early Days

Last to speak, John Schilstra told the audience, "I was impressed by the facility at Biebesheim—" Instantly, the sound of angry voices filled the hall. Schilstra raised his hands in an effort to calm the crowd. "—but I don't think the OWMC facility should be located in West Lincoln or in Niagara."

Before turning to Dennis Wood, Mayor Colyn again emphasized the above-ground landfills. He also noted that at Eibenhousen, tests had revealed heavy mercury in the soil around the stack.

To Marilyn's surprise, no one alluded to Tricil's announcement of a possible rotary kiln for Sarnia.

Dennis Wood opened the floor for questions on this portion of the meeting. Max Woerlen, a farm owner close to OWMC's proposed site, raised the question of the crops grown around the Biebesheim facility.

"German farmers have no choice," Mayor Colyn informed him. "For every acre of available land, there are forty farmers who want it."

For the second segment of the meeting, David McCallum introduced Cameron Watson, one of the Township's consultants. "This week," Mr. Watson reported, "I interviewed about six real estate agencies. There is no doubt that residential and farm properties…within a few kilometres of…the site…have become difficult to sell." He waited for the irate rumblings to die down. "A chunk of the market is saying they don't even want to come anywhere near the area…" He went on to quote the probable statistics: "We expect a five- to fifteen-percent drop in property values. But," he added, "values would probably recover after about five years."

He then noted that he could not be more precise about the proposal's impact on property values or how large an area would be affected. "The consultant team has not yet completed a survey it is conducting to back up any compensation demands the Township makes at the future hearings."

The uproar became deafening at the mention of compensation. "It's a defeatist attitude," one resident shouted angrily.

"How can the Township fight the OWMC," Marilyn demanded to know, "when they are busy negotiating their own defeat?"

Dennis Wood rose and attempted to quiet the crowd. "This discussion on compensation...is not an admission by the Township that they want the OWMC facility. But you must be realistic," he stressed. "If the unthinkable occurs, we have to make sure people are treated as fairly as possible...We can't bury our heads in the sand."

Marilyn turned around to John and Mary in the row behind her. "Is he saying that, if we lose, we can't expect to be treated fairly, even though it's the decent thing for OWMC to do?"

"Sounds like it," John said.

"Shhh!" Mary cautioned. "Listen to what he's saying!"

"The Township has decided," said Dennis Wood, "it's better to address the compensation issue at the hearings because the OWMC is developing its own compensation scheme...If residents decide the Corporation's scheme is unsatisfactory, it may be too late to change it after the hearings are over."

He then held up OWMC's thin, forty-six-page report, called "Managing Change." Ever since it had arrived at the Gracey house, Marilyn had pored over every word, highlighting until the pages looked like a rainbow of colours. Dennis Wood told the residents, "OWMC sent 'Managing Change' to 650 households plus those who live along the access route. It's based on the much larger report titled 'Background Material for the Ontario Waste Management Corporation Phase 4B Public Consultation Site Assessment Meetings.' Anyone who wants either of these reports will have to go to the OWMC office and ask for it."

Dennis Wood handed the meeting back to Cam Watson. Watson held up his copy of the hefty 4B report. "Around the office, we refer to the 4Bs as 'bash, burn, boil, and bury.'" Once the laughter died down, he attacked the OWMC's proposed Nuisance Impact Zone. "I don't believe the zone is large enough," he told his listeners. "Other factors should be considered...including stress..., increased truck traffic to and from the facility, and the potential drop in value of the area's farm produce." He cited additional impacts such as added power lines, noise in decibels, and the glare from nighttime lighting. For the first time, Marilyn realized she had an extremely limited idea of the impacts involved.

Watson cited Ontario Hydro's agreement to compensate property owners in Bruce County when it built a nuclear power plant there thirteen years before. "Hydro had an eight-kilometre Nuisance Impact Zone. That's a five mile radius, as opposed to OWMC's 1.6-kilometre—or one-mile—radius."

Marilyn leaned over to Clifford and whispered, "Hydro's impact zone might be bigger, but OWMC's follows fence lines." She made a raggedly jogged shape in the air with her hands. "And surprise, surprise, it leaves us out." Ever since the first maps appeared, showing the exclusionary lines, Marilyn had simmered with rage over the injustice. "Since when has noise, dust, and ugliness stopped at property lines?" she had demanded repeatedly of anyone who would listen.

Cam Watson told the audience that the Township would likely fight for a compensation area similar to the one in Bruce County. Such an area would take in a much larger portion of West Lincoln and would also cross municipal borders into neighbouring Pelham, Lincoln, and Wainfleet.

"The Corporation should be required to offer generous compensation to the community," he argued. "It is unfair that a few hundred property owners should be affected in a negative way so that others, mainly industry, should benefit."

Lawyer Dennis Wood again opened the floor for questions, and residents hastily queued at the microphone positioned in the centre aisle.

"I've been postponing making some improvements to my farm," one man said. "I don't want to do the work and then have OWMC refuse to acknowledge my work or the increased market value. But I can't wait much longer. Farming is operated on a planned, long-term basis."

Dennis Wood replied, "OWMC says you should do your improvements. Keep the bills and a record of your labour."

"But who's going to verify that I actually did the work—or how long it took me?"

Dennis Wood shrugged and turned his hands palms up.

Paul Balint came to the microphone and held up a map of the aquifer under the proposed site and the surrounding area. "I-It's inevitable that our farm ponds, and our wells and cisterns for our drinking water w-would be polluted by OWMC. How can they compensate us for that?" No one had an answer.

Dieter Schuender spoke next, in an exasperated voice. "I want to move, but I can't sell my property. I'm tired of waiting. It's unfair!"

Dennis Wood could offer no solution.

Another neighbour questioned the meaning of "Fair Market Value," fearing it wouldn't permit the purchase of equivalent property.

"Shouldn't we get industrial prices for our land?" shouted a man from the fifth row. "After all, it'll be used for industry."

Cam Watson rose to respond. "Land value is determined by its present use. As for market value, since the OWMC chose LF-9C as the preferred site, prices in West Lincoln have declined generally in 1985 and 1986. However, data from the regional assessment office suggests an increase elsewhere of six to seven percent. Clearly," he stated emphatically, "OWMC's plans are casting a stigma over the Township and turning people off in this area."

"I think," said a young woman, her voice trembling, "we should be compensated for all the worry we have right now. Every day, every night, I worry." Her voice broke. "I think, what will happen to us, to our children, if no one will buy our milk?" She turned suddenly and sat down.

Marilyn hesitated and then stepped to the microphone. "For us and for a lot of people here, no amount of compensation can ever make up for severing our ties with this community. People's families have lived here for generations. How could the OWMC compensate us for the impact the facility would have on our quiet, rural way of life? Or on our health?"

Not waiting for an answer, Marilyn sat down. John Dykstra took her place at the microphone. "I've been thinking," he began, and the Concerned Citizens burst into smothered laughter.

"Famous last words," Marilyn whispered to Clifford.

Grinning broadly, John glanced over at the group, then started again. "Seriously, I've been thinking: if the OWMC had to compensate everyone within a five-mile radius, wouldn't it make it too expensive for them? Wouldn't they have to give up and leave?"

Dennis Wood, solemn as always, said, "Unfortunately, by law, the OWMC is not required to compensate any residents beyond its boundaries. But yes, if such an agreement could be struck, it would make this area much too expensive."

The public meeting ended shortly afterward. John's "thinking," however, had only just begun.

– 24 –

By the time the Concerned Citizens met a few days later in the Dykstra home, John had done considerable "thinking" about Hydro's five-mile impact zone. "We need to get up a petition," he said, amid groans from the others.

Marilyn felt herself wilt. Not another petition! she begged silently. Not another nightmare marathon of knocking on doors.

John detailed his plan. They would collect signatures within a three-mile radius. If they gained support, they would enlist the help of other residents. One by one, the others conceded that, like it or not, a petition was their next best step. "John Jackson has agreed to come to our next meeting and look over the wording," he told them. While the group discussed the petition, Marilyn sat in an easy chair with her hands in her lap, twisting her pen between her fingers. If only I didn't have to go knocking on doors, she thought. If only—

* * *

Over the next week, in between raking mountains of leaves and hauling them from the front lawn, Marilyn worked on letter number thirty. The Swan Hills toxic waste treatment facility in Alberta had opened officially on September 11, 1987. Like Dr. Chant, who insisted that the proposed OWMC facility was intended only for Ontario's waste, Premier Don Getty had repeatedly assured the residents of Swan Hills that the hazardous waste facility proposed for Alberta would accept only Alberta's waste.

Yet, on the day of the grand opening, Premier Getty had barely pushed the bright red button before he said in his speech that

OWMC—The "Wasted" Years: The Early Days

Alberta might eventually lift its moratorium on handling waste from outside the province. "In other words," Marilyn warned in her letter, "once the facility was in operation, they were free to change the rules."

*　　　*　　　*

Meanwhile, Dr. Joseph Highland, a prominent U.S. scientist, hired by the OWMC to study the potential health and environmental risks associated with its proposed facility, met with the Niagara Regional Health Inspectors in mid-October to discuss his findings. "OWMC scientist says health risk 'acceptable,'" declared the write-up on the front-page of the *Standard*. Long-term exposure to the OWMC facility would pose a marginal but "acceptable" health risk to the surrounding population, with less than one additional cancer in a population of a million.

"Toxic wastes are 'a fact of life,'" Highland told the health inspectors. "The benefits to society of having proper disposal facilities outweigh the risks to those living near them... They may shoulder more of the risk for the common benefit... That," he added, "is a common problem."

Instant rage swept Marilyn like a searing flame. "We're *people!*" she said through clenched teeth. "Not just statistics!" She jumped up from the chair and raced upstairs. Her jaw jutting in angry determination, she yanked the chair out from her father's rolltop desk and seated herself in front of her old typewriter. Letter thirty-one, she jotted down on a new file folder.

First, she quoted Dr. Highland's statement calling the risk to residents to be marginal but "acceptable." To show that his view was typical of the OWMC's attitude toward the people in the area, she reminded readers of statements made to her daughter Kathy by Dick Griffiths at the Open House in Wellandport a year and a half earlier: "I know you can't understand it. And it is unfortunate. But sometimes a few have to suffer for the good of society as a whole." Lastly, she again quoted Dr. Highland: "[The people living near the proposed West Lincoln site] may shoulder more of the risk for the common benefit."

"As far as we know,' she concluded, "no resident of West Lincoln or the surrounding area has volunteered for the job of sacrificial lamb. But we would be willing to accept applications from the OWMC."

* * *

During the fall of 1987, the relentless struggle between the OWMC and the residents was not so much a battle as a series of small skirmishes. John Jackson sent an article to the *Niagara Farmers' Monthly*, challenging Dr. Highland's risk assessment as "acceptable." In it, John quoted from a paper co-authored by Dr. Chant and Dr. Ross Hall:

> The present attitude assumes that the environment can carry a certain amount of contamination...This is a false assumption because we do not know to what degree contamination can be tolerated.

Dr. Chant promptly claimed he had been quoted out of context. Marilyn wrote to the paper, labelling Dr. Chant's response as an example of "OWMC's famous half-twist."

* * *

On November 11, Remembrance Day, Clifford, Marilyn, and their daughter Kathy set off for the Royal Winter Fair in Toronto. In the trunk, resting in a cradle Clifford had made for it, rode Kathy's latest and most meticulously constructed wheat sheaf ever.

The following morning, the telephone rang. A woman's voice informed Marilyn that Miss Kathleen Gracey had won first prize with her wheat sheaf, and had taken World Championship over all the sheaves in every category—wheat, oats, barley.

Within the hour, John Fox from the *Hamilton Spectator* called Marilyn for an interview. "Where exactly is the family farm?"

"It's about a mile from the proposed site for the OWMC's toxic waste facility." Marilyn felt a surge of triumph. "The OWMC is trying to make this area sound like marginal land. But this shows—" She stopped suddenly, afraid that she would steal Kathy's glory. "Of course, the sheaf is judged on how well it is formed."

OWMC—The "Wasted" Years: The Early Days

That same day, a huge colour photo of Kathy and her wheat sheaf, complete with an interview and write-up of the award, hit the front page of the *Spectator*.

* * *

The community had only hours to bask in the fame brought by Kathy's achievement before disaster struck. At three o'clock in the morning, Friday, November 13, 1987, a car blasting its horn woke Stan and Helen Kszan to find their dairy barn totally engulfed in flames. Within hours, radio station CFRB reported that from a herd of sixty-seven, only thirteen young cattle penned outside in a small enclosure had survived. In the morning, as soon as Clifford finished his chores, he rushed over to view the smoking ruins. When he came back, he said that most of the men in the community had gone to see if they could help in any way.

"Stan says he'll probably rebuild, and I told him I'd help as much as I could. John Dykstra was there. He said he and his boys would help when the time came."

"How's Helen?"

"She looks terrible."

Marilyn felt heartsick. "How much more does that family have go through?"

* * *

In mid-November, the *Tribune* featured Don Austin with his fiddle. In the 1950s, Don had played on the radio with a number of well-known local groups including the Lincoln County Peach Pickers.

"Did your wife, Marie, travel with you?" asked reporter Shirley May.

"No," he replied, with a huge grin. "There was no room for her in the trunk."

* * *

True to his word, once the Concerned Citizens had drawn up their petition, John Jackson met with the group in the Gracey home. The same night, two more residents joined the group—

Jean Trudell and Reggie Kuchyt.

Uppermost on the agenda came the wording of the petition. "If we can expand OWMC's one-mile Nuisance Impact Zone to a five-mile compensation circle, it'll make this area too expensive," John Dykstra explained to Jean and Reggie.

John Jackson read over the draft of the petition. "I suggest you change the word compensation to restitution," he said. "Otherwise, it looks good to me."

* * *

Early in December, the *West Lincoln Review* reported that while the Region and the Township would manage the study on waste quantities, it would be done in "full co-operation" with the Ontario Toxic Waste Research Coalition.

In the *Tribune*, the regional councillor for Welland cautioned the two bodies to avoid duplication, with the reminder, "Donald Chant clearly stated that the Corporation will not pay for a duplication in studies."

The newspapers reported that at the December meeting of Regional Council, Chief Administrative Officer Michael Boggs revealed, "The OWMC hearings could start as early as the spring of '89, but they could be delayed until the fall of '89."

Stan Pettit, Chairman of the Region's Steering Committee, stated, "We could very well be looking at a budget of one million dollars, and that doesn't include the hearings."

"One reason for the delays," Regional Chairman Dick reported, "has been a lack of staff in the Environment Ministry...to analyze all the OWMC reports."

"The whole process is a consultant's dream!" declared Gladys Huffman, Regional councillor for Lincoln.

Grimsby Mayor Ross Hall grumbled, "If Dr. Chant is going to take forever and a day to build his mouse trap, we'll never get to an Environmental Assessment Hearing!"

Bill Smeaton, the mayor of Niagara Falls, wanted to hurry things along: "OWMC has already spent $55 million. I can't understand why we want to lengthen the process any further."

Regional Council decided to ask Chairman Dick to set up a meeting with Premier Peterson to try and speed up the process.

Speed up the process! Marilyn didn't want them to speed up the process. The sooner the hearings started, the sooner her life would be shattered. And the less time she and the others would have to scour documents and fight back.

The following Monday, however, West Lincoln council rejected the idea of a meeting with David Peterson and Jim Bradley. John Schilstra, Chairman of the Township's Toxic Waste Committee, told council, "I think the Region's push for faster hearings is premature."

Mayor Allard Colyn concurred. "I don't believe you can speed up the Environmental Assessment process."

Good! Marilyn thought, astonished that, for once, Colyn and Schilstra had managed to agree on something. Every day that it wasn't a done deal, she could summon up hope.

* * *

Four days before Christmas, the *Standard* ran a front-page story, complete with photos, showing Helen Kszan preparing a turkey that had been donated and homemade cabbage rolls for those helping to rebuild the barn. Only five weeks after the fire, the men were already on the roof, nailing metal sheeting to the rafters. It had taken about four days to remove and bury the remains of the cows. Then, once Helen and Stan had made the arrangements for supplies and so on, volunteers had showed up almost daily for the last three weeks. For many of the men, including thirty-six year-old neighbour Ross Allen, the barn raising had been a first.

Stan told reporter Paul Halliman, "To us, it is a Christmas present... There is no way we could have done that by ourselves... With luck, it could be finished by the end of January."

Helen confided to Marilyn, "I called the papers. I thought it was a perfect example for OWMC to see how the neighbours in this area pull together."

* * *

On December 29, the *Pelham Herald* reported that the OWMC had published a manual to help manufacturing companies reduce their wastes. Instantly Marilyn felt consumed by anger and resentment. The reaction startled and puzzled her. Wasn't the Crown corporation doing precisely what she and the residents had lobbied for—placing a greater emphasis on the Four Rs?

The article cited an example from the manual—a steel pickling plant that had installed an acid recovery system. The system had saved the company $379,000 in operating costs, eliminated waste, and improved pickling efficiency. As Marilyn put the paper aside, she realized it wasn't anger she felt, but fear—fear that by doing something right for a change, the corporation was guaranteeing its success.

She folded the *Herald* and picked up the *Review*. As she leafed through the last issue of the paper for 1987, her eye caught the letters OWMC on the editorial page. The editor said she had turned to her crystal ball:

> It isn't difficult to see West Lincoln's fate in '88. This newspaper predicts continued pressure...from such quarters as Ontario Waste Management Corporation and its proposed facility, the PCB cleanup, [and] the planned four-municipality dump...We predict the OWMC will have its way with Wellandport...

— 25 —

Nineteen eighty-eight promised to be as hectic as the previous year. The Concerned Citizens met at Dykstra's on January 4, followed three days later by the monthly Coalition meeting at the Vineland Experimental Station. "I can't find time to take the Christmas tree down!" Marilyn told the Coalition members. "But I did find a great quotation by Joseph Conrad, dated 1900:

"Each blade of grass has its spot on earth whence it draws its life, its strength—" Unexpectedly, her voice faltered. *"—And—and so is man rooted to the land from—from which he draws his faith together with his life."*

She had meant it to be a rallying statement, words of encouragement. Instead, in the silence that followed, she saw stark pain surfacing in the faces around the table, swiftly hidden by downcast eyes and lowered heads.

John Dykstra gave a grim, mirthless laugh. "Yeah, well, this blade of grass has his own spot on earth, and he doesn't intend to be mowed down by the OWMC."

*   *   *

That same week, Shirley May, the reporter who contributed human-interest articles to the *Tribune*, featured Marie Austin's life as an organic farmer. It had begun when their only son, Wayde, was a baby.

"Wayde had so many allergies," Marie said, "he had to have needles for them…Our doctor told us he had chemical allergies and that he was receiving these chemicals in his food and water."

Soon afterward, the Austins had traded in their well-kept house and lot in the city for an older home on a quarter-acre of land in North Pelham. From then on, they had grown everything—chemical-free—and rejoiced as they watched their son restored to health. Totally opposed to the OWMC and its emissions, Marie explained, "We have seen first-hand how chemical additives in food and water created life-threatening allergic reactions in our son."

*   *   *

During the rush of Christmas and the festivities at New Year's, Marilyn had fallen behind with her letters to the editor. Now, she pulled out the file containing an interview in the *Tribune* with John Jackson, printed in November 1987. At the time, John had admitted, "There is inevitably the need for destruction facilities... Some wastes cannot be recycled or avoided." For these, he urged the use of smaller, regional plants.

Dr. Chant had responded in a letter circulated to papers across the peninsula, claiming that Mr. Jackson was advocating hundreds, perhaps thousands, of little treatment plants scattered across the province, apparently without any commitment to the best technology, perhaps sited in unsuitable locations, virtually impossible to police effectively, and presumably without any public participation in a lengthy, careful, and extensive hearing and approval process.

Marilyn had been infuriated when she first read the letter back in November, and she was no less infuriated now. John Jackson and the Coalition had favoured the idea of five or six smaller regional facilities for some time, and Donald Chant darned well knew it. She ran an orange marker over Chant's ridiculous claim of "hundreds, perhaps thousands, of little treatment plants scattered across the province." She would definitely attack that statement.

In November, the day after Donald Chant's letter appeared, the *Tribune* had featured a second interview with John Jackson. A photo accompanied the interview. In it, with his features good-natured and relaxed, and his high, wide brow unfurrowed, John had been laughing with boyish abandon. In the interview, he had criticized the OWMC's plans, which called for only twenty-eight percent of Ontario's hazardous wastes to be disposed of by reduction and recycling in ten years' time.

"That's a pitifully low objective," Jackson said. "Instead of shipping wastes from throughout all of Ontario to West Lincoln..., five or six smaller plants should be set up around the province. This system," he had stressed, "would reduce the inevitable negative impacts of a large facility on any one community."

Marilyn paused as she peered into space at the snowflakes drifting past the living room window. John's statement made her think of her father's lifelong philosophy: never do anything to nature from which it cannot recover. Once the OWMC's mega-facility reached full capacity, it would put out two and a half times the emissions of Biebesheim's twin incinerators.

Her thoughts shifted to Rudi Ohler, who had recently made a fact-finding trip to Germany at his own expense. A professional engineer, he spoke German fluently. Upon his return, he had revealed that agricultural production around Biebesheim was restricted. He had shown the Coalition photos of fields planted with sugar beets. "The tops may not be used for animal feed. They must be destroyed as hazardous waste," he told them. "No grains can be grown there for human or animal consumption, only for use as seed. As for the crop pickers shown in Draper's photo, they were not from that area, or even from Germany. They were unaware of any possible dangers."

Marilyn turned her attention back to the write-up in the *Tribune*. In the interview, John Jackson had also slammed OWMC's proposal to bury between two and three million cubic metres of contaminated material in the ground in West Lincoln over the next twenty years: "Burial of hazardous wastes is unacceptable," Jackson insisted. "Regardless of the depth of the clay the waste is buried in, eventually all landfills leak. The result is the permanent destruction of essential water supplies."

Although fixed on the clipping in her hands, Marilyn's eyes did not see the words. When it came to the landfill, no matter how hard she tried or how much she had learned, she could not totally dismiss the expertise expressed by the OWMC's consultants. Her lingering uncertainty stemmed, she knew, from the letter written more than a year earlier by Dr. Chant. In it, despite the discrepancies over the trucking of chlorine and the high-low water content of the waste, he had spoken with authority when he insisted that the OWMC's engineered landfill would prevent contaminants from entering the underground water supply.

She glanced at the photo of John Jackson's laughing face. He seemed so sure of his declarations, her doubts made her feel like a traitor. Things were never as black and white for her as they were for Clifford. Or for Marie. Or Marija. For any of them, she decided, except me. She expelled a long, soul-searching sigh as she continued to read John's condemnation of all landfills, and OWMC's landfill in particular.

OWMC claimed that its landfill would pose no serious risks because it would put only treated, solidified wastes in the ground. According to John, this had misled many people into thinking it was safe, which was not true. The wastes left over after treatment were often the most dangerous ones, because they were the ones for which destruction methods were not available. Solidification was simply a method to slow down the movement of toxic substances—not to stop the movement. That was why the Coalition believed they should be stored above ground, to enable better monitoring.

"In addition," he said, "the visible presence of warehoused wastes...will keep people aware that the wastes have not magically disappeared, and will maintain public pressure to develop new waste reduction and destruction methods."

With the clippings in her hand, Marilyn climbed the stairs and sat down at the typewriter. As she rolled a sheet of paper into position, she reminded herself of her own rule: deal only with those issues where she knew she stood on solid ground. Therefore, first and foremost, she poked holes in Dr. Chant's ridiculous claim that the Coalition advocated hundreds, perhaps thousands, of little treatment plants. She then focused on the negative impact a huge facility would have on any one community. Finally, she attacked, with all guns blazing, Dr. Chant's assertion that OWMC's proposal had been "developed and approved with full public participation."

OWMC—The "Wasted" Years: The Early Days

"That's a lot of rubbish!" she declared.

The infamous questionnaire remains incomplete, lacking full input from the most informed of West Lincoln's residents. The concerns raised by those same residents have not, to this day, been addressed...

The following morning, she mailed letter number thirty-two.

* * *

At the end of the first week in January, Marie Austin called Marilyn, her voice loaded with doom. "There's an article in the *Standard* tonight," she declared, "and it *ain't* good news. The OWMC is going to release six more volumes of final documents."

As Marilyn screwed up her face at the thought of more reading, Marie read a statement by OWMC spokesman Murray Creed. "It is coming down to the wire. The OWMC is only about a month away from submitting the final documents."

Marilyn experienced a spasm of panic, the same kind she had felt, as a child, hopping from railroad tie to railroad tie as she took a forbidden shortcut home from school along the train track. She had listened to the train whistle far off in the distance as it sounded periodically at level crossings to signal its impending approach, only to look up at the sudden shrill scream of the whistle and see the freight train bearing down upon her.

"How many more 'final documents' can they release?" Marie questioned.

"I don't know. I have boxes filled with them. What else does Mr. Creed have to say?"

"Only that it will take the Ministry of the Environment months to go over them. He thinks the public hearings could begin before the end of this year."

Marilyn tried to sound as though the reports were no big deal. "Oh well, you know the OWMC. Their hidden agenda is to cut down a whole forest and turn it into OWMC documents."

Marie gave a half-hearted chuckle. In an attempt to lighten Marie's spirits, Marilyn joked, "Hey! They're only halfway through the forest. You can't expect them to quit now!"

After Marie hung up, Marilyn let her shoulders sag. The hearings could begin within months.

* * *

Before the month of January ran its course, the Concerned Citizens held five meetings. With the bulk of their hats sold, they now had enough money to reimburse Steve Dinga for the $1,250 he had loaned the group, and John Dykstra for the liability insurance he had placed on his tractor and wagon for the parades. From Carruther's Printing, they ordered another run of petition cards, duplicates of the ones they had presented to Jim Bradley at Queen's Park. "We'll keep bombarding the Premier's office!" John declared, with his familiar gleeful grin. "Just make sure to get new names. We don't want to get caught having the same people sign over and over again." With all of their bills paid, they donated another $200 to the Coalition.

However, the Concerned Citizens' main priority at each meeting remained the development of John Dykstra's strategy for a five-mile-radius Impact Zone surrounding the proposed site.

"When we go door to door with our petition, we'll need name tags." Marilyn held up a sample. "I photocopied a picture of our banner and then reduced it to fit inside one of these plastic thingies to pin to our coats."

Gradually, out of the planning emerged the conviction that they needed to hold a public meeting. Pointing to OWMC's latest mammoth-sized document, Marijah Balint, stated with venom, "I no like this report."

Reggie Kuchyt agreed. "People need to know the other side of the story."

"Well," John Dykstra stated firmly, "any meeting's got to be as early in March as possible, before everyone gets too busy on the land." He gave a beleaguered little laugh. "I've got a lot of clover seed to sow before the frost goes out of the ground."

* * *

Meanwhile, the Township's Toxic Waste Committee brought in Denys Reades, of Golder Associates, to give his assessment of the OWMC's 4A documents. "I am concerned with OWMC's overkill interest in West Lincoln's clay," he told the committee and the handful of residents. "Even if the OWMC declares West Lincoln as a landfill site," he continued, "the whole facility need not be located here, especially since more land will be needed for burial than what the OWMC has designated."

So! thought Marilyn, I was right about their dinky little landfill!

Following the meeting, the residents congregated on the sidewalk outside. "Just think!" Jean said. "He gets paid ninety-five dollars an hour plus expenses to tell us nothing we didn't already know."

* * *

Within days of the Concerned Citizens' decision to hold a public meeting, Reggie Kuchyt, who looked after the hall, had checked her records and booked the Concerned Citizens for Monday, March 7. At the next meeting, as they discussed topics for the public meeting, Marilyn hauled out her big, thick, well-marked copy of "Background Material for the Ontario Waste Management Corporation Phase 4B Public Consultation Site Assessment Meetings," dubbed "Background Material," for short. She read off the topics she had listed on the cover.

"Our priority," John reminded the group, "has to be the unrealistically restricted Nuisance Impact Zone—"

"Just a little dust, a little noise, a little ugliness," Marilyn interjected playfully.

"We have to make sure people understand what the *real* impacts are. And that we're not looking for compensation—we're lobbying for restitution over a five-mile radius to emphasize the seriousness of the proposal, and to make the facility too expensive, even for the OWMC and its bottomless pockets."

Marilyn urged the others to search their documents. "Read, read, read! We *all* have to be prepared to answer questions."

Clifford closed with prayer, and the ladies took the coverings from the goodies they had brought. "Isn't this a party!" sang

Marilyn, pirouetting around John and Mary's dining room. But it was no party when, night after night, she was still sitting up in her old orange chair at three-thirty in the morning, poring over reports and scribbling down page numbers.

As Marilyn and the others began to collect data for the presentations, Paul Balint, John Dykstra, and Clifford worked on letters to lobby their respective farm marketing boards—beef, milk, and pork. Between times, the group conscripted Marilyn to attend an ecumenical meeting, organized by Doris Migus, at the Hagerman home overlooking Lake Ontario. "John Jackson is going to discuss strategies for organizing church congregations," John Dykstra told her.

"I'll go," she said reluctantly, "because it'll give me a chance to go over the wording with John Jackson for the petition for the five-mile radius and the information sheet before we print them up. But I don't agree with the purpose of the meeting. I make it my business not to mention the OWMC at church. It's one place where people should not have to duck divisive issues."

"But lots of church people don't want OWMC," asserted Marija.

"That's true," Marilyn admitted. "But I'm telling you ahead of time—I won't implement it at our church. I teach the adult Bible class, and it makes those in my class a captive audience. Clifford and I have agreed that people have a right to their own opinion on this issue. I don't intend to make anyone at church feel uncomfortable."

The group quickly roughed out its agenda for the public meeting. Paul would open with prayer. Jean Trudell, one of the newest members, would welcome the audience and take part in the more straightforward presentations. Marilyn would chair the meeting and, along with John Dykstra, would handle the more complex presentations.

Edith Hallas, from Citizens For Modern Waste Management, had agreed to speak on the impacts suffered by those living in close proximity to Tricil's toxic waste facility in Moore Township, near Sarnia. John Jackson, as guest speaker, would zero in on the need for increased waste abatement and Four-Rs programs. He would also tie up loose ends and act as a resource person for any questions the residents found difficult to answer.

Less and less able to sleep, Marilyn stayed up all hours of the night, laboriously organizing the material for the public meeting. She insisted upon page numbers and exact quotes. Often, to be sure in her own mind that the quotes were word-perfect, she dug through the piles of documents and searched her boxes filled with newspaper clippings. Her home had become a poorly-run office—poorly run because she had no filing cabinets, and had to search through boxes stacked on top of boxes for every reference. As the mounds of documents grew, the task had become increasingly onerous.

Kathy's bedroom had become the repository for data—piles of reports and documents, boxes of newspaper clippings randomly filed, and numerous little piles of data she had not quite had time to deal with. Judy's bedroom housed the typewriter, buried under work in progress on her father's huge rolltop desk.

As Marilyn typed and retyped the drafts, adding, subtracting, and organizing in logical sequences what the group would present, it became clear that it would be a long, jam-packed meeting.

"We'll never hold the audience's attention," she said flatly. "We need something to break up the flow and keep them from falling asleep." She had watched the hard-working men of the farming community at other meetings. Like Clifford, the minute they lost interest, they folded their arms over their chests, slid down in their chairs, and closed their eyes.

"We need visual aids," Marilyn told them. "How can we get hold of a projector, the kind that projects a page from a book onto a screen?"

Clifford called Etta Lane, a teacher at Bismark Public School, and enlisted her help. "Our projector only uses transparencies," Etta told Marilyn, "but we have equipment to copy a page onto a transparency." Etta then gained permission for Marilyn to reproduce transparencies at the school. The school agreed to lend the projector and screen to the group on the night of the public meeting, provided that Etta ran the equipment.

"Perfect!" Marilyn exclaimed. "That'll leave the Concerned Citizens free to do other stuff."

To keep to the distinctive and now-familiar theme of the town crier, Marilyn produced fliers using the same design as for the trip to Queen's Park: the outline of a parchment scroll, edged with drawings of smokestacks and trucks. In old-style script, she wrote,

> Hear Ye, Hear Ye, Public Meeting...to discuss the IMPACTS the OWMC's proposed facility could have on YOU.

She included the date, time, and place. "Sponsored by the Concerned Citizens," she added as an afterthought, to make sure no one mistook the meeting for another OWMC Open House. She ran off copies at Wellandport Library, and plastered one to the glass in her back door. It made a perfect bulletin board, and it gave whoever hammered on her door something to hold their attention while she ran down the stairs from the typewriter in Judy's bedroom, through the entire house, to reach the door before the caller became impatient and drove away.

At the end of January, the Concerned Citizens began knocking on doors within a three-mile radius of LF-9C, the official name for OWMC's proposed site. To Marilyn's relief, Clifford canvassed a portion of their neighbourhood, leaving her free to work on material for the public meeting.

− 26 −

The month of February quickly became crammed with meetings, sometimes two and three in a week. Marie's minutes, when she read them aloud, became a series of terse directives—John will hire a sound system, Marie will notify newspapers and radio stations, John will get permission to use Rudi Ohler's report, the Concerned Citizens will attend West Lincoln's council meeting on February fifteenth to request support for the petition.

At every meeting, new issues came to light. Reggie had been reading a report that told of hydroelectric lines to be installed along Silverdale Road to service the OWMC facility. Albert Vanderliek suggested that a brief question-and-answer period should follow each speaker.

John Dykstra, as always, had been thinking. "West Lincoln Council meets a week from tonight. I think we should ask them to take our group and our petition to Regional Council as a follow-up to our public meeting. Then we should go to Lincoln, Pelham, and Wainfleet, maybe all on the same night."

When everyone agreed, Marilyn said, "Then I move that John gets on with making the arrangements." Reggie seconded it.

John consulted his notes. "I've talked to John Jackson. He agrees that we need to finalize our presentations. John says he can come for a luncheon meeting on Tuesday, February 23, since he'll be speaking that night at Vineland United Church."

"Okay," said Marilyn, "you can have the luncheon meeting at our place. I'll make a pot of soup."

"We bring everything else," Marija stated.

*   *   *

A week later, the Concerned Citizens received immediate support for the petition from West Lincoln Council. Afterward, they met at Jean Trudell's. John told the group, "I hate to say this, but I've set up a meeting with Harry Pelissero for nine-thirty in the morning on February twenty-ninth. The timing's bad, I know, but..." With a shrug, he left the sentence hanging.

The meeting had been set, with all in agreement, to initiate the process of visiting Queen's Park for a second time. "I told him we want to see Premier Petersen this time. We'll only take one bus and just enough passengers to meet our expenses. I told him that our objective was to bring the true impacts to the premier's attention and that Marilyn would be making a presentation, for example, on stress in the community."

The luncheon meeting with John Jackson the following week went more like a friendly get-together than a business meeting. As the discussion drew to a close, Marilyn made a suggestion. "After we've highlighted all the problems associated with the proposed facility, I think we need to draw people's attention to the OWMC's callous attitude toward how deeply we feel about their dump."

She hefted her well-thumbed copy of "Background Material" onto her lap and flipped to the last page. "I say, let's hang them with their own words." She picked up her granny glasses and settled them low down on her nose. "The area is expected to remain relatively stable—" Marilyn paused for effect. "—and residents' satisfaction with the community is likely to remain high."

"I don't believe it!" declared Marie, as groans circled the room. Marilyn scolded the groaners with mock indignation. "What's the matter with you people?" she queried. "It's only a little dust, a little noise, a little ugliness..."

*   *   *

As Marilyn watched the calendar fill up, she wondered if she would ever survive the month. A concert at Hamilton Place, a meeting with the bank manager, prayer meeting every other

OWMC—The "Wasted" Years: The Early Days

Wednesday, and a hay ride for the young adults from Judy's church in Burlington.

As February drew to a close, she finally gave in and violated the last bastion of tidiness in their home. She removed from the dining room table her grandmother's wedding present to her and Clifford, a lovingly hand-crocheted tablecloth, and replaced it with a silence pad and an old plastic cloth. Then, every time she completed a section of their presentation or thought of something they would need for the meeting, she placed it on the table.

"This place looks more like a dump every day!" she complained to Clifford. "We dump stuff here, we dump stuff there! I *hate* clutter! And I *hate* the OWMC for making me live like this."

"Now, now, now," Clifford said, pretending to scold. "You know what you always told the children—you mustn't hate people."

"I don't hate the people—I hate the Corporation, I hate the proposal, and I hate them for making me even think about hate."

"Now, now, now," he said again. "You know what you've been told: 'You people are creating your own stress by opposing the facility.'"

Marilyn snatched a pillow from the sectional chesterfield and hurled it at her husband. "When my surroundings are in a muddle, my brain's in a muddle!" Her shoulders suddenly sagged. "Clifford, I'm scared to death we're going to forget something excruciatingly important for the public meeting."

*　　　*　　　*

During the meeting held in the Dykstra home on March 2, just five days before the public meeting, it suddenly occurred to the group that they needed a portable lectern to place on the speaker's table. "How big do you want it?" Clifford asked. "I think I've got enough lumber at the house to build one."

Once they had settled on the size of the lectern, Marilyn confided, "At night, the minute I lay my head on the pillow, my brain kicks into high gear. I've been picturing our presentation. We need someone to point things out on the screen as the speakers go through their presentations."

It took a little coaxing, but Marie Austin finally agreed to do it.

* * *

Marilyn waited until just a few days before the public meeting to make up the transparencies at the school. Meanwhile, she had obtained blank flimsies and special coloured pens to trace any of the illustrations that could be done at home.

The group pasted informational news clippings onto coloured paper, made up signs and posters, and mounted photographs of the Biebesheim plant taken by Rudi Ohler. Marilyn, with Rudi's permission, took excerpts from his report on his trip and added them to the presentation.

* * *

At ten a.m. on March 7, 1988, the day of the public meeting, the Concerned Citizens met at the Wellandport Hall. The soundman arrived to set up the microphones and speakers. The residents put up tables near the front for the media, and one for the projector. They arranged tables at the door where Mary Dykstra and Paul Balint would pass out literature. On the walls behind them, they taped their informational display.

They arranged the chairs with a very wide centre aisle and kept them well away from the walls. "That way," John Dykstra said, "it looks as though we have lots of chairs put out when, in fact, we don't. It'll look better to see us rushing around, putting chairs up the aisles than to have everyone looking at a lot of empty seats."

Finally, they set up a long table across the front of the raised platform and covered it with white tablecloths. "Let the cloths come almost to the floor in front," Mary Dykstra suggested. "Then the ladies won't have to worry about their skirts."

"In front, where the speaker stands," added Marija Balint, "we should hang our banner."

Etta Lane begged time off school and came over after lunch to set up the equipment. Marilyn produced the presentations and the residents rehearsed them like a script, with Marilyn directing. "Okay," she'd say to a presenter as she stood at the back of the room, "speak a little louder. I can't hear you clearly enough."

"Speak a little slower," she would say to another.

As they moved through the presentation, Etta Lane needed little coaching. Marilyn had clearly marked, on Etta's copy of the presentation, when to turn on the projector, what flimsy or flimsies to use, and when to turn off the projector.

"We could just leave it on," Etta suggested.

"I'd rather not," Marilyn said. "Even the blank screen, lit up, will act as a distraction. Besides, each time it lights up, it signals a change, which jerks the audience back to attention."

Sometimes, they ran over a section a second and a third time. "Marie," she would coach, "move your pointer from this spot to that spot as John talks."

When they had finally gone through it once, Marilyn said, "Okay, I think we should do it again."

The suggestion met with groans and complaints. "Look," she told them, "we don't want to look like a bunch of amateurs."

"But that's what we are!" joked Don Austin in his carrying voice.

"But we don't have to *look* like amateurs," Marilyn insisted. "If we fumble and bumble around, we'll lose the audience. Once we lose them, we're done. We won't get them back. This is our one chance to win their support and cooperation."

After more good-natured muttering, the residents took their places. Those not presenting stood along the back wall and acted as critics. The second run-through went off with scarcely a hitch.

As they surveyed the auditorium for any last-minute details, Marie said, "I have to admit, I feel a lot better after doing it the second time." She turned to Marilyn and gave her arm an affectionate squeeze. "Thank you for making us do it."

"And for all your work," Paul said.

Etta Lane added, "I've never seen a better-orchestrated presentation. You'll all do just fine."

* * *

That night, Marilyn watched the hall quickly fill with friends and neighbours, Coalition members, and a number of people she had not seen before. A group of nine young men, students from the Robert Land Academy, dressed in full uniform, arrived with their instructor. Clifford, Don Austin, and the other men began adding rows of chairs across the back and up the aisles.

Shirley May was the first member of the media covering the event to arrive. Marilyn took candid shots right up until it was time for the meeting to begin. Then she handed the camera to Clifford, who sat in the front row with the rest of the Concerned Citizens. Marilyn took her place at the table, next to the portable lectern.

Paul stood and, once the room had fallen quiet, opened the meeting with prayer. The minute he finished, one of the reporters jumped to his feet, slapped his notebook closed, and strode from the hall. A strained moment of silence followed. Then Jean Trudell launched into her words of welcome. Before she sat down, she drew attention to the maps and pictures on the walls, including an article on above-ground storage written just two days earlier by Doug Draper. "Please look at them before you go home," she urged. Then she introduced Marilyn.

Marilyn stressed the farming nature of the area and the long-term residency of the people. The quiet yet intense tone of her voice, her manner, her message, all projected the determination to fight. "The purpose of our meeting tonight is to discuss impacts. There are so many, we could easily keep you here until midnight. But one thing we want to make clear—we have not folded! We are simply calling a spade a spade."

At Marilyn's nod, Etta Lane flipped on the overhead. A map from OWMC's documents, delineating the Nuisance Impact Zone, filled the screen. Utter silence filled the hall. "As you can see," said Marilyn, as Marie used her pointer to trace the outline, "the impact zones zigzag along existing property lines and extend out anywhere from 500 to 1500 metres from the site." Marilyn looked up, distracted by whispering about five rows from the back of the hall. She spotted Mary Lou Garr, the local Communications Officer from the OWMC's Smithville office, and another woman carrying on an intense conversation.

Marilyn looked down again at her notes. "By keeping the

Impact Zones small," she said, "and labelled *Nuisance*—" Her voice fairly dripped with sarcasm. "—the OWMC thought to convince the public that the dangers were insignificant." She then cited examples of OWMC's attitude toward the residents, always harping on their "fear," as they called it. Then she explained the petition in detail, and urged any resident who had not yet signed to please do so. "At present, we have over 500 signatures."

Steve Dinga, Don Austin, and other men from the group acted as ushers. They started copies of the petition circulating from six points in the room.

John Dykstra then made a lengthy presentation of the report by Rudi Ohler. As the minutes ticked by, Marilyn watched the audience. Scarcely a sound reached her ears, except for the persistent whispering by the two women from the OWMC's Smithville office.

Next, Jean gave a brief rundown on the effect of airborne pollutants on the food chain. Then John Dykstra provided information on the increased truck traffic. Not only would toxic wastes, chemical additives, and reagents for the physical/chemical treatment plant be trucked in, there would be trucks carrying clay away from the landfill along with employee traffic both during and after construction. Some trucks, he noted, would pass the local public school, and many would travel the same routes as the school buses.

Then Marilyn, who treasured silence, tackled her pet peeve—the noise impacts. Etta Lane placed a map taken straight from OWMC's documents on the projector, and Marie traced the irregular boundary lines of the Noise Impact Zone. "The noise does not stop at the dark line," Marilyn told her listeners. "People outside the noise impact zone will still be subject to noise levels in varying degrees, all day long, day in and day out. But never mind. The OWMC has kindly offered to make people virtual prisoners in their own homes by sealing up the windows and installing air conditioning."

"That may be fine for some, but what about the farmer working in the field all day long? What about the recent concern that tightly sealed buildings may have serious air-pollution problems? What about the family who likes to have the windows up? Or wants to spend time outdoors?"

She motioned for Etta Lane to change the transparency on the overhead. Onto the screen flashed three diagrams, side by side. "These are noise barriers, enclosures measuring roughly twenty-four and a half feet by twenty-four and a half feet. These enclosures would be six or seven feet high. They're called the 'outdoor living area,' and as you can see—" At the pause, Marie tapped her pointer on each of the diagrams, "—they come in three designs." Marilyn waited for the laughter to die down, the only sound the audience had made in nearly two hours. She turned serious and her voice hardened. "I believe it's referred to as a 'place of refuge.' For example, if I wanted a few minutes of quiet, I could go out and sit in this little cubicle—"

A commotion near the rear of the hall made Marilyn look out over the crowd. The OWMC Communications Officer was shaking her head vehemently and denying Marilyn's statements in an audible voice.

Marilyn's temper soared and, without thinking, she snatched up her copy of "Background Material" from the table beside her. Waving it in the air, open to the diagrams on the screen, she cried, "This is your report, not ours!"

The anger in her voice made the rest of her presentation more forceful than ever. Afterward, John Jackson once again made a case for, as he called it, a reasoned solution to the toxic waste problem—recycling, reclaiming, and above-ground storage, as opposed to landfill and incineration. "But the OWMC dismisses all our solutions," he declared. "We are described as silly, and told that if we are not in favour of the proposed facility, then we are the cause of the problem."

As the meeting drew to a close, Marilyn appealed to the audience for financial support. She outlined the expenses they had incurred by holding the meeting. Then those acting as ushers passed containers for donations.

Jean Trudell concluded the meeting by touching upon the effects on municipal finance, air quality, and decommissioning. From beginning to end, the Concerned Citizens had made a strong case for resident dissatisfaction. Now, Jean stressed OWMC's lack

OWMC—The "Wasted" Years: The Early Days

of understanding of the community by reading the statement on the last page of "Background Material": "The area is expected to remain relatively stable and residents' satisfaction with the community is likely to remain high."

Paul rose to close with prayer, and Marilyn felt a rush of exultation. We did it! she thought. We held this audience for two solid hours, with no breaks or disruptions except for time out to sign the petitions and later to pass the hat. Both exhausted and exhilarated, she felt weak in the knees as she descended the stairs. She had no doubts that the meeting had been a smashing success. She made a point of thanking Etta Lane and inviting her to the celebration afterward at the home of Jean and Frank Trudell.

As the crowd thinned, a slender young man with a serious expression approached Marilyn and pumped her hand. "That was an excellent meeting," he assured her, "except for one thing. This was a public meeting. You should not have opened and closed with prayer."

Marilyn felt as though she had just been punched in the stomach. Dumbfounded, she stared at him, unable to utter a reply.

"As you saw," he explained, "people got up and left."

"One reporter left, that was all," she reminded him. "Anyway, at our meetings, we always open and close by asking God for help and wisdom."

"Well," he said, giving her the benefit of a thin smile as he turned to leave, "I just thought I'd give you my opinion."

The Concerned Citizens quickly pulled the display down from the wall, gathered up their tablecloths and literature, and headed for Trudells'.

As Marilyn walked in the door, she exclaimed, "Oh! Look at the cake!" A slab cake, iced in white, declared in green lettering, *"The Concerned Citizens, There is a better way."* Eyeing the platters of food, she declared, "I'm starved! I haven't eaten properly for days."

The next day, Doug Draper reported the public meeting at length in the *Standard*. "It's a good article," Marie told Marilyn over the telephone, "right up to the last paragraph. Then he quotes OWMC Communications Officer Mary Lou Garr, who said that the Concerned Citizens group is basing many of its charges on information it obtained from the Corporation's own reports, which contain the good side and the bad side of everything. 'We have provided (the ammunition) for them by putting everything in that environmental assessment.'"

Edith Hallas' report on the impacts of the Tricil operation near Sarnia made most of the papers. In an attempt to downplay her remarks, Leslie Daniels, OWMC's Regional Communications Officer, told the *West Lincoln Review* that, "the Tricil facility is old and frequently operates in violation of its certificate..."

On March 16, 1988, just a week later, the *Review* published a letter to the editor written by John Jackson.

> Tricil's so-called "old" incinerator was opened in 1983. Ms. Daniels should have no problem remembering its grand opening; she attended as a guest of Tricil. This incinerator was then—and continues to be—"state of the art," a phrase the OWMC loves to use to describe its proposed facilities.

He revealed that an OWMC report comparing incineration technologies had concluded that liquid-injection incineration was the highest-ranking technology. The OWMC had chosen to build a rotary kiln, not for environmental reasons or because it was a superior technology, but because it could burn solids as well as liquids.

In an editor's footnote, Leslie Daniels defended her statement. She contended that her information was correct and concerned the general history of Tricil, which went back to the 1970s.

"Why is it," Marilyn asked Marie, "that the OWMC always gets the last word?"

— 27 —

The first three months of 1988 had seen no let-up between the two opposing sides in the battle for media attention. In January, OWMC Director of Marketing Bill Lightowlers told the *West Lincoln Review*, "OWMC will accept toxic waste from the U.S. We can't stop wastes from coming across the border, as they do now, but OWMC's proposed facility will not be catering to an American market."

The article drew a flurry of letters from outraged residents, which in turn led to cries of NIMBY. To Marilyn's surprise, the strongest words of defence came in an editorial in the *West Lincoln Review*. It was easy, the editor said, to brand OWMC's protesters as NIMBYs, and maybe they were. But they would have to leave the neighbourhood and continue protesting to prove otherwise. Either that, or put out the welcome mat.

"Unfortunately," the editor wrote, "those many others heaving a collective sigh of relief can all wear the halo of caring for the environment without ever being branded as THINIMBY (Thank-Heaven-It's-Not-In-My-Back-Yard)..."

Toward the end of March, almost as if the OWMC had just come across Marilyn's letter to the editor decrying the Crown corporation's intention to use chlorine in its processing of waste, Communications Officer Murray Creed informed the *Lincoln Post Express* that OWMC had chosen reagents based on their "lack of hazard," and would not be using chlorine.

During this time, Marilyn sent a steady stream of letters to the editor. She asked why the OWMC didn't allot some of its sixty-eight million dollars to help waste generators buy equipment to cut their wastes instead of building a huge facility. She wrote an open letter to the Ontario Federation of Agriculture decrying the OWMC's restricted Nuisance Impact Zone. She pointed to the unfair advantage the OWMC gained from its publication, *Update*, paid for out of the public purse. It had a circulation of 80,000, while residents had only letters to the editor and what limited coverage they could coax from the media.

At the monthly Coalition meetings, attention focused on funding, what consultants to hire, and the unhappy realization that the OWMC had long ago cornered a large percentage of the most highly regarded people in the business. Or it had, in some way, connected itself to so many legal and consulting firms that the opponents were finding it almost impossible to secure expert witnesses. In some cases, the search had already extended beyond Canadian borders into the United States.

Inevitably, at some point during each meeting, came the reminder that Coalition members should be working on presentations to make at the Environmental Assessment hearing. Marilyn felt overwhelmed by panic at the thought of giving public testimony. Her cousin, a lawyer, had told her flatly that OWMC's lawyers would tear her to pieces—and she believed it. As a result, between her natural reluctance and her newborn fear, she invariably slid lower in her chair, fixed her gaze firmly on her notebook, and gave a definite, although almost imperceptible, shake of her head. Let Edith give evidence, she thought. Edith, outgoing and confident, welcomed opportunities to give voice to the wealth of information she had acquired.

Time and events telescoped into a tangle of rushing, always rushing somewhere. The group barely had time to collect their thoughts after the public meeting in Wellandport Hall before they were off to attend a mock hearing at Brock University in St. Catharines. Marilyn could never quite remember how the Concerned Citizens had learned of the hearing, but she thanked God that they had.

The organizers had brought in an expert on landfill to add authenticity to the event. His testimony revealed that, unless the treated waste slated for the OWMC's landfill had been encased in ceramics—a horrendously expensive procedure—the chlorides contained in the waste could not be kept out of the groundwater.

Marilyn furrowed her brow as her mind raced in circles. What were chlorides? And why hadn't she heard of them until now?

Eleanor Lancaster, who chaired the mock hearing, recognized the residents' confusion and asked the witness to clarify. Chlorides, he told the audience, were salts produced in high quantities by the treatment of chemicals—impure salts, which retained a measure of toxicity. Salts that would, in time, enter the drinking water used by humans and livestock alike.

Stunned, Marilyn looked from Clifford to the others in the group. Shock and outrage marked their features. In an atmosphere of intense silence, the landfill expert explained that the Crown corporation planned to dump the waste in slurry form from trucks much like cement mixers. But, in his opinion, the liquid content in the cement would be too high to make good-quality cement. Plus, he said, the salts themselves would be deadly to the concrete.

Afterward, the Concerned Citizens met at a nearby restaurant for coffee.

"Why haven't we come across chlorides in the documents we've read?" Reggie questioned.

Marie turned to Marilyn. "You have to write a letter on this!"

Marilyn shook her head. "Not without documentation. We have to search through everything on landfill until we find the information about chlorides, or salts, or whatever they call them." She made a gesture of frustration. "I don't know how we could have missed it."

"Maybe they didn't want us to find it!" Don Austin declared in his deep, penetrating voice.

John Dykstra gave a mocking laugh. "You mean, the way school bus routes aren't under transportation, they're under finance."

"And driveways onto the truck route in Vineland aren't under transportation, they're under social impacts," Reggie added.

Don Austin muttered, "Or maybe it's not there at all."

Marilyn reflected on Dr. Chant's strong letter to the editor written nearly a year and a half earlier. In it, he had belittled the delegation to Pelham Council led by Mr. Edward Heuniken for the concerns expressed over the possible contamination of the groundwater under the area surrounding the proposed site. In that letter, Dr. Chant had assured the delegation that any contaminants remaining in the solidified waste residues would be isolated from the environment.

"And to think I halfways believed that letter," Marilyn said, in a low, sad voice.

* * *

The Concerned Citizens searched the OWMC documents in vain. They could find no references to chlorides or salts.

"Obviously we don't have the right report," Marilyn told Clifford. "I guess I should go to the OWMC office in Smithville—but…" She knew she never would. The minute she stepped through the door, she would be under attack. The encounter last time had left her shaken for days. And in the end, she had failed to get the data she had gone for.

* * *

On March 21, 1988, at seven-thirty p.m., in their continuing effort to force the OWMC to expand its impact zone to a five-mile radius, the Concerned Citizens took their petition with more than a thousand signatures to the town council in the neighbouring municipality of Lincoln. The petition, amended by West Lincoln's council to read a "minimum" of five miles, included those residents who lived along Regional Road 24, otherwise known as Victoria Avenue, the proposed toxic waste transportation route. After a brief period of questioning, Lincoln's council adopted the proposed minimum radius and accepted a photocopy of the signatures.

At eight-thirty, they made the same presentation to Pelham council. "Extending the boundaries," explained John Dykstra when questioned, "would show good faith on the part of the OWMC. If the Crown corporation stands firmly behind its claim that residents and their property would suffer no adverse affects, then the OWMC

OWMC—The "Wasted" Years: The Early Days

should have no trouble getting rid of property vacated by those who don't want to live in the shadow of OWMC's smokestack."

Pelham council asked for time to examine it. "I can't picture the OWMC wanting to buy the whole village," said Alderman Doug Beamer, "nor the whole village wanting to sell."

In exasperation, Marilyn told reporters afterward, "Rather than allowing the OWMC to first build its plant and then force residents to try to prove harmful side effects..., the Concerned Citizens' suggestion would guarantee nearby homeowners the chance to move away, no matter what. Sometimes it's impossible to prove harmful side effects. We don't want people trapped in an area where they can't move out because there's no policy for them. If it's left unchallenged, people will be stuck with it."

Jean Trudell further explained, "By lobbying for a minimum five-mile impact zone, our group thinks this strategy will keep the OWMC out by placing a formidable financial barrier in front of the Crown corporation."

When the Concerned Citizens had split up, Lynda and Bob Bradley, Ziggy Soyka, and Mike Kicul had formed their own group, known as the Niagara Residents for Safe Toxic Waste Disposal. The Niagara Residents had quickly recruited additional members, including the slim young man with the serious demeanour who had faulted the Concerned Citizens because they opened and closed their public meeting with prayer.

Without consulting either the Concerned Citizens or John Jackson, without fully investigating the strategy behind the minimum five-mile radius, the Niagara Residents had decided to publicly oppose the petition. Despite the explanations quoted clearly and accurately in the newspapers, the Niagara Residents for Safe Toxic Waste Disposal viewed the strategy as nothing more than a demand for compensation and issued a press release: "While we do not deny the West Lincoln property owners the right to demand compensation, we believe that such demands are premature at this time."

Marilyn felt sick with dismay. We're all on the same side! she thought, in frustration, as she pictured the weeks they had knocked on doors all for nothing. OWMC personnel would be laughing up their sleeves as one Coalition group shot the other in the back. Supporters would begin to have doubts. Politicians might withdraw or withhold their support. Marilyn laid her glasses aside and massaged her closed eyes with her fingertips. Did Bob and Lynda's group really believe that a grudge against the Concerned Citizens was worth undermining the fight?

Meanwhile, the newspapers had a heyday with both the petition and the apparent rift between member groups of the Coalition. "OWMC told to buy Fenwick," screamed the headline in the *Guardian Express*. The *Lincoln Post Express* took a less inflammatory position with its front-page banner: "Larger OWMC compensation area backed by Lincoln town council."

Norman Nelson, editor of the *Pelham Herald*, asked, "Are cracks appearing in the unity of groups opposed to the proposed toxic waste dump?"

Marilyn called John Jackson and apprised him of the situation. Over the next few days, the telephone lines hummed. John speedily prepared and issued a press release to clarify the position of the Coalition and its seven member groups. He stressed the total opposition of the groups to the OWMC's proposed facility. He explained that the Concerned Citizens had used the petition as a means to stress the serious ways in which the OWMC was understating the impacts.

Then he said, "Some people have misinterpreted this petition as a plea for compensation...The seven member groups agree that no compensation and no mitigation measures can make up for the serious impacts of this facility."

In an effort to solidify the Coalition, John Jackson arranged to meet with the two groups an hour before the regular monthly Coalition meeting. Right from the start, the meeting went badly. Mike Jones, the young man who had objected to prayer at the public meeting, had assumed the role of spokesman for the Niagara Residents, and neither he nor the rest of his group, except, perhaps, for Ziggy, exhibited the slightest desire to resolve the misunderstanding. Instead, the group's strong political leanings quickly surfaced, and Marilyn sensed a determination to wield power.

Marilyn expected John Jackson to take a strong stand on behalf of the Concerned Citizens. Instead, in the face of the implacable attitude of the Niagara Residents and their powder-keg personalities, she watched as, with a calm, mild-mannered voice, he distanced himself from the issue and struggled to negotiate a truce. Although she felt hurt and betrayed, she recognized his reason. He had determined, at all costs, to hold the Coalition together.

"You should have included us in your plans!" Lynda charged unfairly.

"Why should we?" John Dykstra replied. "You don't include us in yours. You don't even include John Jackson. If you had—"

Able to hold back no longer, Marilyn snapped, "It was absolutely stupid to air our differences in public!" She narrowed her eyes. "Lynda, you signed the petition. Why didn't you or your group object then?"

Before Lynda could respond, voices sounded in the hallway. John Jackson looked at the clock on the wall and called a halt to the discussions. Angry and resentful, Marilyn kept silent throughout the Coalition meeting.

I've half a mind to quit! she thought in rebellion, then dismissed the idea. She would not quit. Nor would the rest of the Concerned Citizens. And John Jackson had no doubt gambled on that.

* * *

The Concerned Citizens took their petition and their request for support to Wainfleet council. Because of the wild headlines in the papers and the suspicion of a rift among Coalition members, Wainfleet gave the group a cool reception.

"We've received almost one-hundred percent support," John Dykstra told Stan Pettit, mayor of Wainfleet and chairman of the Region's OWMC steering committee.

"But not from all of your own groups," muttered one councillor.

"What sort of impacts are you referring to?" asked Mayor Pettit.

"Well, for example," John Dykstra replied, his head tipped to one side as he organized his thoughts, "they'll be mixing grain from this area with other grain to reduce the level of contamination in the feed for livestock."

"Oh c'mon," scoffed Mayor Pettit. "Where did you read that?"

John looked anxiously at Marilyn. She shrugged. She had read it, but in a momentary lapse of efficiency, she had failed to write down either the name or the page number of the report.

Pettit shook his head. "When you are in an adversarial position, it's necessary to have something specific that can be put before the OWMC, that is worthy of further examination."

"We'll get it for you," John promised.

Pettit thanked the delegation for their presentation. "We'll discuss it and let you know."

Although the group hunted through document after document, no one found the reference.[3]

The next day, Marilyn learned that the OWMC had finally published its information on Waste Quantities. However, before the Township's consultants could review the material, they had to wait for the Crown corporation to approve the funding.

"That," Marilyn grumbled to Clifford, "could take forever!"

* * *

As spring approached, the Concerned Citizens explored the possibility of the group applying for intervener funding as a means of funnelling funds into the Coalition. At the next meeting, John Dykstra read a letter he had received, which sounded favourable. Despite repeated attempts to contact John Jackson, he had not heard back from him. Therefore, based on the letter, the group decided to pursue the idea. "We can find out the details," suggested John. "If we don't like the strings attached, we can refuse."

The next day, John Jackson called Marilyn. Any funding to the Coalition would be granted on behalf of the member groups, which received no funding. To apply separately for money would splinter the Coalition and siphon off money earmarked for the organization. "We won't do it, then," Marilyn assured John.

---

[3] Marilyn and John Dykstra stumbled across the statement several years later, when it was too late to be of use.

OWMC—The "Wasted" Years: The Early Days

Later the same day, John Jackson called Marilyn again. Mary Lou Garr had contacted Helen Zimmerman, secretary for the Coalition, to tell her that the Concerned Citizens were breaking away from the Coalition and going after their own funding. It was rumoured, John told Marilyn, that Ms. Garr had been contacting other member groups with the same information, and suggesting that the Concerned Citizens thought they were the only group doing anything.

Then John told her, "I had a call today from Tricil as well." Marilyn could hear the amusement in his voice. "You remember, after the public meeting, how Leslie Daniels tried to downplay Edith's report on Tricil's impacts by telling the *West Lincoln Review* that the Tricil facility was old and frequently operated in violation of its certificate?"

"Yes."

"Well, she phoned Tricil and told them she had not called their facility 'old.' She had been misquoted." John laughed. "I told Tricil that she didn't claim to be misquoted when the *Review* gave her an opportunity to respond." His voice lost its laughter. "I notice, though, that the papers don't contact me before press time for last-minute comments when the OWMC takes issue with any of *my* letters." Marilyn could sense the injury behind the words. Neither the press nor the politicians took the Coalition seriously or gave it the same respect they gave the Region or the Township.

*   *   *

During the first week of April, the Concerned Citizens received a file box bulging with OWMC's Draft Environmental Assessment reports. The group divided up the reports and took them home to study.

By mid-April, between the money left over after they paid the expenses for the public meeting and the donations that continued to trickle in, the Concerned Citizens had a balance of $414.36. They decided to refurbish their float to enter it in Dunnville's Mudcat Parade and in Smithville's Canada Day Parade.

On April 12, 1988, consultants for West Lincoln, along with lawyer Dennis Wood, made a presentation in the council chambers.

Janet Lees of the *Spectator* reported that the consultants "outlined fourteen areas where the OWMC has been found lacking in choosing West Lincoln as the site for its disposal plant." The *Standard* noted that the report charged, "there is little evidence West Lincoln residents were meaningfully involved in the corporation's site selection process…"

At the next meeting, held at the home of Kathy and Albert Vanderliek, the group decided to add the statement to their expanded presentation intended for Regional Council ten days later.

− 28 −

On Thursday, April 21, 1988, John Dykstra and the Concerned Citizens took their 1,200-name petition backing the five-mile radius for OWMC's impact zones to Regional Council. Right from the first, Marilyn sensed a lack of support from the politicians.

John used material from both the public meeting and the Township's recently released report. "It lists a number of alternatives not considered by the Corporation," John told the politicians, and proceeded to read the list. "Decentralized facilities, mobile treatment units, above-ground landfill, emphasis on Four Rs, consider siting in lands zoned industrial, or consider the environmental uniqueness of the Niagara Peninsula."

Thorold mayor William Longo stood up and undermined support for the residents' concerns by asserting that his municipality would welcome the OWMC facility. "We have areas zoned for such a facility and I'm disappointed the OWMC did not seriously consider any possible sites in Thorold." A seasoned politician, he played to the media and the television cameras, as well as to his fellow councillors. "I don't think...we can say we don't want it. We have to work along with the OWMC...There's no question we have to have it, and if it has to come to West Lincoln, so be it."

The statement drew a round of boos from the fifty-or-more residents in the gallery. "He's sick!" declared LaReine Foden.

Regional Chairman Wilbert Dick remarked acidly, "I'm sure Dr. Chant would be glad to get a letter from you volunteering your community."

After more heated debate, Regional Council voted to send the residents' request for support for the petition to its OWMC steering committee for consideration.

Alderman John Schilstra, chairman of West Lincoln's Toxic Waste Committee, had been among the spectators. Afterward, the Concerned Citizens invited him to join them at Jean and Frank Trudell's home for a brief meeting and refreshments.

During the meeting he said, "I personally feel that your petition should have had spontaneous support at Regional Council."

"We thought so, too," said Steve Dinga, with a quiet chuckle.

Schilstra went on. "I am very impressed with the attitude and commitment your group has...in opposing the OWMC facility." He leaned forward, resting his elbows on his knees. He looked down at the floor for a moment and rubbed his hands together, as though reluctant to say what was on his mind. Then he looked around at them. "I'm sure sometimes it may seem to be a waste of time and effort, since there is another group trying to undermine your good work...Do not let this discourage you. Keep up the good fight."

Again he gazed around the room, his glance touching for a moment on each face. "I was particularly pleased to see that you opened your meeting here tonight with prayer. As long as we all seek guidance and wisdom from above in this very difficult task, and seek His blessing in whatever you are doing, I'm sure this community will be rewarded for it."

John Dykstra looked at Marilyn. She gave a slight nod. He glanced at the others, his face solemn. "I'd like you to know—and I know I speak for the group—how sorry we are for the presentation we made that first night at council, when we demanded your resignation. It was a cruel and improper thing to do—not so much what we did, but the way we did it."

The group then explained how they had been taken in and manipulated by Alasse Plaines. "With being a new group," Marilyn explained, "we didn't know one another well enough to call a halt. But I can honestly tell you, nobody felt good about it."

"W-we'd like to ask you to forgive us," said Paul.

Too overwhelmed to speak, Alderman Schilstra nodded and Jean promptly served the refreshments. Then, as Marilyn drank her tea, Jean appeared at Marilyn's shoulder with a long, flat cardboard carton. "The Concerned Citizens have decided that you can't keep working out of cardboard boxes, so we've bought you a filing cabinet."

Flabbergasted, Marilyn opened her mouth and then closed it, unable to speak. Jean patted her shoulder. "Now, don't you start crying! And just so that Clifford doesn't feel left out, we've bought him a screwdriver so that he can put the filing cabinet together."

As laughter filled the room, Marilyn turned to John Schilstra, her eyes crimped in merriment. "No matter how bad things get, this group always finds something to laugh about."

*   *   *

The next morning, Mayor Longo's disastrous statement to Regional Council hit the airwaves. Radio stations CHOW in Welland, CHSC and CKTB in St. Catharines, CJRN Niagara Falls, and CJFT Fort Erie all carried the startling announcement that Thorold would happily accept the OWMC's proposed toxic waste facility. By night time, the newspapers blazed with similar headlines: "Thorold mayor would welcome OWMC plant." "'I am disappointed OWMC never asked,' says Longo."

Over the next few days, the story spread from the *Hamilton Spectator* and the *St. Catharine's Standard* to the *Tribune* and the *Guardian Express* in Welland, and, finally, to the *Toronto Star*.

Marilyn could have done damage to Mayor Longo. Nothing travelled as fast or as far as negative, demeaning words pronounced by a politician.

In a follow-up story, Janet Lees of the *Spectator* contacted the Smithville office. Leslie Daniels, the Regional Coordinator for the OWMC, said, "There's no chance the OWMC will even consider taking another look at Thorold...We're convinced that West Lincoln is the best site."

"Ironically," Ms. Lees wrote, "Ms. Daniels—who lives in Thorold—led the local opposition attack against a 1980 federal government proposal to locate a similar facility in Thorold."

Leslie Daniels, however, was quick to assure Ms. Lees that her reaction would not be the same today: "If OWMC had found a piece of land in Thorold, I certainly wouldn't be leading the opposition group in Thorold knowing what I know about the OWMC."

When Thorold's Town Council met during the first week in May, Mayor Longo's inflammatory statement hit the headlines once more. Thorold's Regional Councillor, Mal Woodhouse, refuted the claim that Thorold would welcome the toxic waste facility. "A municipal referendum taken two years ago showed an overwhelming majority were opposed to waste treatment in Thorold." He then referred to the vacant and highly controversial Walker Brothers' Quarry. "I will continue to oppose a site in Thorold. Storage of toxic chemicals underground can leak and we find out twenty years later…"

Thorold councillor Mike Strilliski said, "Why welcome it here when it is going to West Lincoln?"

In the midst of all the controversy, Clifford and Marilyn received a letter written in a shaky hand from Ella Krick, an elderly resident whose family had lived in the area for generations. Marilyn removed the ragged clippings featuring Mayor Longo, cut with an unsteady hand from the *Spectator*. In a letter that encouraged Marilyn greatly, Mrs. Krick wrote, "If the OWMC is put here, it will not be because your husband and you and a lot of others didn't try to stop it."

*　　　*　　　*

Meanwhile, Marilyn buried herself in research. The Ministry of Natural Resources had requested that the Concerned Citizens make a presentation to its organization on May 24. As well, she and the others had been gathering material for their second presentation to Queen's Park.

On May 11, 1988, the Concerned Citizens gathered outside the Regional Offices. After counting heads, John led the way into the room where the Regional Steering Committee would meet. It quickly became clear that the Committee would not support their petition for a minimum of a five-mile radius. Committee members insisted the residents did not have a strong technical case on the possible long-range impacts of the facility.

"Without data," Stan Pettit told them, "you can't make a case

OWMC—The "Wasted" Years: The Early Days

to the province for a larger compensation area." Marilyn knew that Pettit's remarks referred obliquely to the missing data on blending corn to reduce the risk of toxicity. "It's not that we don't sympathize with what you're saying," he added, "but if you're going to play hardball with the big boys, you have to have ammunition."

Afterward, when interviewed by reporter Doug Draper, John Dykstra told him, "We're disappointed the committee won't support the larger area now, in time for our meeting with Environment Minister Bradley."

"Where have you received support?" Draper asked, his pen poised over his notepad.

"West Lincoln and Lincoln councils have already supported us." John hesitated, then admitted with reluctance, "Pelham and Wainfleet are still reviewing the matter."

Without the full support of Regional Council, Pelham, or Wainfleet, the petition requesting an extended impact zone all but died. Still, they had the appointment with James Bradley, and they intended to raise a multitude of issues.

*   *   *

No matter what the month, Marilyn looked at a calendar jam-packed with obligations. In May alone, she and Clifford had fourteen events, including two weddings, a fiftieth wedding anniversary, a baby shower, and a graduation. In addition, Clifford and Judy constantly worked on restoring the MGA. Marilyn hated to admit it, but the little car looked pretty snappy. With the bodywork completed, it had been painted a pale butter-yellow. The chrome had been polished and, in some cases, re-chromed.

The *Review* printed the salaries paid to some of the OWMC's staff and consultants. "It boggles the mind," Marilyn declared, hit with a rare case of envy. "We work all ends and hours of the day, as hard or harder than they do—for nothing! Instead, we are out of pocket. And all we're doing is trying to protect what we'd worked for—before the OWMC came along."

Dr. Chant, Murray Creed, and Richard Szudy declined to divulge their earnings.

"You can just imagine what *they* take home," Clifford grumbled, "if OWMC's salaries and benefits for 1986-1987 totalled nearly three million dollars."

"If Dr. Chant and his buddies are such great environmentalists," Marilyn asked Clifford, "why don't they take just enough to live on—and no more?"

*   *   *

On May 19, the Concerned Citizens met at Clifford and Marilyn's home to work on the presentation to Queen's Park in June. Harry Pelissero still had not contacted John with a date or time. As a result, no transportation plans could be made, no press release could be issued.

Marie read a letter from the Coalition, asking each group for $500 to help with expenses for the hearings.

On Tuesday May 24, John and Marilyn made their presentation to the Ministry of Natural Resources in Fonthill. Marilyn almost lost her nerve when she discovered Leslie Daniels had also been asked to speak. But just as the group had unexpectedly learned of the chlorides problem at the mock hearing at Brock University, so they learned from Ms. Daniel's remarks that the OWMC had set up test plots around the proposed site.

The next day, at a meeting at Vanderlieks', John informed the group that they would not be allowed to meet with Jim Bradley. John, Paul, Clifford, and Jean volunteered to go to see Pelissero one last time.

A week later, the decision had been reversed and a date set—Thursday, June 9, from three-thirty to four p.m. The group closed their meeting by singing, "*Those Who Wait Upon the Lord.*"

Marie added, humorously, "And wait, and wait, and wait!"

*   *   *

Marilyn dreamed she could hear the school bell ringing. The insistent sound filled her, as a past public schoolteacher, with panic. I'm not prepared! she thought, with the frantic helplessness of nightmares. As the bell continued to ring, she ran down the hall toward her classroom. With every step, the deserted hallway stretched longer and longer before her, and the door to the classroom receded. As she thrashed in distress, the ringing became more distinct, and she realized suddenly that the telephone was ringing. She leaped out of bed and dashed downstairs to the kitchen.

"Hello?" she gasped. She untangled the long telephone cord and sank into the easy chair around the corner in the living room.

"I was about to give up," John Jackson said. Marilyn could hear the amusement in his voice. "I know you're a late sleeper, but I thought you'd be up by now."

"I was so tired, I decided to sleep in. I don't have to be *anywhere* today." She glanced at the clock on the wall. "It's only seven-thirty!" She folded her legs under her and covered her bare feet with her granny gown. "I suppose you've been up for hours."

"As a matter of fact, yes. I wanted to catch you before I left for the day. I am to be a moderator in October for one of the sessions for HazMat—"

"What's HazMat?"

"It's a conference on hazardous materials. I'm lining up speakers for the session on public consultation. I'd like you to give a short presentation on—"

"Oh John, I couldn't! I'm no public speaker!"

"You'd be giving a resident's view of the OWMC's consultation program."

Despite her dread of anything public, she felt the hook set. She had plenty to say about the OWMC's open houses and all the rest. "Who would be at this conference?"

She expected him to say environmentalists, other citizens' groups, people who would be friendly to her remarks. "It's an industry conference, held at the Hamilton Convention Centre. It's for companies involved in dealing with their own industrial wastes or the wastes of others."

Marilyn felt the bottom fall out of her stomach. "You mean corporation executives?"

"Anyone involved in the waste industry—"

"Oh no, John! No! I couldn't do that. What do I know—"

"You wouldn't be discussing the technical side. You'd be telling industry what's wrong with OWMC's public consultation process and what they should do to try to fix it."

Part of her felt awash in panic with the kind of terror that shrieked a continuous warning for self-preservation. Yet another part of her felt a stirring of excitement as her mind leaped to the question of what to say and how to say it. She exhaled heavily. "I'll have to think about it." Caution urged, "Stay safe, stay at home. Refuse right now. Get it over with."

Meanwhile, her mind busied itself by resurrecting the problems with the questionnaire, the complaints other residents had expressed to her, her own experiences with the superior attitude of OWMC personnel. "How long should the presentation be?"

"Twenty minutes to half an hour. You won't have any problems. You and the others held a two-hour public meeting."

"That was different. That was a friendly audience—mainly—except for our friends from the OWMC. And we're getting ready to go to Queen's Park."

"You've got four months to work on it. Just tell them what it feels like."

Marilyn hesitated, wishing John had never called. She exhaled again in contemplation. "I won't promise anything. I'll see what I come up with."

"Well," said John, "I don't want to push you—"

"Like fun you don't! You're like a mother hen, constantly urging the chicks from the nest." A thought struck her. "I won't go without Clifford."

"I should be able to arrange that," John assured her.

"Well, then, like I said, I'll let you know in a few days."

– 29 –

The sun shone brightly on Thursday, June 9, 1988, as the Concerned Citizens boarded the bus to Queen's Park in Toronto to meet, for the second time, with the Provincial Minister of the Environment, the Honourable James Bradley. Marilyn paused beside the driver and surveyed the other passengers. Not all of the Concerned Citizens had chosen to make the trip, so Jenny Dinga, Frankie Ozog, and Shirley and Jeannie May had been included. Even so, with only thirteen adults and six young people scattered throughout the seats, the bus looked empty. She picked a seat as isolated from the others as possible and sat alone, with her head back and her eyes closed. She had awakened this morning with a headache that had rapidly developed into a crushing migraine.

The group arrived in Toronto with little time to spare before they met with Harry Pelissero at twelve-thirty. With John in the lead, they took over the benches outside the legislative building and gobbled their lunches. Despite the sunshine and clear skies, a cool breeze blew.

They had barely finished their lunches when Harry Pelissero dashed outside long enough to advise the group that Mr. Bradley would be unable to spare the delegation more than fifteen or twenty minutes, if that. With a hurried apology, he rushed back inside. Stunned, Marilyn looked from Clifford to John, to Marie, to Marija, to Steve. The brilliance of the sunlight drilled through her eyes and into her tortured skull. Carefully, she lowered her face into her hands. She wanted to give up. She wanted to fling her arms into the air and shout, "I can't do this anymore!"

Instead, she raised her head, and rummaged in her satchel for a pencil and her copy of the presentation. Sentence by sentence, she evaluated the worth of it. What they would keep, she left plain. What they could skip, she bracketed and labelled omit. Then she marked John's copy and Jean's copy to match her own, as well as copies for Harry Pelissero and Jim Bradley.

She found that the altered presentation shook her confidence. It no longer flowed as smoothly as before, and she feared the logic of the arguments or the clear train of thought had been lost. In the middle of her worries, word came that it was time to go inside.

The group gathered at the foot of the Grand Staircase and Marilyn took several snapshots. Then Mr. Pelissero escorted them up the stairs and into one of the visitors' galleries overlooking the Legislative Chamber. Huge panels of carved mahogany and Canadian sycamore adorned the walls. From the lofty ceiling, four magnificent chandeliers hung suspended on long chains. A wide aisle divided the Legislative House, with three rows of padded blue seats on either side for the 125 Members of Provincial Parliament.

As the debates raged on with voices raised in rhetoric, Marilyn closed her eyes and touched her fingertips to her throbbing temples. She knew she looked ashen, she always did with these headaches, and she feared the waves of nausea would get the best of her. A faint rustling made her turn toward the door. Relief spread over her at the sight of John Jackson. If worst comes to worst, she thought, maybe John can step in for me.

Near the end of Question and Answer Period, Harry Pelissero rose and read the residents' petition to the House. Minutes later, he appeared at the door of the gallery and beckoned the residents to follow. As they trailed after him along the gallery of the West Wing, constructed of Italian marble, Marilyn took a hurried picture of Jean and Mary admiring the marble railings and pillars.

The delegation crowded into a small room. Marilyn had barely seated herself when Mr. Bradley came striding in, accompanied by his aides, whom she could hear muttering and complaining: "You don't have time to meet with these people. You're going to be late for..." Marilyn didn't catch the rest, but she could sense the Minister's restrained urgency.

Marilyn handed Mr. Bradley an amended copy of the presentation. Jean Trudell opened with the group's request for the expanded Nuisance Impact Zone, and John presented him with copies of the more than 1,300 signatures in support of their request.

Marilyn skipped the first page of her portion of the presentation and plunged into the issues of stress and social injustice. "When we speak out in opposition to the OWMC proposal, we are labelled a 'self-interest' group. Yet, no one points to the OWMC personnel or the consulting firms hired by OWMC as 'self-interest' groups, even though their professional lives hang in the balance. Careers will either advance or take a step backward based on the outcome of this project."

Marilyn felt out of breath as she raced from topic to topic, touching only on the high points. She drew attention to the controversy over the third incinerator planned for the Biebesheim facility: "The building of a third rotary kiln became an issue during the state elections in Germany in 1987. At that time, all three political parties promised not to expand the Biebesheim incineration plant." She looked Mr. Bradley straight in the eye.

When he had been in the Opposition, during the time the Conservatives were still in power, he had spoken out strongly against the OWMC. Since the Liberals had come to power, however, and he had become Minister of the Environment, he had consistently claimed he couldn't interfere. She went on. "Shortly after being elected, the party now in power in Germany broke its promise by instructing HIM to proceed with the installation of a third incinerator.

"Mr. Minister, every time we raise some issue, we are told to wait for 'The Hearings.'" She bracketed the two-word title with her fingers. "Yet, these hearings will be based on decisions and judgments made by the OWMC *now*."

Next, she attacked the Nuisance Impact Zone in detail. John Dykstra followed by outlining the possible problems a farm family would face, whether it moved or whether it stayed.

He referred to the findings in an official government report from the state of Hesse. "During the year 1984, a test grid around the emitting plant was established. Test results based on a three-year comparison show that concentration variations [of heavy metals] are becoming noticeable—two, three, four, and five kilometres from the object plant."

He then cited both real and perceived negative impacts, complete with documentation, on agricultural products. "PCBs found in milk caused a drop in sales, resulting in a four-percent cutback in quota to dairy farmers for a year. Recent deaths due to contaminated mussels affected sales of all shellfish. A shipment of hogs contaminated with sulphamethazine caused pork to drop twelve dollars per hundredweight the day the residues were detected." John went on for three uncut pages. "Forage crops such as alfalfa may build up levels of fluoride, toxic to grazing cattle, without showing visible injury to the plants. Sulphur dioxide inhibits photosynthesis...Heavy metals cause leaf injury and can accumulate in toxic levels in plants, animals, and human tissue...Plants injured by air pollution are more susceptible to disease...Increased spraying to combat increased susceptibility to disease would drive up costs to the farmer, add to environmental pollution, as well as increase risk to the food chain..." He then backed up his claims by quoting similar statements from OWMC's "Site Assessment—Phase 4B— Agriculture," page xiii.

Despite his obvious need to hurry, Minister Bradley took additional time to ask questions and make comments. Marilyn tried to impress upon him the strain they all suffered day in and day out. "The residents of West Lincoln feel like pawns in a chess game..."

John Dykstra emphasized the reason behind the request for a minimum five-mile impact zone. "OWMC personnel have told us that the vapour plume would circle in a seven-mile radius...Why did the OWMC bother asking for community input up to three and four miles away from the proposed site if, as they claim, the impacts to residents will be restricted to a mere 500 to 1,500 metres?"

With her head still throbbing and the light from the overhead fixtures piercing her eyes like a knife, Marilyn could not keep the bitterness from her voice. "The OWMC's inky-dinky little Impact Zones do not *begin* to deal with the negative impacts of such a facility."

In parting, Jim Bradley promised the group, "The environmental hearings will be as complete and thorough as possible." With his aides agitating at his side, he scooped up the papers on the table before him and strode from the room.

Outside, in the late-afternoon sunshine, Marilyn shook hands with John Jackson. "Thank you for coming. But you never said a word."

His wide smile resembled that of a proud parent. "I didn't need to. You all did a great job!"

On the trip home, Shirley May interviewed John Dykstra and Marilyn. "Do you think your trip to Queen's Park was worthwhile?"

"I think so," Marilyn said. "At least Mr. Bradley didn't misconstrue our request as a plea for compensation, like other politicians have pretended to do over the past months. He pegged it for what it was—an issue in our continued opposition—and I respect him for that."

She looked over at John Dykstra for affirmation. He nodded. "We presented the signatures and our point of view," he told Shirley. "And Bradley listened and took notes." He shrugged. "What more could we realistically expect to achieve?"

* * *

As Marilyn pushed open the back door, the shrill ring of the telephone felt like a dagger slicing through her pounding head. Carefully she lifted the receiver and placed it gingerly to her ear. Ella Krick's voice quavered with age. "Have you seen today's *Spectator*?" she asked slowly and with difficulty.

"No, Mrs. Krick, I haven't." Marilyn sat her purse and her dog-eared copy of the presentation on the kitchen counter.

"Mayor Colyn is threatening to pull the Township out of the battle against the OWMC."

"Why, for Pete's sake?" Marilyn exclaimed, holding her throbbing head.

"Apparently he met with Dr. Chant on Tuesday. Dr. Chant's only giving the Township half as much funding as the Township thinks it needs to build a case. So, Mr. Colyn says he's not willing to do only half a job."

Marilyn leaned against the door jamb and closed her eyes. The struggle was never-ending. She thought of a news clip she had once watched on television, of a community trying to hold back floodwaters. While the sodden workers frantically piled sandbags in one place, the raging river churned and boiled as it broke through somewhere else.

Her elderly neighbour went on. "Alderman Schilstra, Chairman of the Township's Toxic Waste Committee, agrees that the $445,000 Chant wants to give the Township isn't enough. But he thinks council has to accept the money and do the best it can. He says, "We cannot go into the environmental hearings with emotions, we must go in with arguments backed up with facts."

As Marilyn stroked the right side of her face, which had gone numb, she tried to sort out what she had been told. As usual, Colyn and Schilstra had taken opposing positions on the issue.

Ella Krick's labouring voice then turned heavy with scorn. "Of course, the newspaper interviewed that Leslie Daniels. She says if the Township pulls out, it'll only stall the hearings."

Minutes later, just as Marilyn hung up the telephone, Clifford walked in from checking on the pigs and picking up the mail. "Here's a letter and a flyer from the OWMC. They're holding another Open House in Smithville and Vineland—right in the middle of haying."

Marilyn took the pages from Clifford, plugged in the teakettle, and retreated to the living room, where she lowered herself carefully into her old orange recliner. The words blurred and spun on the page. With her face screwed up in pain, she squinted her eyes in an effort to bring the letters into focus. "Consultants and staff will be available from 4:00 p.m. to 9:00 p.m…"

"That's chore time," Marilyn murmured. "We hope that you will be able to take advantage of this opportunity since you are the public most directly affected by OWMC's proposal." This time, OWMC's agricultural consultants would be present.

At the sound of the kettle boiling, Marilyn started to get up.

OWMC—The "Wasted" Years: The Early Days

"Stay there," Clifford commanded. "I'll make your coffee."

"Could you bring me one of my pills from the bottle marked 'head'?" she asked as she settled back and closed her eyes. "I wonder how long it took the OWMC to figure out a date and time when none of us could come to their Open House," she murmured bitterly.

*   *   *

Marilyn wrote a letter on behalf of the Concerned Citizens, protesting the timing of the Open House. Mary Lou Garr replied with a letter titled, "You can't please everyone." The Niagara Residents, non-farmers, picketed the event. The following week, the *West Lincoln Review* reported that only about fifteen people attended the Open House at the Old Farm Inn in Smithville and only ten at Prudhommes' in Vineland.

In mid-June, the Concerned Citizens entered their float in Dunnville's Mudcat Festival Parade. Two weeks later, they took part in the Kinsmen's Canada Day Parade in Smithville.

"That's it for parades!" John declared emphatically. "I've cancelled our invitations to both the Labour Day Parade and the Grape and Wine Festival Parade."

"Good," murmured Marilyn, in unconcealed relief.

− 30 −

One morning during the final week of June, more than a month after the group's presentation to the Ministry of Natural Resources, John Dykstra called Clifford and Marilyn. "Have you got time to go looking for those test plots that Leslie Daniels talked about?"

"We'll just take the time," Marilyn told him. Inwardly, she gave a weary sigh. She already had more work piled up ahead of her than the day could hold.

As Marilyn understood it, the test plots were intended to form a basis for checking possible future contamination. Various sample crops would be tested for residues present now, and the data checked against the effect of OWMC's emissions once it was in operation.

By the time she had combed her hair and changed her clothes, John and Mary pulled into the driveway. John handed Clifford a photocopy of a map with the location of the test plots marked on it. "The girl in the OWMC office didn't really want to let me have it." His face broke into an impish grin. "I wonder why not?"

They stopped first at Stan and Helen Kszans'. "Is it okay if we go back along your fenceline and look at the test plot?" John asked.

"Sure," said Helen. "But there's not much to see."

A few minutes later, all four leaned on the fence and shook their heads in disbelief. The main plot, which was on the site, had obviously been spring ploughed, something no right-thinking farmer in West Lincoln would do.

"It looks like pictures I've seen of the moon!" Marilyn exclaimed, as she and John took pictures.

"You know," said John, "we should have been notified about

these test plots so that we could have observed what was done, when and how they did it." He jabbed his thumb in the direction of the spindly plants and trees. "That stuff's never going to grow. Then the OWMC will say, 'See? We told you this ground's no good for agriculture!'"

"And since when do we grow tobacco and gladiolas in this area as crops?" Mary questioned.

After taking more snapshots, they got back into the car and tracked down one of the off-site plots across the road, a metre and a half square and overrun with weeds. After driving around until they had viewed three or four of them, all in the same condition, they decided to head for home.

As they drove past the proposed site and turned the corner where Heaslip Road met Vaughn, John pulled around a parked van. Marilyn caught a glimpse of a man standing on the site at the edge of a swampy area near the roadside, waist-high in weeds. He was hammering something to the largest tree in a small stand of trees. She twisted around in the seat and strained to see out the back window, exclaiming, "Hey! What's that guy doing?"

John put on the brakes and backed up. "I don't know, but we'll soon find out."

The car stopped and all four got out. They walked over to where the man was packing up his tools. Marilyn stared at the white box nailed to the tree. From the road, it looked about the size of a shoe box, but wider. Large white letters printed on a scarlet oval proclaimed, "Danger." Beneath it, in black print, she read, "High voltage, keep away." A small padlock dangled from the box.

After exchanging a few pleasantries, John asked the man, "What are you doing?"

Instantly he tensed, his expression guarded. In a mild voice, Marilyn asked, "Don't you think those boxes are too easy for kids to reach if they are as dangerous as they say?"

"They're not dangerous," he said defensively. "The high voltage signs are just to keep people from tampering with them."

"Then what are they?" she asked.

"Noise monitors," he answered.

"Why are they here?"

"To measure the ambient noise levels of the area."

"I know I sound like a dummy," John said, with a short, choppy laugh, "but what does ambient mean?"

The consultant licked his lips, and Marilyn saw a flicker of anxiety in his eyes as he gazed around at his isolation. "Look!" he exclaimed in sudden agitation, "I'm just doing my job."

"Your job could affect the rest of my life," John retorted. "So, what does ambient mean?"

"It means the normal daily background noise for the area. This data will then be compared to the expected noise levels from OWMC's proposed toxic waste facility to determine the amount of impact on the area."

"Is this the only monitor?"

"N—no." Marilyn could see the nervous pulse in his neck jump and race. "This is the fourth one."

"How long will they be here?"

"Two days."

"Are you from around here?" asked Clifford.

"No."

Clifford gave a sarcastic laugh. "So much for OWMC's claims that it would give employment to local people."

The consultant snapped, "They hired people from outside the area so that people like you couldn't intimidate them or influence the results."

Marilyn squinted in concentrated thought as she pondered each fragment of information. "Have I got this right—the less noise these monitors register, the greater the impact of the proposed facility on the surrounding area?"

"Yes, basically." He looked uneasy with the direction her question had taken.

"Where are the rest of the monitors?" asked John.

The consultant hesitated before answering. "There's one on each of the four sides of the site."

"*Proposed* site," Clifford corrected.

Marilyn, her brows still furrowed in thought, said, "Now let me get this straight. The higher the ambient noise levels, the better for the OWMC. The quieter it is, the worse for the OWMC."

The man hesitated to consider. "I suppose you could put it that way, yes."

Still looking puzzled, Marilyn asked, "So, is that why you've placed this monitor at a stop sign, so that vehicles stopping and then accelerating will make more noise than a vehicle just going along at a normal speed?"

The man's face turned a deep, angry red. "It's here because this is the halfway point." Without another word, he snatched up his tool case and marched to his vehicle. He slammed the door and roared off down the road.

"Well," said John, "I think we'd better take a look at the other monitors."

"Just a minute," Marilyn told him. She took out her camera and took close-up pictures of the monitor. Then she moved back until she could include the road sign and the stop/yield sign.

Once inside the car, they drove to the corner by the old school house and turned up Schram Road. "There it is!" exclaimed Mary. She had to shout to be heard. The consultant had placed the monitor on a tree in the field where Ed Comfort was working with his tractor and baler. Each time the baler arm packed a clump of hay into the chamber, the governors opened up. Then, as the arm pulled back, the governors eased off. The steady rhythm altered only slightly when the machine spit out a finished bale.

When Ed saw John and Mary, and Clifford and Marilyn get out of the car, he stopped to visit. "How much longer will you be in this field?" Clifford asked.

"I'll be done tomorrow. Why?"

"Isn't that a beggar!" Marilyn exclaimed, shaking her head.

"Were you baling hay when that consultant put up his noise monitor?"

"So, that's what that is," Ed said, nodding his head. "Yeah, I was here. Why?"

Between the four of them, they filled him in. Ed grinned, the kind of resigned grin that Marilyn recognized. It said, "They're a crafty lot, aren't they?"

"Three hundred and sixty days of the year, this area is as quiet as a tomb. But today and tomorrow?" He shrugged and left the sentence unfinished.

With a wave, the four of them returned to the car and went in search of the other two monitors. One had been placed on a hydro pole along Highway 20. Marilyn got out and placed her hand on the pole. Sure enough, it hummed and vibrated under her touch. "That should boost the ambient noise level."

The fourth monitor had been placed in the middle of a secluded swamp, where tree frogs chorused constantly.

"He picked the noisiest spots possible," Marilyn said, enunciating each word carefully as she tried to contain her frustration. "Noise meters can't tell the difference between tree frogs and earth movers. To a monitor, it's all noise."

Minutes later, John dropped Clifford and Marilyn off in their driveway. As Marilyn turned toward the house, she glanced over at her well-tilled garden and then at the field of thriving corn west of the house. Thinking back to the hard, lumpy ground of the OWMC's test plot, Marilyn told Clifford, "I'm taking pictures for comparison. That way, if the issue of the test plots ever comes up, it's not just our word against the OWMC's *experts*." The disparagement in her voice showed how greatly her respect for experts had diminished.

* * *

At their next meeting, the Concerned Citizens decided to send a letter to the editor, objecting because they had not been notified in advance about either the test plots or the additional drilling which had taken place on the site the previous year, during the summer of 1987.

"OWMC brags that it has developed such an excellent public consultation strategy that they're thinking of selling it to other jurisdictions," John said. "So, why do we only find out about things after the fact?"

Marilyn wrote in John's name, on behalf of the Concerned Citizens. Leslie Daniels responded to the complaint by sending a photocopied log to both John Dykstra and John Jackson, listing all the residents who had observed the drilling in 1984, and a letter to the editor. "The fact that the Concerned Citizens were not formed at the time," she stated, "does not alter the fact that citizens from the community…did indeed observe the drilling."

Marilyn responded to the papers on behalf of the group. "As Ms. Daniels very well knows, the drilling in question took place on LF-9C in 1987, not in 1984… Ms. Daniels also suggests that in future it would be better if we raised our concerns directly with the OWMC. Why? Ms. Daniels' letter is a perfect example of the convoluted run-around we would get…"

Within days, John Dykstra received a letter from the OWMC's Smithville office, signed by Leslie Daniels. In it, she complained that the Concerned Citizens had made a number of presentations to various groups in the Region of Niagara:

> In most cases, a formal written brief was distributed to the participants and the media. OWMC, however, has never been sent copies of these position papers. OWMC has always made every effort to keep all interests informed of its activities. We would appreciate it if, in the future, you would send us copies of information you distribute.

Marilyn promptly responded in a letter to the editor. "…We would like to point out that the OWMC has been making public presentations all over the province for years, none of which has ever been sent to the Concerned Citizens."

A week later, John Dykstra received a formal request for copies of all written presentations made by the Concerned Citizens. Marilyn called John Jackson. "Is there some rule in the Environmental Assessment that requires us to send them copies of everything we do?"

John Jackson laughed. "No. That's just an intimidation tactic."

Marilyn read from the second paragraph of the letter:

OWMC would be happy to supply the Concerned Citizens with copies of written speeches...However, presentations made by the regional office staff are not done in this formal fashion, but we attempt to react and address current issues and questions raised during presentations in an informal atmosphere. In other words, we try to keep our presentations as flexible as possible and not second guess an audience's interests before the fact.

She laid the letter aside. "John, that's just another of OWMC's half-twists. I've seen her speak several times. No, she doesn't have notes, but she most certainly has her speech lined up in her head. She speaks first, and then responds to questions. She's second-guessing, just the same as we are, if that's the way you want to put it. But she's smart—she commits nothing to paper."

"Don't let this letter worry you," John advised. "She's just trying scare you into backing off."

Marilyn let out huge sigh. "What a relief! I don't want to have to deal with that office—ever—if I can help it!"

* * *

Marilyn stared at the newspaper clipping in her hand. Why did Donald Chant always manage to overshadow their efforts? Last year, right before the Concerned Citizens made their first and historic trip to Queen's Park, he had stolen some of their thunder by being appointed to the prestigious panel of the International Joint Commission, the official watchdog on Great Lakes pollution. Then, barely a month after their trip, Environment Minister James Bradley had announced that Dr. Chant had been re-appointed as chairman of the OWMC.

Now, just a month after the Concerned Citizens' second foray to Queen's Park, Dr. Chant had been awarded the Order of Canada by the Governor General of Canada, Jeanne Sauvé. The high level of public esteem attached to these appointments scared Marilyn. A person had to move in the right circles just to be considered for such honours.

She felt a black cloud of hopelessness settle over her. We're just a bunch of nobodies, she thought. Not like Dr. Chant, who had been an invited guest recently at a dinner given in April by Premier Peterson for the governor of Michigan. The faces of Environment Minister Jim Bradley and Premier David Peterson crowded into her mind. "In their place," she admitted glumly, "I wouldn't listen to us. I'd listen to Dr. Chant."

- 31 -

Marilyn rubbed her aching back as she glanced at the clock. Only four-thirty. She turned the potatoes from high to simmer and flipped the pork chops. I don't care what else needs doing, she thought, I'm going to sit down and put my feet up. As she turned to plug in the teakettle, the shrill ring of the telephone cut into her thoughts.

"Get the *Standard*," Marie snapped without preamble. "Read Edgar Lemon's letter to the editor." Her normally cheerful voice vibrated with fury. "I can't talk now," she stated abruptly. "I've got supper on the barbeque."

The phone clacked in Marilyn's ear and she hung up. With a quick flick, she turned off the burners on the stove, snatched the car keys from the hook on the wall, and headed to the corner store for the paper. Ten minutes later, she dashed up the back steps, turned the elements on under the potatoes and the pork chops, and spread the newspaper out on the kitchen table. With a weary sigh, she sank onto one of the chairs as she leafed to the editorial page. Between the exhausting work of summer and the never-ending demands made upon her time and energy by the OWMC mess, she felt constantly worn out.

The title, "Fleecing the taxpayer," made her heart sink. Edgar Lemon, from Niagara-on-the-Lake, had once told Marie Austin that if she didn't like the OWMC, she should move. Now he proposed that Canada should institute the "Golden Fleece Award" for the costliest boondoggle in government. He did not blast the OWMC for its scandalous spending. He blasted the opponents:

The Golden Fleece Award…honours those whose hands are deepest in the public till. The OWMC opposition merits such an award—that flock of lawyers, consultants, activists, and…politicians who…want more money, having spent over 2.3 million in the past three years with little to show for it…

To be accused of fleecing the public purse hit Marilyn hard. She felt as though she'd had the wind kicked out of her. Even worse, Edgar Lemon had timed his attack perfectly—the next round of funding hearings would take place in two weeks. As she rose slowly from the table and walked into the living room, she tried to judge the level of damage he had done. Public support was crucial for the opponents. Stir up enough pressure on the politicians over spending and they would cut funding to a pittance.

She dropped into her old orange recliner and stared sightlessly at the ceiling. No one understands, she thought. No one cares. In the end, the only thing that matters is money. Not what is right or just. Only money. She thought of all the hardship and work the citizens groups had endured, all the long hours, the fundraising, the reports they had ploughed through. Then, one letter, ignorant of the facts, and decrying their efforts as a waste of money, could undo the years of struggle.

She felt tears build behind her eyelids, but she could not weep. She was too weary to weep. The crushing injustice of Edgar Lemon's accusations sapped her already flagging will to go on. Let someone else raise up the residents' voice to the world, she thought.

At last, the shifting shadows on the ceiling reminded her of supper. She pushed herself to her feet and walked listlessly to the kitchen. She supposed she should write a letter in rebuttal. But at the moment, her mind refused to function and, quite frankly, she just didn't care.

In the days that followed, she tried to find the motivation to compose a letter refuting Mr. Lemon's remarks. But the sun shone and the grass needed mowing. She carried water to the garden and pulled weeds. She spent her time canning, freezing, pickling—physical work that drained her strength, but required little thought. Each day after lunch, she took her pillow and stretched out on the picnic table in the hot sun and watched the small tufts of white clouds drift and change shape in the cobalt-blue sky. Like the clouds, her mind drifted aimlessly, focusing on nothing but the rustle of the leaves in the maple tree, the cardinal whistling from the top branch of the dead tree in the bush across the road, the constant throb of crickets and frogs and insects.

A week went by before she turned to the Scripture in Chronicles: "O our God…we have no might against this great company that cometh against us; neither know we what to do: but our eyes are upon thee." The passage did not quicken her hope as in the past. She seemed numbed to everything, too weary even to rise from the chair and put the breakfast dishes in the dishwasher.

Instead, she reached over to the tape player and loaded the new tape Pastor Snyder had given her the Sunday before. It had been recorded by a group of street kids to raise money for a Christian group home where they had found refuge and hope. Though the lives of these teenagers had been turned around, the struggles with addiction and abuse were far from over.

She settled back in her chair, her eyes watching the sunlight dapple the leaves on the trees outside the window. The untutored young voices rang with a sincerity no polished choir could attain, and she felt her chest constrict. As the words of the last song flowed quietly into the room, an ocean of tears rose in her eyes and rolled down her cheeks. She covered her face with her hands and sobbed, the hard wrenching cries of intolerable pain.

During the days that followed, she played *Press On* over and over again. Each time, as the young voices rose in the strains of struggle and hard choices, she would feel her cheeks wet with tears. "I've given my all," she would whisper, her chest heaving with emotion. "I've nothing left." Then would come the words of courage and determination.

"In Jesus' name, we press on…"

✻ ✻ ✻

Despite the long, work-filled days, Marilyn found herself gradually gaining back her spunk. She had agreed to speak at HazMat

OWMC—The "Wasted" Years: The Early Days

before Edgar Lemon's letter had laid her low, and her name had subsequently been added to the official list. She couldn't back out. In a half-hearted piecemeal fashion, she began to work on her presentation—while rotor-tilling the garden, while chopping up tomatoes for chili sauce, while washing dishes.

She kept paper and the stub of a pencil in her pocket at all times, ready to jot down thoughts and better ways to express them. Ideas for cartoon-like sketches began popping into her mind, to emphasize the various points, and she roughed them out wherever she happened to be.

Finally, she decided to call John Jackson. She needed to know if a projector would be available. "There's no use going to all the trouble of preparing these drawings if there is no way to project them," she told him.

There was a moment of silence. "What kind of drawings?" he asked. Marilyn heard a high degree of uncertainty in his voice.

"Oh, cartoon-type drawings—something to nail down the points and hold the attention of the audience."

"I'd—like to see them before you use them."

"You don't trust my judgment."

"It's not that…" His voice trailed away.

"You don't trust the quality of my work."

"It—it's not that, either…" She heard a touch of exasperation in his voice. He didn't want to be questioned. "I'd just like to see them first, that's all." Nevertheless, despite his obvious reservations, he gave her the number to call.

Marilyn had hoped John would make the arrangements. She hated making phone calls almost as much as she hated knocking on doors. She dialed the number, hoping, with a total lack of logic, that no one would answer. A woman picked up the receiver and Marilyn gave her name. "I'm—um—s-supposed to speak at—uh—HazMat," she explained. She suddenly felt short of breath. "I—um—need a projector for my presentation, the kind that takes an image from a page and throws it onto a screen."

"I'll make certain there's one available for you," the young woman assured her.

After Marilyn hung up, she leaned on the kitchen counter and stared out of the window onto the yard. Something about the conversation left her with an uneasy feeling. But she couldn't decide what it was.

* * *

At every meeting of the Coalition, funding—or rather, the lack of it—entered into the discussions. The groups held garage sales, walk-a-thons, and collected donations. The Niagara Residents for Safe Toxic Waste Disposal moulded palm-sized decorative blocks of West Lincoln's famous clay, baked them, and sold them for two dollars apiece. John Jackson made the rounds of the municipal councils, requesting financial support to research alternative methods of dealing with Ontario's hazardous waste. Once again, the Concerned Citizens, along with Coalition members, drove to council meetings across the Niagara Peninsula. The effort met with little success.

St. Catharines' aldermen gave the group a scathing denunciation. "It's a well-formed political pressure group," stated Alderman Al Unwin. Alderman Bill Wiley added, "It appears that the bulk of the group's money covers salaries." A groan of disgust arose from the spectators.

"What an idiot!" Lynda declared.

However, Alderman Don Barber argued, "We are downstream of the emissions... I don't think it's unreasonable for the city of St. Catharines, with a budget of fifty million, to spend two thousand dollars to ask a couple of questions."

Alderman Joe LaPlant agreed. "This is a twenty-four-hour continuous use of a smokestack of the most toxic substances known to man. Two thousand dollars is a drop in the bucket, but we'd be saying a lot more... It's a small gesture to say we're concerned with what's going on."

In the end, the council turned down the Coalition's request. Marilyn felt as though the small group of residents had been cut adrift on an ice floe in the middle the ocean, with no help in sight.

At the next meeting, Edith Hallas came up with the idea of

putting on a supper. "It would not only be a fund-raiser, but it's a way to raise public awareness."

Marilyn, who had been involved in numerous church suppers over the years, predicted too many problems for a group like the Coalition. "I hate to be a wet blanket," she said, "but people in the Wellandport area have big families and not much money. They won't fork out nine dollars a ticket for a meal."

Yet, that night, after Clifford had gone to bed, Marilyn couldn't rid herself of the belief that, as a public awareness project, Edith's idea had enormous potential. Even after she went upstairs, she lay awake for hours as the idea tumbled around in her mind, shifting, forming, changing, until, in the morning, she felt convinced she had a workable plan.

"A fundraising barbeque," she told Clifford over breakfast. "That way, people can spend as little or as much as they want. Maybe have a reduced price for kids. And this way, we won't have a lot of salads and stuff left over."

That day, she spent hours on the telephone. First, she called Edith, to make certain she didn't mind Marilyn stealing her idea. Then she posed the project to the Concerned Citizens. Before the week was out, plans had been put into motion. John Dykstra called an impromptu meeting. Reggie brought a list of available dates for the hall. Marie and Don had obtained prices and the promise of a discount for the buns from their favourite bakery. Clifford had quotes for meat.

"We need to keep the price as low as possible," Marilyn urged. "Make it affordable for any and all. First and foremost, let's regard it as a means to generate community support."

"What'll we serve?" Marie asked. "Hamburgers, hot dogs—"

"Clifford wondered about back-bacon-on-a-bun. Get it sliced thick, five to a pound. He's already inquired about renting the Pork Producers' big barbeque. We can cook five or six pounds at a time—"

"More than that," Clifford interjected. "And what about some of our farmer's sausage, made from the whole pig? We could provide it at cost."

Mary spoke up. "I think we need something like hot dogs for the kids. And pop. John and I have already bought six cases. We saw it on sale."

"I think we should get *Snowbird Recycling* to put in containers for the pop cans," John suggested. "Let him put up a sign, advertising his service. We need to make people aware that they can recycle in this area. Maybe he could be there in person to sign up customers."

Next, they settled on the prices. Two-fifty for bacon-on-a-bun or sausage-on-a-bun. Coffee, tea, or pop would be extra. Two-fifty for a hot dog *and* a drink for the kids.

"What about a bake table?" Marija asked.

Marilyn felt skeptical. "Well, I suppose we could offer desserts. A piece of pie for a dollar…"

"I no mean that!" Marija insisted. "I mean sell—to take home."

Marilyn shrugged. "If you think…"

Marija patted her chest emphatically. "I *know*!"

John laughed. "Then I think Marija should be in charge of the bake table. All in favour?"

All hands went up. Reggie volunteered, "I'll make Rice Krispie squares for the kids."

Marija looked over at Steve Dinga. "I get your Emilia to help."

As the women volunteered their specialties, the men decided who would help Clifford pick up the Pork Producers' huge barbeque, buy charcoal, and help cook.

Marilyn volunteered to make flyers and design the tickets.

"I think we should keep to the town crier's scroll design we used to advertise the rally at Queen's Park and the public meeting. Everyone knows it by now," John reminded them.

For food tickets to sell, Marie suggested using a miniaturized photocopy of their banner with its slogan, "There is a better way!"

Marilyn jotted down the rest of the wording with care. She would need to go to town and buy a new supply of rub-on letters to make the masters.

Just before the meeting closed with prayer, Marilyn said, "I have a clipping to read. It's a quote by a famous general: '*He suc-*

ceeds who believes he can!'" She suddenly discovered she believed the general's words. Elated, she raised her fists and gave two thumbs up. "The way we've got this set up, we can't fail! And the food'll be so good, people will just *have* to buy seconds!"

*   *   *

By this time, the Concerned Citizens knew the routine necessary to noise their latest project throughout the community and beyond. Marilyn's letters to the editor now ended with a plea to readers to add their moral and financial support by attending the Concerned Citizens' barbeque. Flyers went up, Marie contacted radio stations and newspapers, and everyone sold tickets.

Once, and sometimes twice a week, they met to turn in their ticket money to Marilyn, tally ticket sales, and add to the growing list of items for each member to do or bring.

On August 3 and 4, with the barbeque just two weeks away, Mary Munro chaired another round of funding hearings in Smithville in the Legion Hall.

The first day of the hearings dawned hot and muggy. The sunshine poured down out of the heavens with a brassy glare, giving the impression it had already scorched the fabric of the day.

Inside the Legion Hall, the heat was oppressive. A decrepit air conditioner crammed into a small window rattled and roared, but gave out little relief. Marilyn surveyed the room, with the large rectangle of tables set up as before. Until these hearings, she had not fully appreciated the placing of parties around a negotiating table. Now she did. Last time, the OWMC had claimed the chairs at the foot of the table, giving Ian Blue and his associates direct eye contact with Mary Munro. Those seated along the sides had automatically fallen into the category of lesser beings.

Among the first to arrive, Clifford and Marilyn took seats in the front row. Soon afterward, John Jackson, with his usual smile, came through the door. John and Mary, Don and Marie, and Paul and Marija followed right behind him. Marilyn indicated the vacant chairs at the foot of the tables set up for the hearing. "Why does Ian Blue get to sit at the foot of the table every time?"

John Jackson looked taken aback. "I don't know why." He scanned the room. "He's not here yet." Decisively, he walked to the end of the tables, and placed his satchel on the chair Ian Blue had used the last time.

Doug Draper from the *Standard* stood on the landing. Unsmiling, he squeezed behind a table set in a small alcove. Marie shielded her mouth with her hand. "He hates me. I tore a strip off him the last time we were here."

Marilyn said nothing, but her instant response was disapproval. Nothing could be gained by alienating the press.

Just then, Ian Blue and his retinue of assistants clattered up the half-dozen steps and strode to the end of the table. Seeing John's satchel, Blue grabbed it by the handles and pitched it up against the wall. Astounded, Marilyn stood with her mouth open, then turned to John Jackson.

John exclaimed, "Can you believe that?" His face had reddened at the affront, and he wagged his head in disbelief as a slow grin spread over his features. "What arrogance!"

Just then the Hearings Officer came into the room and stated, "All stand." Residents who had congregated in groups hurried to their places and the room fell silent. Herman Turkstra, the lawyer advising Mrs. Munro, led the way to the head of the table. Mary Munro, tiny beside the tall rangy lawyer, took her place at the head of the table. "Be seated," instructed the Hearings Officer.

Following Mrs. Munro's opening remarks and instructions to the applicants, a representative of the Ontario Liquid Waste Haulers Association made a plea for $20,000. Next, the lawyer for Ministry of the Environment gave an update on the Ministry's study of OWMC's Draft Environmental Assessment documents. "It will take another four to seven months to conclude our review," he stated as he sat down.

Ian Blue for the OWMC rose. White-haired and lean, he spoke with his head back and his eyes focused on the wall above Mary Munro's head as he launched into his presentation. "OWMC plans to have fifty witness panels," he announced. A look of shock rippled over the faces at the table.

"Fifty!" Edith Hallas gasped behind Marilyn.

Mr. Blue waited for the murmurs to die down. "Each panel will have at least two members on it, some will have four or five...Each witness will testify for probably half-a-day to a day, and some possibly for a week." He swung on his heel to view the audience. "Surely, since everything has been done carefully, opponents will be able to accept some data without challenging it." He swung back to face Mrs. Munro. "We suggest that funds to the interveners begin at the time of public notice, when the Final Environmental Assessment is received."

Mrs. Munro asked, "Mr. Blue, can you give an estimate of how long you expect the OWMC's case to last? Are you suggesting fifty weeks? Have you done any serious scoping?"

As Mr. Blue embarked upon a detailed response, Marilyn puzzled over the word scoping. She found it difficult to understand the proceedings when she didn't understand the lingo.

Next, Stan Stein, the lawyer for the Regional Municipality of Niagara, gave a detailed background of expenses. "The taxpayers of the Region should not have to pay to examine the proposal for an undesired and uninvited facility, designed to serve the whole of the province. Therefore, we request full funding." He glanced down at his notes and adjusted his glasses. "Also, we had anticipated a hearing which would last one to one-and-a-half years. Now," he flung an accusing glance at Ian Blue, "we hear that the OWMC could take as long as fifty weeks to present its case. This hearing could take five years to complete."

Marilyn's heart seemed to stand still and she heard voices all around her murmur in dismay, "Five years!"

Five years, she thought. Five more years. As she tried to grapple with the mind-numbing prospect of a future filled with unending hearings, the lawyer for the Region droned on in the background. "The Region needs to address all the major issues—Need, Alternatives, Site Selection, and both the impacts and full mitigation in the event we lose the case. The Region is therefore requesting $4.38 million." He spoke of the Region's need for adequate work and storage space. "The OWMC proposes joint facilities. It won't work... OWMC's time frame far exceeds anything previous. We need transcripts available on a daily basis. OWMC's provision for computer hardware and software is inadequate. We also need access to the OWMC database. Access is limited at this time, and some access is denied."

Stan Stein then argued for a staged hearing. If the OWMC was unable to prove "Need," for example, it could be the end of the hearing.

Mary Munro then posed a series of questions to him. Finally she asked, "Why can't the Township of West Lincoln and the Region of Niagara see things in the same way? What makes them distinctly different?"

"Without going into endless detail," answered the lawyer for the Region, "the OWMC has recognized the Region and West Lincoln as separate entities."

"Thank you. We'll adjourn for lunch and reconvene at two-ten."

*   *   *

After lunch, Mrs. Munro called upon Joe Castrilli, the lawyer representing the Coalition. Dark-haired and slight of build, he resettled his glasses on his nose and shuffled nervously through his notes. "The Coalition has been incorporated since August 1987. But the group has been in existence since the early eighties. We intend to present a complementary case to the Region's and to West Lincoln's. A supplement, if you like. Therefore, the timing of the funding is crucial." In seemingly endless detail, he spelled out the Coalition's request. He wanted the funding to begin as soon as the OWMC filed its Draft E.A. If the parties were forced to wait for the Final E.A., the funding wouldn't arrive until just before the hearings began, leaving almost no time to study the thousands of pages in the final documents.

He looked up at Mrs. Munro. "Mr. Blue tells us today that he will present fifty expert panels of witnesses. We have seen no witness statements, nor the data involved."

As Mr. Castrilli continued to spell out the problems, Ian Blue interrupted. "We have seen much speculation in Mr. Castrilli's data." As he continued to expand his criticism, Marilyn felt her

OWMC—The "Wasted" Years: The Early Days

stomach knot. Was Blue right? Did they have faulty data that wouldn't stand up to scrutiny?

Joe Castrilli twisted his hands together in a gesture of nervousness. His voice lacked assurance as he continued to list the impediments facing the Coalition. "OWMC's suggestion of one set of transcripts for all interveners is ludicrous." He glanced at Ian Blue. "As for the legal fees, if Mr. Blue will work for the legal tariff fee, I will accept the same."

As he moved into Hearing and Pre-Hearing procedures, he repeatedly used the term "evidence-in-sheaf."[4] Marilyn wished she knew what it meant. "I would like to remind the Board that the mere existence of the OWMC's data base is of no value to the Coalition without access to that data."

Blue rose to contest the Coalition's right to funding. As he started to quote figures, Joe Castrilli jumped to his feet. "The Coalition has never received a dime for legal fees."

At that point, Mary Munro adjourned the Funding Hearing until the next morning.

* * *

On Thursday morning, Dr. P. Boldrini made the first plea for funding. Although Marilyn had never heard of him, apparently he was well known to those involved in environmental hearings in numerous jurisdictions. He applied for $20,160 on the grounds that he would bring an unbiased view for the board. "Opinions at hearings such as this are bought. Each side shops for experts who will promote a particular point of view." With a slight, old-world bow, he sat down.

Roger Cotton, on behalf of West Lincoln Township, took up the rest of the morning. Marilyn scribbled like mad. He didn't just say, "We want money." He spelled out the flaws in the OWMC documents, areas that needed more study, emphasizing concerns specific to the Township and residents living near the proposed site. He informed

---

[4] The correct term is "evidence-in-chief"—the initial evidence of a witness before any cross-examination takes place.

Mary Munro that if the Township did not receive funding to cover its work on *Phase 1: Need for the Facility*, the residents would be facing a tax increase of 6.8 percent for 1988 and 121 percent for 1989. He then pointed out that, although some funding had been approved, the OWMC had so far made only a partial payment to the Township.

An hour before noon, the question arose as to the best location in which to hold the hearings. Mr. Cotton insisted that they should be held in a place convenient to the residents.

Marilyn felt a frown form around her eyes. All along, she had assumed that the hearings would be held locally. She listened closely as the lawyers tossed around ideas for location. Then, as quickly as the flip of a page in a book, the topic reverted once again to the lawyers' pet complaint—the hourly rate proposed for the interveners' lawyers, known as the Legal Aid Tariff.

As the debate raged, Marilyn could feel her temper rise. They cared more about how much money they would make than what the OWMC's proposal would do to the surrounding countryside, the groundwater, or the residents' lives. Yet, part of her agreed with the arguments. Why should the OWMC's lawyers earn the going rate, while others worked for much less?

The afternoon session began with Mrs. Munro posing questions to Roger Cotton, the lawyer acting for the Township. Marilyn continued to jot down phrases and sentences, but in truth, with no background understanding of the process, she could not discern the significance of many of the issues.

Mrs. Munro looked at the clock on the wall. "We'll take a short recess. Then we'll hear closing remarks."

Speculation among residents ran all the way from no money and no intervener status to all they needed and wanted because they had a right to it. Although John Jackson had his hands pushed into his pockets in his usual casual manner, Marilyn could sense the tension in him. Everything rested on Mary Munro's decision. The Coalition must gain the right to argue issues at the hearings and the funding that went with intervener status. The last months had proven what Marilyn and the others had long known—they could not possibly raise the necessary funds on their own.

Ian Blue opened the final session. He suggested that the lawyers for the opponents should receive market-value payment. Marilyn wrote down the words, but had no idea what the term meant in dollars and cents.

The lawyer for the Ministry of the Environment recommended that funding be made available when the Final Environmental Assessment was received, as Mr. Blue had outlined, not sooner, as Castrilli and the others had requested. Mr. Stein rose, spit out some venomous remarks about other minor applicants such as Mr. Boldrini, and sat down.

John Jackson spoke on behalf of the Coalition. "The Coalition crosses the lines of municipal, regional, and provincial politics. It is citizens, speaking out responsibly. They are not NIMBYs; they simply have different questions. For example, the Township asks, on behalf of the residents, 'Will my water be affected?' The Coalition asks, 'What other way can toxic waste be dealt with?'"

Roger Cotton raised a final complaint about deadlines. Ian Blue interrupted him to assert that nothing was carved in stone. The applicants completed their arguments and a tense silence fell over the room as Mary Munro spelled out her recommendations. She named the Region of Niagara, the Township of West Lincoln, and the Ontario Toxic Waste Research Coalition as the official interveners. The hall rang with cheers and wild applause. John Jackson's normally calm façade fell away, and Marilyn thought he would explode with excitement.

"However," Mary Munro cautioned when the noise died down, "we would like the Coalition to co-operate with the other parties." Then she fastened her gaze upon Ian Blue. "I am assuming as well that the Ontario Waste Management Corporation will continue to fund administrative structures up to the end of '88."

She then recommended to the Ministry of the Environment that fees for legal representation be at the provincial rate. Thus, instead of the Legal Aid Tariff rate, set at $67 to $83 per hour, the lawyers would receive $100 to $150 per hour.

Marilyn turned to Edith Hallas. "What's Blue's salary?" she asked in a whisper.

Edith whispered back, "He gets the high end of market rate, which ranges from $175 to $225 per hour."

"Figures," Marilyn remarked. Still, she thought, Blue had made an effort to put his fellow lawyers on a more even footing. She narrowed her eyes in thought. Or, had he figured he could afford to be magnanimous because he knew the opposing lawyers wouldn't get market rate? Suddenly, Marilyn thought of the Coalition's lawyer. She asked Edith, "Will Joe work for the provincial rate?"

Edith looked dubious. "I don't know. It depends on what his law firm decides."

Marilyn sighed. Even if she and Clifford had to take a second mortgage on the farm, perhaps it would be for a lesser amount. But on second thought, with five years of legal fees accumulating hourly, she and the others would most certainly face bankruptcy. Even if we win, she thought suddenly, we could still lose everything.

Mrs. Munro then outlined the stages of the hearing. "Stage One: Determination of the Issues and Position on the Issues. Stage Two:…" Inwardly, Marilyn shrugged as she scribbled. Hadn't they already determined the issues and taken a stand? "…And Stage Three: Conducting of the Hearings." Mrs. Munro surveyed the array of lawyers. "If there is nothing more, we are adjourned until Wednesday, September 14, 1988, at ten a.m."

John Jackson was ecstatic. "I can't believe it!" he kept saying. "I can't *believe* it!" He could hardly speak for grinning. "Do you know what this means? We convinced Mrs. Munro that we have legitimate concerns and that we represent a large segment of ordinary people." His face sobered, and he fixed his gaze on Marilyn. "It was crucial for us to present a unified group—no splits for Blue to use against us." She knew he referred to the conciliatory position he had taken in favour of the Niagara Residents for Safe Toxic Waste Disposal. She understood it, but she still thought it unfair.

Instead of answering, she changed the topic. "Does this mean we can cancel our fundraising barbeque?"

"Oh no!" John exclaimed. "We don't know yet how much funding we will receive. Also, we need to raise money to pay the

Coalition's own expenses. The battle is not over. We have to win at the hearings, but we still need to stage public events to keep up public support." Then he broke again into a contagious grin. "We gained official intervener status! I just can't *believe* it!"

– 32 –

The day of the barbeque dawned hot and sultry. Marilyn's clothing clung to her like wet rags. Fortunately, the group had set up the tables and chairs the day before, and organized the kitchen. They had loaded the refrigerators with cases of soft drinks and piled the rest in tubs filled with ice. Three signs placed over the pass-through window of the kitchen indicated the serving lines—bacon-on-a-bun in line one, sausage-on-a-bun in line two, hot dog plus a drink in line three. As at the public meeting, they used the walls to display the latest information on OWMC's proposal.

Family and friends had been conscripted to help. Judy and young Henry Dykstra blew up green and white balloons, and fastened them around the hall. Kathy and her friends Lauri and Blair Lagdon from Burlington sliced dozens of rolls, still warm from the bakery. Kim Dykstra and young Maija Balint filled sugar bowls and milk pitchers and set out relish, ketchup, mustard, and barbeque sauce. Reggie took care of setting up the two big coffee makers and boiling the teakettles for tea. The group had designated Marilyn as hostess. "This barbeque was your idea," Marie said with a merry laugh when Marilyn protested. "So you get to greet the people and take care of any problems."

As the public began to stream into the hall, Mary sold additional food tickets at the door, while Paul looked after tickets for drinks. For a long while, Marie and Marija senior did little or no business at the tables lined with pies, tarts, squares, and Marija's Polish pastries. Marilyn had expected people to come, eat, perhaps buy a piece of pie to eat with their supper, and then leave. But in the true rural tradition, they came and stayed. When they had to abdicate table space, they stood in groups along the walls to visit with neighbours and friends.

As Marilyn circulated, trying to make sure everyone felt welcome, she would find small cash donations and cheques pressed into her hand. Not until the guests finally decided to leave did they hit the bake tables. Then, Marie and Marija did a huge business. John jumped up on a chair and announced, "Before you folks go, we have a lot of food left over—buns, boxes of sausage and back bacon. So, if anyone wants to buy some, we'll sell it at cost."

Marilyn pulled her blouse loose from her chest. She had never felt as hot and tired before in her life. The pain in her back seared like a hot knife. She looked at the monumental mess and wanted to fall in a heap somewhere. Clifford and John joked incessantly, keeping up a stream of laughter that buoyed spirits. Reggie ploughed through the cleanup with astonishing efficiency. "I'm used to it," she laughed. "We have events like this at the Legion all the time."

It was after midnight before they wiped the counters for the last time and cleaned the sinks. Marilyn practically collapsed onto the front seat of the car. But tucked away inside her big enamelled roasting pan was the cash box, filled with money.

It took Marilyn nearly ten days to sort everything out. Then, flushed with success, she reported at the next meeting a gross take of almost $2,200. The bake table alone had taken in $225, and donations had amounted to sixty-five dollars.

After they had paid their bills, they donated $800 to the Coalition and $100 to John Jackson. "We run John's phone bill up all the time," John Dykstra pointed out. "The least we can do is help pay for it."

* * *

Two and a half weeks after Mary Munro recommended that the Coalition receive intervener status, John Jackson submitted a request for funds to cover administrative costs. He quoted Mary Munro's statement made at the end of the final day, "I am assuming as well that the Ontario Waste Management Corporation will continue to fund administrative structures up to the end of '88."

John then pointed out that thus far, the Ontario Waste Management Corporation had covered administrative costs for only the Region and the Township. In addition, Mr. H. Turkstra, Counsel to the Environment Assessment Board for this case, said he interpreted Ms Munro's statement to mean that the administrative costs for the Toxic Waste Research Coalition should be covered as well by the OWMC. John then submitted a budget to cover the costs for the latter part of 1988.

Eight days later, on August 30, Michael Scott wrote back to say that the OWMC currently had no authority from the government to include the Toxic Waste Research Coalition's current administrative costs in its funding of the Township's and the Region's OWMC-related administrative costs up to the end of 1988.

* * *

Marilyn had barely turned the calendar to September when the OWMC sent a letter saying the Crown corporation would be excavating two or three test pits on site to observe the shallow fractured zone and confirm the slope stability conditions. Expected to be approximately twelve metres square and six metres deep, the pits would remain open, but fenced, for three or four days before being back-filled.

The fifth of September, Labour Day, turned out to be miserable, cold, and drizzly. Despite the unpleasant weather and the iron-gray sky overhead, the Austins, the Dykstra family, and the Graceys turned out to cheer for the float entered by the Niagara Residents for Safe Toxic Waste Disposal in the annual Labour Day Parade. "We might not see eye-to-eye," John had argued, "but we're all on the same side."

Even at a parade, the Concerned Citizens couldn't spend time together without planning the next step. As John warmed his hands around a cup of coffee bought from a vendor along the parade route, he said, "The OWMC's supposed to start excavating those pits today. I think we should check them out."

"I can't go tomorrow," Clifford stated with finality. He shoved some French fries into his mouth, then talked around them. "With today being a holiday, I don't ship pigs 'til tomorrow. And I'd like to go to the stock sale, if I can."

"We can't go Wednesday," Marie said. "That's the day we take my dad shopping for groceries and pay bills."

They settled for one-thirty p.m. on Thursday afternoon.

*   *   *

By Thursday, the weather had warmed and the sun shone brightly. Don and Marie, John and Mary, and Clifford and Marilyn all arrived at about the same moment. Marilyn looked at the workmen in hard hats, and the heavy equipment, and her stomach felt queasy.

"I'm not cut out for this kind of stuff," she said as John and Clifford stepped over the dry drainage ditch at the side of the road and trudged into the field. Marilyn and the others trailed along behind. The workmen did not look particularly welcoming.

After a few preliminary exchanges, the hydro-geologist, named Andy Cooper, launched into OWMC's now-familiar speech. "We need a site where the clay is deep and uniform—"

Marilyn interrupted with an impatient wave of her hand. "What I want to ask is, how did you know that this spot—" she indicated the proposed site "—was the one where the clay had no layers of sand or shale?"

Mr. Cooper started to give a detailed explanation of the site selection process.

"No, no," Marilyn interrupted. "I mean, how did you know that *this* farm was the right place to drill test holes?"

He plunged into the social factors that played a part in the choice of a site.

"No," Marilyn exclaimed in exasperation. "I want to know how you, as a hydro-geologist, determined that *this* farm, not that one—" She pointed to the Schuender farm. "Or *that* one—" She swung ninety degrees and pointed. "Or *that* one. I want to know how you knew *this* was the right farm?"

Mr. Cooper's face turned an angry red, and a vein throbbed in his neck. For a second time, he attempted to explain to Marilyn the criteria used by the OWMC in their site selection process.

"I know all that!" she exclaimed in exasperation. "You know, you never quite get around to answering my question." She drew a square in the air with her two forefingers. "You just keep circumventing it." Before he could speak, she asked, "Did you, for example, use infra-red photography from an airplane?"

Mr. Cooper threw up his hands. "It's impossible for me to talk to you people!" he snapped. "I have four years of university. You've only recently started to take an interest in hydrogeology."

Marilyn gaped at him in astonishment.

"That's funny," Clifford retorted, his bushy eyebrows knit together in a fierce frown. "Both of our daughters went to university. The one took seven years of chemistry, but I can talk right along to her!"

Without another word, Andy Cooper stomped off toward his truck. Marilyn watched him jerk open the door and pick up a telephone. Meanwhile, a younger man approached the group of residents. "Maybe I can be of some help," he said in a mild tone as he handed them hard hats.

Using steps cut into the clay, he took the residents, one by one, down the twenty feet into the central pit. When Marilyn's turn came, she pointed to a six-foot vertical seam in the clay that was oozing water.

"We only opened this pit yesterday," he explained "It took all night for that small amount of water to accumulate. That bluish colour in the clay indicates water action on the soil in the absence of free oxygen."

Marilyn noted that the seam started at the level where the fractured upper level of soil stopped and the plasticine-like clay began. "That seam goes right to the bottom here. If you kept digging, how much farther down would it go?" She could picture leachate from the landfill following such a seam straight to the bedrock.

The young man gave no answer. He had already turned away and stood with studied patience, waiting for Marilyn to mount the earthen stair steps.

As she reached the top of the pit, she saw a vehicle crossing the field toward the pit area. A minute later, Mary Lou Garr from OWMC's Smithville office stepped down from the vehicle and approached the residents. "You should have let us know you were coming," she said.

"How long will this pit be open?" John asked.

Mary Lou Garr looked at Mr. Cooper, who had rejoined the group and stood at Ms. Garr's side. "We're closing it tomorrow."

"I thought it was supposed to stay open for three or four days," Clifford said.

Mr. Cooper pursed his lips in annoyance. Keeping his voice level, he said, "We've seen all we need to see."

"What about the seepage in the bottom?" John inquired. "Don't you need to see how much?—"

"I said, we've seen all we need to see." He cast a murderous look at the younger engineer. "That seepage is insignificant."

"Seems quite a bit to me," John said, "considering what a dry year it's been." He looked at the others. "Ready to go?" But as they turned toward the cars, he motioned with his arm. "C'mon," he said.

With long strides, he marched into the middle of the deepest swamp. In seconds he was lost to sight as the dry bulrushes, rattling in the breeze, waved above his head and closed around him. "Dry as a bone," he called back. Marilyn watched as one by one, the others pushed through the bulrushes and disappeared from view. John, his camera to his eye, snapped photos from two or three vantage points.

"Just for the record," he declared, with a mischievous grin.

* * *

On Wednesday, September 14, Clifford and Marilyn climbed the half-dozen stairs into the Legion Hall for the sixth day of Funding Hearings.

Ian Blue, in a surprise submission, stated bluntly, "The interveners' requests are astonishing. The OWMC is a government agency that operates in the public interest. Consequently, we are concerned that the funding be fair to all the public, including the taxpayers. We are concerned about the flagrant duplication of consulting work.

Come on!" he admonished as he swung around to include the audience as well as the lawyers and consultants. "Let's be practical!"

His submission instructed Mary Munro on the Crown corporation's view of what should and should not be funded. He named nine reports. "They are public documents. OWMC never intended these to require funding." He insisted that the interveners should settle on one set of consultants per issue. "In the public interest, the tri-parties need to do some serious scoping."

Marilyn marked the word scoping with stars and squiggles. I must find out what "scoping" means, she told herself.

Dennis Wood, on behalf of the Township, objected to Blue's submission. "We have been taken by surprise. The OWMC should have provided the three parties with a copy of his position in order to permit us to prepare a rebuttal. [We had expected the corporation to take a position of neutrality in these funding hearings. Instead], the OWMC has taken an extremely adversarial position..." Wood appeared offended. "We also object to Mr. Blue's continued use of the phrase 'in the public interest,' meaning the other three parties are *not* working 'in the public interest.'"

Mr. Crocker, lawyer for the Ministry of the Environment, rose to interject. "Mr. Blue's submission is inappropriate. Mrs. Munro, I respectfully suggest that you should give [it] no consideration..."

Mary Munro replied tartly, "I have been asking for this kind of data for more than a year. I was surprised to learn, when we met here on August third and fourth, that the issues had not yet been defined. Mr. Blue's document is not inappropriate." She gave Mr. Blue a cold stare. "However, the timing is inappropriate."

Blue rose hastily. "We did not receive the documents until Friday. We worked all weekend to get them ready." He surveyed the other lawyers and the audience. "Our analyst worked for no additional cost."

Mrs. Munro called a recess. Marilyn stood up and pressed her hands to her screaming back. Looking at Clifford, she declared, "The chairs in this place are a killer." Suddenly she spotted John Jackson. She pushed through the press and asked, "John, what does 'scoping' mean?"

"It means to narrow the field, to zero in on the main issues. Mary Munro wants the three parties to determine the issues, choose one set of consultants per issue, and avoid the present scattergun approach." He tipped his head in contemplation. "I have to agree with her."

After the recess, Dennis Wood continued with his list of complaints against what he considered to be a highly prejudicial submission, sprung on the interveners without notice. "Are there other documents—like this one—that we don't know about? Certainly the OWMC has documents, which it has promised to publish. They are not yet available. In this [instance], we were not advised that there even *was* a document, never mind what was in it. Also, we object to Mr. Blue's tone—telling you *how* you should use the information."

Kirby, the legal counsel for the Region, echoed Dennis Wood's sentiments. Then he listed a number of documents that OWMC claimed were still in the unfinished draft stage, dating back four years to 1984. "There are also some documents, believed by the interveners to exist, but not declared." He argued that the interveners wanted to cut costs by giving their consultants final documents to study rather than draft reports, which were subject to changes. But it was impossible when so many reports were being kept in the draft stage.

Joe Castrilli rose and clasped his hands together. "These draft reports should be issued in their final stage before the hearings—and not *just* before." He shoved his glasses up on his nose with his finger. "Also, I would like to inform the board that the OWMC has not yet funded the Coalition for administrative costs." He submitted the Coalition's letter of request to the OWMC and Michael Scott's negative response.

For the second time that day, Mr. Blue proceeded to give Mrs. Munro directions on how she should deal with the funding.

Mr. Crocker, representing the Ministry of the Environment, objected to Mr. Blue's tone and attitude, and charged that it was inappropriate for him to instruct Mrs. Munro on the funding.

The hearing adjourned for lunch. While Clifford chatted with a farming acquaintance, Marilyn wandered over to where Doug Draper sat ensconced in his little niche along the wall. "Is your chair any more comfortable than mine?" she asked as she rubbed her back.

He looked wary. "I hadn't noticed." He gathered up his pen and notebook, and prepared to leave.

"Will you be back this afternoon?" she asked.

"Yes. Why?" He glanced at where Marie Austin stood talking to Reggie. "Look, if you're planning to tear strips off me, why don't you do it and get it over with?" As Marilyn gaped at him in astonishment, he snapped, "I suppose you're like Marie Austin. You think I should slant my articles in favour of the residents."

"No," Marilyn answered slowly, "I don't. I think it's your job to report what happens. But I *do* think that the residents should get as much space and as fair a deal as the OWMC gets."

"I write an impartial account—"

"No," Marilyn interrupted, "actually you don't. I'd have no complaint if you did. But you give the OWMC more space and they always have the last word. That's not impartial."

He shoved the table aside and squeezed past. "I need to get some lunch."

"So do I," she replied with a conciliatory smile. "I hope I haven't offended you."

He did not smile in return. "No, but you at least are reasonable, and I respect that."

Marilyn watched him go through the door, a solitary figure who dared not eat lunch with anyone for fear he would be accused of taking sides.

When the session reconvened after lunch, Mary Munro told the parties that they should meet here again in two weeks' time. Then, with her expression as cold and set as granite, she issued a long list of instructions to the interveners. As she spelled out the details, Marilyn sensed the underlying message: Get your act together.

Only when she referred to the Coalition's plight did she show any warmth. "I am sympathetic to the Coalition's concerns regarding its administrative costs. I will make my recommendations after the hearing on September thirtieth." Before she adjourned the hearing, the lawyers asked for clarification on several issues. She told them sternly, "I have not made up my mind on anything yet."

* * *

Before the week was out, the Funding Hearing scheduled for September 30 had been postponed until October 13. The three parties needed more time, and Marilyn breathed a sigh of relief.

Frost three nights in a row brought a flare of red and gold to the leaves on the maple trees. Yet, the mellow, sunshiny days of late September were filled with the last scents and sounds of summer. Insects drilled their dying songs into the afternoon heat. Birds, gathering in chattering clouds, drifted en masse from trees to fields and then vanished into the distance, only to return in a noisy sweeping arc to line the telephone wires overhead.

Like the restless birds, Marilyn could not settle for long. Determined to relax for a few minutes in the waning sun, she fidgeted and shifted until it seemed easier to work. As she pulled up corn stalks and dragged them away, she fretted over the presentation she had agreed to make at HazMat. She resented the fact that because of the OWMC, she no longer had time to wander in the woods across the road or tape record birdcalls.

Even when she stretched out on the picnic table, like Charlie Brown's dog Snoopy on the roof of his doghouse, her mind worried over the latest developments surrounding the OWMC. The Region's Medical Officer of Health, Dr. Joseph Burkholder, had recently announced in the *Welland Tribune* that he supported setting up a database to monitor health effects. Marilyn saw this as another step forward for the OWMC.

"Such monitoring would act, not as an early warning," he said, "but more like a late warning system…" His words of caution had settled over Marilyn like a shroud of menace.

But perhaps the most troubling trend of late had been the lack of financial backing received by the Coalition from the town councils across the Peninsula. Most had refused to donate any money whatever. Others, like Pelham council, had given little more than tacit support, reported in a tiny news item in the *Herald*. "Council has approved a grant of $250 to the Ontario Toxic Waste Research Coalition…at its last…meeting."

She tried to be fair as she sat up and dropped her feet to the ground. These councils had municipalities to run. But the lack of whole-hearted support sapped her courage. "I can't tell this to anyone else," she said to the tabby cat from the barn as it curled itself around her legs, "but sometimes I feel as if I can't go on."

Two days later she learned that Welland council's finance committee had recommended that the city contribute $2,000 to the Ontario Toxic Waste Research Coalition.

- 33 -

Marilyn woke at ten after six to the insistent buzz of the alarm clock. Instantly she felt choked by the sick, hollow feeling of doom. Three hours from now, she thought, I'll be speaking at the Hazmat conference in the Hamilton Convention Centre. She closed her eyes as her stomach gave a threatening lurch. Terror always made her feel queasy.

An hour later, she stood before the mirror in her brand new, super-expensive suit. The long cut and trim fit reduced her weight by a good ten pounds. At least I don't look like some dowdy, know-nothing farm wife, she thought, as she descended the stairs.

Clifford brought the car to the back door and they set off. Marilyn sat with her notes in her lap, rehearsing for one last time what she would say, how she would say it, and making sure she had marked on each page the exact moment when she would project the image of the next illustration onto the screen. She again felt a moment of uneasiness. No two projectors were alike. She hoped she had no trouble operating the thing.

Fifty-five minutes later, Clifford pulled into the underground parking ramp at the Hamilton Convention Centre. "At least no one can see the churning in my stomach," Marilyn told Clifford with a shaky laugh as she stepped out of the car. She threw back her shoulders and lifted her head. Remain calm, she told herself. Walk with confidence. At the entrance, she presented her pass. "This is my husband Clifford."

The woman checked her list. "I don't see his name on here. He'll have to pay the $150 conference fee."

Marilyn lifted her lapel and indicated her nametag. "I'm one of the speakers."

"I'm sorry," the woman insisted, "but—"

"Then I'm not speaking!" Marilyn exclaimed in a sudden panic. "I don't go without my husband."

Clifford tried to calm her. "No!" she insisted, "I'm not going up there like a lamb among the wolves without you, and that's that."

The woman turned away and picked up a telephone. After a hurried conversation, she replaced the receiver. "We'll make an exception today, but it's not common practice to admit spouses." She hastily printed up a nametag for Clifford.

Upstairs, in a room the size and shape of a small classroom, Marilyn looked for the projector. Just then, John Jackson appeared in the door. "There's no projector," she told him.

He left the room and came back several minutes later. "They're getting one. While we wait, maybe I could take a look at…"

Marilyn flipped open her folder and went through the drawings. John began to smile. "They're fine," he admitted.

Just then, the young woman returned with a projector. But not the right kind. Marilyn showed her the pages of drawings. "Paper?" the young woman said in surprise and puzzlement. "Is there such a projector?"

"When I went to high school, the teachers used them all the time."

"Well, we only have the type that use transparencies."

Marilyn threw up her hands in despair. "I should have known!" she all but wailed. "That's the kind Etta Lane borrowed from Gainsboro Public School to use at our public meeting."

John interrupted. "You'll just have to make your points without the drawings."

Marilyn pressed her hands together and released a huge sigh. "You're right. But the drawings would have made the impact stronger." She looked around the small room. "Actually, they may be able to see them if I hold them up. It's worth a try."

John looked at the clock on the wall. "It's almost time." Just then, the other speaker walked in. As John made the introductions, Marilyn watched several men in dark suits enter. Absorbed in discussion, they gestured emphatically with their hands as they took seats at the rear of the room. Behind them came others until the small conference room held about a dozen people.

Suddenly, John leaned closer to Marilyn. "See the guy seated on the right, fourth row down, next to wall?"

Marilyn nodded. "Yes?"

"He's with the OWMC."

Marilyn felt her heart lurch and begin to pound. "I hadn't thought about *them* being here!" she whispered in return.

"Don't think about him," he said. Then, casually dressed as always, with his hands in his pockets, he stepped forward. "Ladies and gentlemen, thank you for coming. As you know, our topic views the public consultation process from the perspective of residents. Sherry Morrison from the Lambton Anti-Pollution Association is unable to be here today, so I will read her presentation."

He then launched into Sherry's experiences regarding Tricil's hazardous-waste treatment and disposal plant near Sarnia. Her anger and frustration, her account of the struggles she and the other residents had faced, made Marilyn go cold.

Then it was Marilyn's turn. She gazed out over the expressionless faces until she found Clifford. Concentrating her attention on him, she said, "I'm not a public person. When I was asked to speak here today, I had to do some serious thinking before I finally accepted. 'Just tell them what it *feels* like,' I was told."

She glanced at her notes, grateful that she had long ago learned to break her sentences into naturally-phrased groups of words, one group per line, each line indented more than the last, and a double space between each sentence. It made the delivery easy, smooth, and she could not lose her place.

She emphasized the overwhelming sense of injustice, both social and personal:

> Social injustice, at having one small community made the repository for all special waste from across the province, even though OWMC had been urged not to do so. Rare, indeed, is the person who reacts well to injustice.

A man's head poked through the open door. He listened for a moment, then slipped into a vacant seat.

> Injustice comes in many forms. Sometimes it is nothing more than the tone of voice used, letting the resident know that he or she is a lesser being. Sometimes it is the photocopy of a resident's letter to the OWMC's office in Smithville, returned anonymously with a spelling error circled in red.

Marilyn noticed two more men step quietly into the room and take seats. So far, she thought, no one has left. She returned to her notes:

> Perhaps it is the indignity of a phone call from the principal of the local school, not because some problem has arisen concerning the children, but because the OWMC has embroiled parents and the principal in what was clearly not a school matter.

She raised her head and stated firmly, "Lesson number one: The end does not justify the means."

Marilyn noticed a number in the audience taking notes.

Next, she tackled the issue of the first public meeting in Wellandport Hall. "Questions had to be submitted in advance. We were given a glossy, one-sided view of the issue, with no opportunity to ask questions." She held up the drawing of a man and a woman, wearing gags. "Lesson number two: Muzzling the community at a public meeting may be an accepted method of crowd control, but it will only kindle a sense of injustice."

Several more people poked their heads through the door, listened intently, then sought chairs. By this time the room had nearly filled. Marilyn turned to the next page of her notes:

> At the first Open House, we were properly impressed and intimidated by the sight of so many experts and consultants—until they started giving answers which sounded good but couldn't possibly be right.

"Lesson number three." She held up the picture of an artist's paintbrush, dripping with rose-coloured paint. "Painting a rosy picture may be considered a good way to push a project through, but don't count on the public to be gullible enough to believe it."

The room remained absolutely silent, except for the rustle of pages turned on notepads. Suddenly a loud clack made Marilyn jump. Her head swung to the OWMC representative. He leaned over and Marilyn realized he had been taping her speech. As always, she felt a shiver of intimidation and uncertainty. Then she calmed herself. She had said nothing that wasn't true. She had nothing to fear.

She took a deep breath. She had devoted the next four pages to the now-infamous questionnaire. She condemned the need to give spur-of-the-minute opinions and judgments on complex issues, using the "yes-no-don't know" format.

> They asked us to rank our concerns, thereby relegating some concerns to a lesser position of importance. A number of questions were offensive. How much do you earn? Have you attended meetings held by the West Lincoln Task Force, a citizen group opposing the OWMC? What's your ethnic background?

She paused to gaze out over the crowded room. Men and women, who had quietly entered while she spoke, stood along the back wall.

> During one of the OWMC Open Houses, I asked one of the consultants his ethnic origins. Since all four of my grandparents had emigrated from the land of his birth, I knew I could not be accused of racial prejudice. His response was interesting, to say the least. He became quite huffy and defensive, and demanded to know what his ethnic origins had to do with the issue.

Marilyn couldn't help chuckling. "Same shoe, different foot, but it pinched just the same." She picked up her page and read,

> Lesson number four: If you are going to pretend that your public consultation arises out of genuine concern for the

residents affected, at least make every effort to satisfy legitimate questions and concerns.

She held up the drawing of a set of scales, loaded on one side with pages. "Lesson number five: Don't count on getting away with a questionnaire that is psychologically loaded. The OWMC failed to address the residents' concerns over the questionnaire. At that point, the OWMC consultation process went down the tubes." By this time, the doorway to the room was packed with people straining to hear. Others stood in the hall.

As a result, forty-four percent of farmers in the local community declined to answer the questionnaire. To fill in the gaps in primary data, the OWMC consultants did a "windshield survey." That means they drove past, counted the silos, checked to see if the buildings were painted, guessed at the acreage in a field.

Marilyn's voice suddenly became hard with anger. "They gathered data from other sources, made up reports, and submitted them *to other residents* for correction and validation." She could feel her eyes turn steely. "Lesson number six: Most citizens view this form of public consultation as questionable at best, and a moral, if not legal, invasion of privacy."

She moved on to the topic of the kitchen table meetings. "This is a method by which the OWMC can choose *not to meet* with those most actively opposed to the facility." She pulled out a photocopy of Dr. Chant's remarks made at a five-day conference of public health inspectors: "I have been meeting with small groups of West Lincoln property owners in their homes...successful farmers, with roots in their community that go back several generations. *They are not interested in joining active protest movements against our proposal.*"

Marilyn paused and let her gaze fall upon the man from the OWMC. "Lesson number seven: A policy of divide-and-conquer is not well received in a community which traditionally functions within a network of mutual aid and support." She noted the time and speeded up her presentation.

Lesson number eight: In an agricultural community, the busy season is no time to carry out public consultation. Exceptions, for example, would be fieldwork in biology, et cetera. This past June, with only two weeks' notice, the OWMC scheduled open houses with their agricultural consultant. Now, if the citizens attended simply to see the colour of his hair, or to get his autograph, they could probably have squeezed out a few minutes in between baling loads of hay.

She waited for the ripple of snickers to die down.

OWMC does not conduct its open houses out of the kindness of its heart or out of concern for the residents. The Crown corporation is required, under the Environmental Assessment Act of Ontario, to solicit public input.

She cast her gaze over those in the seats, those lining the back and the sidewalls, and those standing in the doorway. "Need I say, almost no one attended the open houses in June?"

She held up a diagram. She had written "public" vertically and "consultation" horizontally, in the form of a crossword, with the u in both words as the common letter. "Consultation" had a hand attached to either end of the word. The left hand held the word "law." The right hand released the word "spirit."

"Lesson number nine: Public Consultation, which abides by the letter of the law, but frequently fails to catch the spirit of the law, becomes little more than a formality."

A loud clack again broke the silence, and every head swivelled toward the sound. Annoyed by the distraction, Marilyn blurted out, "At OWMC functions, I ask for permission to tape. I don't just—" She had been about to say "sneak," then thought better of it.

"Lesson number ten:" Marilyn held up a cartoon, which showed a man's head with huge ears, plugged by two old-fashioned thermos corks. "It isn't enough to print in newspaper ads and on flyers, 'Drop in—voice your concerns.' You need to listen."

Lesson number eleven: Bad manners is bad public relations. For example, when Great Lakes United took a position in favour of above-ground storage, as opposed to landfill, OWMC pestered the office in Buffalo, trying to persuade personnel there to reverse their decision. The reporter who wrote up the story also received a phone call urging him to change what he had reported.

"In March 1988, the Concerned Citizens held a public meeting." Marilyn held up a simple cartoon, showing a podium, a speaker, the backs of many heads, and two in the back row making a disturbance. "Present were two representatives from the OWMC who distracted those around them by talking all through it."

Her next illustration showed, in stick form, a giant-sized figure intimidating a much smaller figure. On the ground in the foreground, she had drawn a slingshot.

Recently, the Concerned Citizens Group has come under pressure because of formal presentations which members have made around the area. In letters to us and to the editor, the message has been two-fold: One, if you have something to say, say it only to the OWMC. And two, copies of all presentations should be sent to the OWMC. Lesson number twelve: Trying to strong-arm unpaid citizen groups only emphasizes the injustice intrinsic in the classic David-and-Goliath struggle.

She touched on the inequities in the funding. "Even as the Township of West Lincoln and the Region of Niagara were being cut back on funding to study the OWMC's proposal, the OWMC was busy making another promotional video for public distribution."

Marilyn then turned to the issue of the corporation's attitude toward the residents. She cited two incidents, quoted from the *Niagara Farmers' Monthly* and from the OWMC's own publication, *Update*, in which OWMC personnel demonstrated a low opinion of the residents' knowledge.

Citizens like myself have assimilated a great deal of data. I feel as though I'm taking a university degree in something I hate and, as one OWMC consultant noted, we have been obliged to become knowledgeable in a number of disciplines. So lesson number thirteen is: If members of a corporation consider themselves to be more enlightened than the citizens they are dealing with, that sense of superiority *will* show.

She no longer spoke in generalities but directed her words to every person in the audience.

Members of the general public may not be experts in your field. But, then, you don't shine in theirs, either. Lesson number fourteen: Don't underestimate the public's ability to learn.

Marilyn turned to her last page of notes and gazed intently from face to face.

As part of a corporation, you may find yourself at some time in the future dealing with the rural community. Before you fall prey to the common misconception of the farmer as a straw-chewing hayseed, a simple-minded clod with little education, keep in mind that you are, in fact, dealing with the head of a corporation—a hard-headed survivor in an industry chronically plagued by slim profit margins, while faced with ever-changing technology.

She held up her hand, as though to quell objections. "Oh, the farm may not be legally incorporated, but in essence, it is a corporation, all the same." She felt her eyes take on a steely look as she drove her point home:

The farm business probably won't have nearly the assets of the corporation that employs you. However, you should keep in mind—you don't own your company's assets, but the farmer owns *his*!"

She held up her final illustration, showing a balanced scale. On one end stood the word "citizens," written vertically and wearing a peaked cap. On the other end, stood the word "corporations," also written vertically, and wearing a businessman's hat. Between the two words, and precisely over the fulcrum, she had drawn an equal sign. "Lesson number fifteen: Give citizens the respect they deserve."

She paused, then closed her folder. "Thank you."

Not until she sat down did she realize her knees were weak and trembling. John introduced Cameron Wright, the young man from the Citizens' Network on Waste Management. His topic was funding for environmental hearings. By the time Marilyn stopped speaking, the small conference room had been filled to capacity and beyond. Now, some of the listeners began to drift into the hall. Even before Cameron Wright had finished his presentation, small groups had formed in the hall and the voices elevated in discussion proved a distraction. Marilyn glanced at John Jackson. He looked extremely pleased.

Even when they left the room to find some lunch, groups still clustered in the hall, intent upon debate.

Over lunch, John beamed. "I checked around. No other seminar sparked the discussion afterward that ours did. We were a huge success." He looked like someone who had just run the one hundred-metre dash—and won.

– 34 –

Marilyn had taken it personally when Edgar Lemon wrote his devastating letter to the editor claiming that the OWMC's opponents deserved the Golden Fleece Award for dipping their hands deepest into the public purse. Her spirit crushed, it had taken her three months to respond. Not until October 11, 1988, two days before the funding hearing, did she finally type the finished copy of a letter to the editor.

> If anyone is a prime contender for the Golden Fleece Award, it is the OWMC. It has cost the taxpayers $69 million to date—$30 million on site selection, $15 million on preliminary engineering design, and the remaining $24 million on creating and staffing the corporation.
> As of August 3, 1988, OWMC was seeking "$14.8 million to keep it running for the year ending March 31, 1989." Of that $14.8 million, only slightly more than $.8 million is to be divided between the Region and the Township.
> So far, the Toxic Waste Research Coalition has received no funding for administrative costs.
> When is the last time the OWMC went door-to-door, looking for money to pay the phone bill? Or put on a fundraising picnic or a community barbeque or sold T-shirts? So don't talk about nominating us for "The Golden Fleece Award," not when we are the ones digging deeper and deeper every day into our own pockets.

Marilyn read the letter aloud one last time, then picked the car keys off the hook at the back door. She drove to the library, made twelve photocopies to send to the twelve newspapers around the peninsula that printed the letters on a regular basis, and rushed home. In less than two hours, she, Clifford, and Reggie were off to Brock University to attend a debate involving the OWMC.

* * *

Seven months had passed since the mock hearing, which had been held at Brock University in March. Once again the Concerned Citizens congregated in the parking lot outside one of the university's main doors. Marilyn hunched her jacket around her ears as the mid-October air sent a chill through her. "Who's taking part in this debate?"

"Leslie Daniels, for one," Marie told her. Marilyn nodded and shouldered the door open, arms loaded with reports and notebooks.

John Dykstra pushed the button for the elevator and, minutes later, they settled into chairs in the designated room. Scarcely had the debate begun before one of the participants, a professor from Brock, directed a ticklish question to Leslie Daniels, OWMC's Coordinator of Regional Communications from the Smithville office. "Last spring, when Dr. Ray Durham was here at the mock hearing, he told us that when you attempt to solidify substances with high levels of salt, the cement doesn't form. As he put it, 'you can't get it to set up.' Has that situation changed?"

Leslie Daniels looked uncomfortable. "Salt *is* a problem," she admitted. "We're in the process of developing alternate strategies concerning the salt."

"That was over half a year ago," the professor retorted. "I was hoping that by now you would have some answers."

Clearly flustered, Mrs. Daniels replied, "I don't know what the answers are. I really don't know what they are. I don't know where the emphasis is. It's a problem of long standing."

John Dykstra glanced at Clifford and Marilyn and lifted his eyebrows the merest fraction. Marilyn returned the look. OWMC's wonderful state-of-the-art landfill had a major problem.

Later, over coffee in the restaurant, Reggie asked Marilyn, "Are you going to write a letter to the editor about the OW?" As usual, Reggie had shortened the name to two letters.

"They've sneered at our concerns long enough," Marie snapped, her usually merry blue eyes fierce under eyebrows knit together in indignation. "And now we find out that those baskets have been struggling for years with a problem they can't solve." "Baskets" was Marie's way of calling them foul names in thought, if not in deed.

Marilyn remained firm. "Without the facts to quote, chapter and verse, I'm not going public. One mistake and the OWMC would make Clifford and me look like fools. I'd rather say nothing than say something I can't prove."

Reluctantly, the group agreed. "Okay," said John, "then I suggest we keep scouring the reports for anything we can find on salts—or chlorides—or whatever it is they call them."

* * *

That night, after Clifford had gone to bed, Marilyn sat in her housecoat and slippers, browsing through the *Shopper's Guide*. The name OWMC leaped off the page at her. The OWMC had announced the creation of a new industrial waste reduction award program. Each year, two awards would be presented to companies or trade organizations that demonstrated successful industrial waste reduction practices.

Her bosom heaved in resentment. The OWMC had money to dole out here, money to dole out there. It bubbled up into the OWMC coffers like oil from the ground. Giving the newspaper a sharp snap that threatened to rip it in half, she held it under the light.

The announcement had been made by OWMC engineer Jody Sabo. Hundreds of companies across Ontario were helping to solve the province's industrial waste problems, she explained, yet nobody ever heard about them. Smart waste reduction programs, which incorporated the Four Rs of reduction, reuse, recycling, and recovery, had diverted millions of gallons of industrial waste from landfill and sewage treatment plants. "We feel it's time to recognize some of these success stories," she said.

Marilyn felt like a stream with its current running in two directions. Less toxic waste meant less need for the OWMC, which was good for the opponents. Yet, the corporation's poor record regarding the Four Rs had been a strong rallying point for the citizens' criticism. She tossed the paper aside and rubbed her eyes in a gesture of weary frustration. She wanted to see industry encouraged to do the right thing. But not by the OWMC.

* * *

For the third time in three months, the Region, the Township, and the Coalition left the funding hearing empty-handed. Marilyn came away from the hearing furious with Stan Stein, the lawyer for the Region. In the other hearings, Marilyn had sensed that the Region wanted to head up the tri-parties' opposition in the upcoming environmental hearings. This time, his tone and manner left no doubt that he thought the Region should take the lead in the case. Without consulting the Township or the Coalition, he had proposed slashing the request for funding from $1.9 million to a crippling $271,000, with the Coalition's share chopped from $42,000 to a mere $27,200.

Joe Castrilli had vigorously opposed the plan. He argued that the province's intervener funding program was designed to encourage members of the public to participate in the decision-making process, not just regional and municipal government. "Such a plan," he said, "would limit the Coalition's role in issues such as alternatives to the OWMC facility, an area in which it expressed an early interest and has already spent considerable time investigating."

Mary Munro, compelled to report to the provincial cabinet, informed the three parties that since they had failed to reach a consensus of agreement she would be preparing some sort of funding requirement. She also scheduled another hearing for October 31, 1988, to be held at the Environmental Assessment Board offices in Toronto. "We need to decide on the funding of project administration," she said, glancing at John Jackson and Joe Castrilli. "We need to ensure that the parties have enough funding to carry on for now."

* * *

OWMC—The "Wasted" Years: The Early Days

In the past, Marilyn had taken only a passing interest in the municipal elections. Now, she avidly read everything that came into her hands. Her concern focused upon one issue. What position did each candidate take on the OWMC's proposed facility? For the first time in their lives, Clifford and Marilyn attended the Candidates' Night held in the auditorium of the West Lincoln High School.

After the meeting, as Clifford drove home, Marilyn broke the silence. "They all talk a good story," she observed and gave a deep sigh. "I just wish I knew which ones to believe."

However, once she arrived home, she had no time to ponder. She had to finish packing the large suitcase and mixing up egg salad filling for sandwiches. She and Clifford, and John and Mary, were setting off for Sarnia first thing in the morning.

To explain away the residents' opposition to OWMC's so-called state-of-the-art facility, representatives of the Crown corporation had coined a stock answer: "If they're afraid…it's because it's an unknown to them. They haven't anything in their lives to relate it to, not even another facility…"

Marilyn, for one, nearly went wild every time she read the statement. "I grew up in a factory town," she declared heatedly during a meeting of the Concerned Citizens. "I know about ugliness and stack emissions and all the rest. And Clifford worked at Atlas Steels in Welland for twenty years. They make us sound like members of some isolated tribe seeing modern man's thunder-stick for the first time!"

As a result, the four of them had decided to see for themselves. On November third, under a sky promising rain, they travelled to Sarnia in Dykstra's van to have a look at Tricil's toxic waste facility. Their first glimpse of the facility was a tall stack with a slender plume of dark-coloured emissions trailing a mile or more over the countryside. After snapping photos, they booked into a motel, made a visit to see Sherry Morrison and her husband Garry, and made impromptu arrangements to tour Tricil's facility the next morning at eleven. Then they went back to the motel for something to eat. To save money, Mary and Marilyn had packed enough food for lunches, supper that night, breakfast in the morning, and snacks along the way.

The next day, under a canopy of sullen cloud, they arrived shortly before eleven for their tour. A steady drizzle fell from heavily overcast skies as Marilyn snapped roll after roll of photos. "It's worse than I thought," she murmured to the others, her head bowed against the cold misty rain as they stared out over the vast acres of landfill. The huge cranes and earth-movers looked like the tiniest of miniature toys as they crawled along the bottom of the pits.

Nearly a year and a half earlier, angered by the dinky little rectangle representing the landfill in OWMC's literature and releases to the public, she had written in a letter to the editor, "We do not believe that the people of Niagara have been given a true picture of the magnitude of the landfill site proposed by the OWMC." She had based her accusation on a statement she had stumbled across in one of the OWMC's documents: "LF-9C is free to expand to the west." I was right, she thought with a sinking heart. More right than I knew.

Their guide then escorted them through the facility itself. "What do you think?" Marilyn whispered behind her hand to Clifford, who, having been a factory worker for more than twenty years, would be better able to judge than she was.

"I'm surprised at how clean it is," he whispered back.

In the parking lot, Marilyn took more pictures of the plume trailing away into the distance, a dark smudge against the gray sky.

Once they were on the highway, headed for home, Marilyn remarked, "I read somewhere that if the emissions from a stack are just steam, they appear as a puff of white that dissipates almost immediately. Tricil's emissions streamed out over the countryside for at least a mile, maybe two."

John expressed his disapproval of the landfill operation. "That stuff went straight into the ground."

"It looked like soggy-wet giant tea-leaves," Marilyn remarked. Then added grudgingly, "I hate to say it, but the kiln area itself looked quite respectable. And it's not as if they had time to clean it up before we came. They didn't know we were coming."

"Well," said Mary with a sarcastic laugh, "at least we don't have to be afraid of the unknown. We've seen it, and we like it even less than before."

The next night, the Coalition held a Christmas potluck supper. Of special interest was the Coalition's video. Some months earlier, the group had received funding to produce a video, demonstrating the ways that industry could reduce or reuse its waste. Although the *Turnaround Decade* had not passed beyond the early stages of development, the producers had set up a viewing area. Marilyn watched in amazement as representatives from a variety of companies took the camera through their plants and explained how they had changed their procedures to cut back on waste and save the company money.

Marilyn felt a surge of guilt. She realized that no matter how many times John Jackson, Edith, or Al DiRamio had described these changes, deep down inside, she had not believed them. Now, her eyes believed what her brain had traitorously condemned as wishful thinking.

* * *

It became increasingly obvious to Marilyn that she would have to find a better way to keep track of information. As the piles of documents grew, she wasted more and more time searching for data she had read, but couldn't find. In frustration, she would try to visualize the paragraph or the chart, visualize the colours she had used to mark it. She pictured it on the left-hand side of the page or the right. But what page? Which report?

"Mom, you need a computer," Kathy insisted.

"Not me! I don't know the first thing about a computer."

"You could learn."

Marilyn shook her head. How could she admit she had an irrational fear of a machine so clever it spoke its own language, a language so technical, she knew she could never learn it? Yet, the time came when she and Clifford scoured the computer stores, and finally brought home equipment they could ill afford. They sat it on the dining room table. It was ten-thirty at night before they finished hooking it up and plugged it in.

"Well?" Clifford urged.

Poised for flight, she reached out at arm's length, her eyes squeezed shut, and pushed the *on* button. The computer whirred and beeped and flashed, and she kept her distance. But once it had calmed down, she declared in triumph, "C-prompt! Just like the salesman said!"

She inserted the program disk and pressed the keys. Clifford went to bed, and Marilyn soon wanted to. But the young man at the store had failed to tell her how to shut the thing down.

One a.m. Two a.m. She poured over the manual. Do this, do that, it instructed. Then exit. "But how?" she cried in frustration, and Clifford called out from upstairs.

"Come to bed," he told her. "It's after three."

"I can't get out of the program."

"Just shut it off."

"You're not supposed to do that," she called up to him. But in the end, that's what she did. Shut it off. The screen burst into a shower of confetti and turned black.

Thus began the struggle that seemed too great for her. Night after night, she stayed up until three and four in the morning. But true to his word, the young man at Radio Shack in Seaway Mall, the only computer salesman who had treated her with respect despite her ignorance, answered her many distress calls with undiminished enthusiasm, and spent two hours with her one morning at the store under the watchful eye of Brian Pychel, the store manager.

Almost immediately, she began putting data into files.

※ ※ ※

On November 14, 1988, the Township of West Lincoln elected a new mayor. Joan Packham, an alderman for the past two terms, replaced Allard Colyn.

– 35 –

The last of the leaves fell and swirled in the cold winds of November. On the thirtieth day of that bleak month, the OWMC released the final Environmental Assessment Reports. They contained approximately 10,000 pages and weighed forty pounds. If Marilyn could have held that day at bay forever, she would have. For, as long as the reports remained pending, she could fool herself into believing that life would go on much as usual, even though "usual" had fallen into a pattern of endless reports, letters to the editor, and meetings.

The night following the release of the Final E.A, as the members of the Coalition assembled for the monthly meeting, an air of doom permeated the second-floor room of the Vineland Agricultural Station. All hope of political intervention had vanished with the announcement that the Preliminary Hearings would begin in January.

"No evidence will be heard at that time," John Jackson explained, with an air of unflagging optimism. "The purpose of the preliminary hearings is to decide where and how the actual E.A. hearings will be conducted." Then he handed out photocopies of a new report. "This is part of the Waste Quantities Study done jointly by the Region and the Township," he announced. He wore the pleased look of someone beaming inwardly. "It shows that the OWMC has greatly exaggerated the amount of waste to be dealt with, just as we've said all along. The corporation has ignored trends in industry and made decisions based on outdated information."

Although the Waste Quantities Study couldn't have come out at a more opportune time, providing a strong basis for the tri-parties' opposition to the OWMC's Final E.A. reports, Marilyn couldn't drum up much enthusiasm. Foolishly, she had pinned her hopes on stopping the hearings before they ever got started. She stuffed her copy of the report in the back of her notebook, thinking, what is the point of reading it? It won't change anything.

After the meeting, the seven regulars from the Concerned Citizens—Reggie Kuchyt, Clifford and Marilyn, John and Mary, and Don and Marie—met at a local restaurant to hash over the latest events. It was after eleven before Marilyn climbed the back stairs and dropped her notebook on the kitchen counter. Clifford checked the barn, then came in and went straight to bed. Marilyn, in her flannelette granny gown and housecoat, curled up in her old orange recliner and reached in anticipation for the latest Mary Higgins Clark mystery. As she tipped the recliner back, a bone-weary sigh escaped her. I'm so exhausted, she thought, I'm almost too tired to read.

Unexpectedly, the familiar voice of God gave her a guilty nudge Read the Waste Quantities Study. With a defiant flick of her hand, she smoothed open the page of her new book to chapter one. Read the report, the voice insisted. Her gaze moved without comprehension over the opening sentences of the mystery. Annoyed, she started at the top of the page again. "Surely I have the right to read a new book without feeling guilty," she grumbled. I need to unwind. I need to escape from the OWMC and everything connected with it. Instead, she felt a steady, uncompromising compulsion to read the Waste Quantities Study. Not until she had struggled through the same paragraph for the third time without knowing what it said did she finally give in.

With markers and pens close at hand, she picked up the Waste Quantities Study. Immediately, she was gripped by the economy of words in the report. As fact followed fact, without the excess verbiage common to the OWMC documents, she found it difficult to keep from highlighting every sentence, every phrase.

The consultants for the Region and the Township had devoted much of their study to private sector waste management companies, focusing on the excess capacity prevalent among them.

Renewed hope and excitement surged through her. Excess capacity diminished the need for the facility, and need had always been a key issue. She glanced at the clock. One-thirty in the morning. Her eyes felt like two hot coals, sunken into her head, but she couldn't put the report down.

As she neared the end, she learned that facilities outside of Ontario, such as the ones in Alberta and Maryland, were also operating well under capacity.

> One reason is that experience demonstrates that industry will select the most cost effective alternatives for waste management, including construction of on-site treatment and waste reduction facilities, even when public facilities are available.

However, the report stated, "The Study Team could not consider the experiences of these and other jurisdictions because this task was deleted [by OWMC] from our original work plan."

Marilyn read the sentence, stopped, then disbelieving, read it carefully a second time. With every word, she felt an almost ungovernable rage. The OWMC had prevented the authors of the report from checking out other facilities, like those in Swan Hills, Alberta, and the one in Maryland, which couldn't find enough toxic industrial waste to run at full capacity.

It was nearly three a.m. before she put the report aside and went upstairs. But the study had clearly spelled out its conclusions. "...The sizing and type of facilities selected by OWMC cannot be justified."

*   *   *

Marilyn had just plugged in the vacuum cleaner the next morning when John Dykstra telephoned. He sounded extremely upset. "Mary Lou Garr just brought me a Summary of the Final Environmental Assessment reports. She says an interview's been set up for twelve o'clock on CBC's *Radio Noon* between Chant and myself. We're supposed to discuss the E.A."

"Oh John!" Marilyn exclaimed, remembering her own frenzy in a similar situation. "Can't you refuse?"

"It's a setup—to make me look like a dummy!"

Marilyn's mind rushed incoherently here and there. "Bring the Summary here. We'll do...something."

Seven minutes later, John and Mary burst through the back door. John's panic was plain to see. They sat down at the table and tried to read through the slim thirty-page document. Marilyn could feel her heart pounding until she could barely breathe, her thoughts so turbulent, she couldn't concentrate. After twenty minutes, they had only read a few pages. She glanced up at the clock. "We can't read this report. We don't have time." Her voice turned harsh. "Chant's had months to become familiar with what's in here."

Mary looked anxiously at the clock. "We don't have much longer."

In the frantic way of nightmares, Marilyn felt incapable of actions or thought. Involuntary words came from her mouth. "We need to pray."

The three of them joined hands and began to ask for wisdom. Suddenly, Marilyn broke off and exclaimed, "I know what to do!" She jumped up from the chair and ran into the living room. She grabbed the Waste Quantities Study from the arm of the chair, along with the notes she had made the night before. Waving them as she rushed back to the kitchen, she exclaimed, "The information in here is as fresh in my mind as the E.A. is in Dr. Chant's mind." She gave a gleeful laugh. "John, you will take charge of the interview. You will talk on how this report shows there is no need for the OWMC facility."

Baffled, John gasped, "But what do I do when I'm asked a question?"

"Don't pay any attention to the questions. Just keep hammering away at this information." Marilyn fetched several sheets of paper and started printing information in capsule form in large letters. As fast as she finished with one page, she shoved it over for John to read and started on the next. Within ten minutes, she had summarized the main points of the report. "Whew," she sighed. "Thank goodness—no, I change that, thank a very persistent God—that I stayed up half the night reading a report I didn't want to read. Now we'll see how Dr. Chant likes being hit with data he's not geared up for."

John had a grin on his face, but Marilyn could tell he was nervous. "You'd better come back to our house," he urged. "I'll get on the upstairs phone. You can listen in on the one downstairs. Mary can run notes upstairs to me if you need to bail me out."

Marilyn picked up her clipboard, with the Waste Quantities Study and her notes from the night before safely secured to it. With her heart still beating rapidly, she followed John and Mary out the door and jumped into her car. At quarter to twelve, they had everything ready. John had set up the tape recorder, Marilyn had run John through the high points of the Waste Quantities Report one last time, and they sat watching the telephone, jumping involuntarily when it shrilled. Suddenly from the radio trumpeted CBC's familiar theme song, signalling the start of the programme.

In a strong, bright voice, a young woman announced, "Hello, I'm Donna Tranquada, and this is *Radio Noon*." After a few introductory remarks, she said, "First of all, Dr. Chant, what sorts of problems did the report turn up for the people of Smithville?"

Dr. Chant sounded calm and confident. "Well, we've done a more extensive risk and impact analysis on that site over the last three years than has ever been done before in Ontario. And when you strip it all away, the residual impacts will be largely dust and noise on about fifty-three families in the zone around the proposed plant."

"Dust and noise?" inquired Donna Tranquada. "Coming from the trucks rumbling through?"

"Largely from excavating the landfill itself, as we move, over the years, with the treated waste residues going into it."

"What about the property values in the area—will they go down?" the host wondered.

"The property values probably will go down over a brief period of time, looking at other similar projects," Dr. Chant replied. "But we've developed a very extensive programme for property value protection. If anybody in that impact zone feels that they want to move away from the plant—ah—we've made provisions for buying them out at a very fair market price."

Marilyn mouthed the word "fair" in exaggerated disbelief.

Almost as if he knew what the residents must be thinking, Dr. Chant suddenly sounded defensive and unsure of himself. "It's entirely voluntary on their part—uh—but, by buying them out at a fair market price, they can go and—uh—establish their farms and their lives in some other community—I guess."

Donna Tranquada turned her next question to John. "Well, John Dykstra, were you surprised at the results, or did you anticipate this all along?"

"Well, the Summary—uh, I just had it delivered to my house about ten-thirty this morning by an OWMC official—after the OWMC knew the CBC would be getting in touch with me. The documents themselves are not yet available to me."

Donna Tranquada sounded surprised. "So, you haven't seen what the impact is then."

"No, I haven't. So—"

Dr. Chant's voice drowned out John's answer. "The documents have gone to—"

Donna Tranquada cut in. "Just a second—uh—"

Dr. Chant paid no attention. "—gone to the headquarters of the Concerned Citizens early this week—"

Marilyn frowned at the misstatement. "Headquarters?" she mouthed to Mary. "What 'headquarters'?"

"—We're sending out 950 of them," Dr. Chant continued insistently. "Mr. Bradley, the Minister, got his copy on Monday, and we've been sending the—uh—950 ever since, but I assure you that the office of the Concerned Citizens *has* their copies already."

Donna Tranquada stated firmly, "Nevertheless, John Dykstra got his copy this morning." To John she said, "You're just aware now that the residents will be facing traffic, noise, and dust problems?"

"No, not really that," John answered. "But aside from anything else, the need for such a facility is a major question. According to a Waste Quantities Study put out by the Region and the Township, the facility cannot be justified."

Caught off-guard by the swift change of topic, Donna Tranquada tried to get the interview back under her control. "Well—uh—Dr. Chant...do you have any other—do you have any alternatives at all to building this facility now that you—"

"There *are* no alternatives!" The exasperation in his voice seemed almost tangible, and for a moment, Marilyn felt sorry for him. Over the years, he had answered this same question every way he knew how, but the badgering over alternatives simply had not gone away. His tone made it clear that his patience had been stretched past the limit by the question of alternatives. "The *whole* province has agreed *for years* that we must move into the modern era of industrialization. We've got to have an industrial waste treatment plant. The questions are, where does it go, and what kinds of treatment technologies will be included in it?...They have dozens of them in Western Europe. There's a new plant in Alberta. Quebec is planning a plant. Manitoba is planning a plant. We simply have to come to grips with the problem of toxic industrial waste in this province."

Donna Tranquada directed her next point to John. "Well, John Dykstra, the province *has* offered to buy the land of residents who won't tolerate this. Isn't that a solution? You just pack up and move?"

John promptly dismissed the idea. "That only applies to residents living from 500 to 1500 metres from the proposed site. There are a lot of people outside of that particular area." With scarcely a breath in between, he launched into the next point from the Waste Quantities report. "And because Ontario is a net importer of wastes, that is, we import more wastes for treatment than we ship out, the OWMC has given little priority to recycling and reclamation of wastes or reuse or reduction at the process level. The private sector facilities are already operating at less than half-capacity, and only about 10,000 tonnes is not being adequately treated; and, to a large degree, those wastes are PCBs, which may soon be dealt with in the future by specialized temporary facilities."

Despite John's infusion of new information, Donna Tranquada did not stray from her prepared questions. "Mr. Dykstra, coming up in June, there are public hearings. What will you be telling the government that they have not heard already?"

John faltered. "Well, that we don't—can't—say at this particular point in time."

Dr. Chant jumped in. "The question of quantities is very interesting. *Our* studies, which are the most extensive ever done in the province, supported by the data from the Ministry of the Environment, show that between 600,000 and 800,000 tonnes of this material are untreated each year. I don't know where Mr. Dykstra gets his 10,000 tonnes, but obviously there's a very great difference in perception there. However, the bottom line is that we have a very demanding and extensive Environmental Assessment and independent hearing process in the province, and it is before that hearing panel that the various viewpoints will have to be resolved."

Marilyn jotted on her notebook and passed it to Mary, "He repeatedly stresses OWMC's work as the most extensive ever done in Ontario."

Donna Tranquada's voice took on the familiar sound of a radio host winding up her interview. "Well, Dr. Chant, is there anything the citizens can tell you or anything that can come forward that would dissuade the province from putting this plant at that site?"

"Not on the basis of the work we've done." Dr. Chant spoke with a confidence that sent a chill through Marilyn. He's convinced he's got the cat in the bag, she thought. All he needs is permission to pull the drawstring. "We've done the most thorough site selection work that has ever been done in Canada...The whole record is there. Mr. Dickstra and his people have had three years to look at it, and—uh—I just don't think it's conceivable that—uh—this will come off the rails."

Marilyn's head jerked up. Why the sudden mispronunciation of John's last name?

Donna Tranquada picked up on Dr. Chant's statement. "Well, John Dykstra, do you think it can come off the rails? Are you giving up hope at all that you'll ever stop it?"

John sounded optimistic. "No, we're not, because the OWMC facility could conceivably treat no more than *half* of its initial start-up capacity—and to treat even that much, it would have to capture a large part of the market share of wastes already going to present commercial facilities. The OWMC decided its facility requirements years before our Waste Quantities Study data was available, like from the Niagara Region and from the Town of West Lincoln. Their study has just been released. And the money spent to date is a drop in the bucket compared to the money, which would be spent putting up a white elephant or subsidizing its operation—"

"But what chances do you see of winning this?—" Donna Tranquada asked.

John had not finished his statement. "—There's no point in sending good money after bad."

A silence fell. Donna Tranquada said, hesitantly, "—Dr. Chant?—"

"Er—yes?—"

"What do you think is...is going to happen, then? The hearings come up in June. They could last another two or three years. Do you see that plant going in, no matter what?"

"Well, I'm confident in the work we've done," he said. Marilyn rolled her eyes in an expression of exaggerated boredom. This is the fourth time he's harped on that! she thought.

"We've had the benefit of—we spent over two million dollars on supporting groups such as Mr. Dickstra's, and the Township and the Region of Niagara in analyzing our work..."

Marilyn knew now that the mispronunciation of John's name was no mistake, but a subtle put-down. John had fared too well in the interview, brought out too many points detrimental to the Crown corporation.

"...But there never *has* been a more complete and detailed Environmental Assessment done in this province, and I'm confident that it will stand the test of the scrutiny at the—uh—independent public hearings."

"And if it does pass the hearings," Donna Tranquada asked, "when will the plant be completed?"

Again, Dr. Chant spoke with the confidence. "Well if the hearings should last about a year and a half, which is the estimate which most people seem to be making, it would take us between two and two and a half years to—to complete the detail design, and to build and commission the plant."

Donna Tranquada spoke quickly. "Dr. Chant, thank you very much. Mr. John Dykstra, thank you as well."

* * *

A week or so later, the Final E.A. arrived at John and Mary's home, a whole file box filled with thick reports. Marilyn made up a detailed checklist, with the subject of salts [chlorides] marked as a priority. At the next meeting, each member of the Concerned Citizens took home a report and a list. Marilyn assigned to each person a specific coloured marker with which to mark the passages. "Then," she explained, "when we trade reports, it'll be easy to identify who marked what."

Thereafter, Marilyn spent hours upon hours setting up files on the computer and inserting her own comments as she compiled the data turned in by the group. But, as she pointed out to Clifford, "It would be so much easier if I knew how to type."

Two weeks before Christmas, the *Niagara Falls Review* reported, "Fires at the plant's 'tank farm'...are another acknowledged problem." Marilyn promptly wrote the paper, challenging the word acknowledged. Fires, she pointed out, were not acknowledged by the OWMC until citizens brought the problem to light and refused to let OWMC push it under the rug as it tried.

She included three statements gleaned from the E.A. documents, then shredded the logic:

OWMC: "Potential risk to livestock and wildlife with an 'upset' or 'accident' [such as a tank farm fire] would be extremely remote after mitigation."

COMMENT: In the event of an accident or upset, the risks would be the same, with or without mitigation.

OWMC: "A tank farm fire would be expected to occur with a frequency of 'once' in 700 to 1,300 years if the tanks are segregated."

COMMENT: Frequency figures mean little, since an upset can occur at any given moment.

OWMC: "The cancer risk associated with facility operations would be many times lower than the current level of cancer risk in Ontario and would fall well within the range that has been judged 'acceptable' by regulatory bodies and decision makers."

COMMENT: The cancer risk would be "in addition to the current level of risk."

*   *   *

Two months had passed since the second debate at Brock University. At that time, one of the professors had quizzed Leslie Daniels regarding the high levels of salts, otherwise known as chlorides, expected to hinder the solidification of residues slated for the proposed landfill.

At the time, clearly flustered, Mrs. Daniels had admitted that the salts were indeed a problem. When pressed by the professor for OWMC's solution to the problem, she had replied, "I don't know what the answers are. I really don't know what they are. It's a problem of long standing."

Now, two months later, almost to the day, OWMC's Murray Creed had given a glowing report to the *Niagara Falls Review* on the proposed solidification process. Marilyn felt consumed by frustration. There was no way the public would know that this statement was the same old lingo the Crown corporation had been spewing out for years, the lingo praising a technology the debates at Brock had shown could not solve the long-standing problem of chlorides.

Nevertheless, Murray Creed had assured the public that the solidification process would "mix the results of the chemical processes with plain old Portland cement, which 'sets up into a mass you could drive on.'"

"Maybe you could *drive* on it," Marilyn stormed to Clifford as she tossed the clipping onto the table, "but it won't keep the chlorides from migrating into the bedrock aquifer!"

*   *   *

In mid-December, John Jackson sent out a letter asking the Coalition members for donations. The past six months had been very costly, what with preparing for and appearing at intervener funding hearings. As a result, the Coalition had run up substantial legal bills, plus other expenses such as travel, copying, faxes, etc.

"Fortunately," John wrote, "our lawyer has worked for us at less than the legal aid rate. His bills to us still totalled $6,000...We have used up our bank account, but still owe half of his billing."

Marilyn had half a mind to send one copy of the letter to Edgar Lemon, another to the newspapers, renouncing all rights to the Golden Fleece Award. Instead, Clifford wrote a cheque on the account that held their operating loan. "So we have to pay a little more interest," Clifford declared with false joviality. "It'll neither make us nor break us—I hope."

* * *

On December 14, 1988, the *West Lincoln Review* carried a demoralizing letter to the editor from Lloyd McKay. While campaigning for alderman a month earlier, he had asked the voters what they considered to be the main issues. Their answers, he wrote, had been the OWMC, the PCBs, and the four-municipality garbage dump. As for the OWMC, the voters said it would be built in Bismark and the money wasted fighting it was just flushing millions of dollars down the drain.

In the same issue, the Review carried an interview with Mark Sheldrake of Smithville. He believed that regional government hadn't done a bad job, but his biggest beef was the money being spent on battling the Ontario Waste Management Corporation. "The fight is a waste of money, because the OWMC facility is a foregone conclusion the province will force upon the municipality. Scrap the battle and spend the money on fixing roads or whatever."

Marilyn's forehead puckered in frowning impatience. Didn't these people have any idea of the harm they were doing by making such reckless statements? Yet, deep down, she knew her impatience was fueled by dread—the dread that there were lots of Lloyd McKays and Mark Sheldrakes out there. Seated in her old orange chair, she could feel the fear start in the pit of her stomach, and travel upward until she was trembling all over.

"It's hopeless," she murmured. "We're outnumbered, outgunned. We're killing ourselves for nothing. I want my life back." She moved her head from side to side in deepening despair. "I can't keep going," she said, knowing she couldn't quit. If they lost and she'd quit before it was over, she'd always wonder what the outcome would have been if only she'd stuck with it. It's like childbirth, she thought. You can't seem to birth the child, but it's impossible to give up and go home.

"Oh, Lord God," she whispered, "how can I stand anymore?"

*   *   *

Three days later, Marilyn opened a second letter from John Jackson. "Congratulations! We have cause to celebrate! Mary Munro has recommended that the Ontario Toxic Waste Research Coalition receive $839,735 (at legal aid rate) or $907,638 (at provincial rate for legal fees) in intervener funding. It is now up to the Minister of the Environment and the provincial cabinet to make a final decision."

And none too soon! Marilyn thought. With the release of the Final E.A., the pressure was building to prepare for the upcoming hearings.

"Mrs. Munro was clearly impressed with our work," John continued. "She made fewer changes in our proposals than she did in the other applications for intervener funding… In her report to the Minister, she describes us as dealing with issues of provincial concern…"

Marilyn felt a surge of pride as she continued to scan the letter. Obviously, Ms. Munro expected the Coalition to play a substantial role at the hearings and she had recommended that the Coalition be funded for every issue it had asked for except one—social impact of the OWMC's proposed facility.

Marilyn stopped reading and frowned. Social impact was one of her pet peeves. It had been assigned to the Township, but she didn't believe that it ranked high on their list of important issues. However, the Coalition had been awarded $55,650 to study specifically the comparative risks and impacts of the alternatives to the OWMC's proposal in the areas of Risk Assessment, Economic and Social Impacts.

"In addition," John wrote, "Mrs. Munro has made us an equal

partner with the Region and the Township on several topics. She has also recommended that we be given $175,000 to cover administration costs over the next two years, including a coordinator."

"Finally!" Marilyn declared as she thrust a closed fist into the air in a gesture of victory. "John Jackson will get paid for his work! And the Coalition members can no longer be looked down on and treated like second-class citizens!"

The postscript, written at the bottom of the page, came as icing on the cake. "Our 15-minute video, *The Turnaround Decade*, has just been completed."

*   *   *

With Christmas only two days away, Marilyn had no time to be polishing a letter to the editor on waste quantities. Yet she needed to strike while the issue remained a hot topic. She simply cited example after example from the Waste Quantities Study, now dog-eared and soiled from constant use:

- Ontario is currently a net importer of incinerable wastes.
- In 1987...Ontario...imported 135,000 tonnes [from the United States] and exported 42,000 tonnes.
- The majority of the waste management companies in Ontario have excess capacity...
- Tricil's...incinerator [in Moore Township] is operating at about 38% capacity...Of this, 25% of the incinerator capacity is used to burn imported waste.
- Quantex in Kitchener..., the recycling operation, is at 40% capacity...The physical/chemical plant is at 60%.
- Breslube in Breslau (which treats oil waste) is at 65% capacity. If the borders were closed to waste transport, the plant throughput would reduce to 25% of design capacity.
- Varniclour in Elmira...the solvent recovery plant...is operating at 55% of its capacity.
- Some...facilities (e.g. Swan Hills, Alberta; and Maryland) are operating well under capacity. The Study Team could not con-

OWMC—The "Wasted" Years: The Early Days

sider the experiences of these and other jurisdictions "*because this task was deleted from our original work plan by OWMC.*"

- According to estimates made by the Ministry of the Environment, only about 72,710 tonnes per year of OWMC's initial capacity of 150,000 tonnes per year "could potentially be managed at an OWMC facility *if* the facility captured a large portion of the [*market share from*] *existing commercial facilities.*"

- 36 -

Marilyn had believed the funding issues were resolved, but in truth, they had barely begun. Mayor Packham, claiming that Mary Munro had awarded the Township only twenty-five percent of the funding while allotting fifty percent to the Region, told the *West Lincoln Review* that she considered the decision a slap in the face. "Donald Chant promised the Township that it would be fully involved in the Environmental Assessment process."

The Region complained that not only had the funding it requested been cut in half, but it had been assigned nine technical areas to investigate, instead of the six it had chosen.

Then, just before Christmas, Mayor Packham threatened that if the Township were forced to take a back seat to the Region, both in funding and areas of study, she was prepared to recommend that the Township drop out of the E.A. Hearings. "With the reduced funding, the hearings will only be a sham," she told the *Review*. "If we're to retain our present lawyers, the town might have to pay up to $200,000 a year to its lawyers out of local tax money. For the nearly ten thousand residents, the burden would be onerous."

Marilyn found these pronouncements extremely worrisome. Nothing could turn the support of Township residents against the fight faster than the threat of higher taxes.

Then, as the parties geared up for the first of the preliminary hearings, scheduled for January 18, 1989, the Township held an orientation workshop on the Friday prior to the hearing to update township politicians on OWMC's activities and to discuss the funding issue. John Dykstra was invited to show the Coalition's video, *The Turnaround Decade*.

OWMC—The "Wasted" Years: The Early Days

* * *

On the morning of the first Preliminary Hearing, Marilyn looked around the public meeting room at the Legion Hall, with the rows of chairs set out for the audience and the tables arranged in a large rectangle for the participants. This place is becoming as familiar to me as my own kitchen, she thought, nodding to friends and acquaintances. As everyone rose for the beginning of the hearing, the murmur of voices softened to a silence.

Marilyn took her seat and the formal introductions began. Hurriedly she jotted down the names of the new faces at the table. Doug Watters, lawyer for the Ministry of the Environment. Philip Morris, exuberant and as green as they come, sat next to Dennis Wood. Marilyn experienced a moment of uneasiness. If he was the replacement for Dennis Wood, who had stated he would not work for the legal aid rate, the Township hadn't a hope of success.

Douglas Hodgson, also a lawyer, accompanied Stan Stein and Dan Kirby, the lawyers for the Region. Hodgson, tall and rangy, with a closely-clipped grey beard and longish greying hair, tended to look shaggy by comparison to the grooming of his spit-and-polish colleagues. Tricil, a new player to enter the fray, had sent jovial Ernie Rovet to act as legal counsel for the company.

As always, Mr. Blue made the opening presentation. He outlined what he considered an appropriate agenda for the day and the legal wrangling commenced.

First and foremost, the parties deliberated over where to hold the hearings. Marilyn had expected the hearings to take place in Smithville. But as the parties outlined their requirements—separate offices and secure storage rooms for files, rooms for the hearing panel, space for copiers and so on, a public area—she realized the problem was more complex than she had anticipated. After considerable discussion, Stein suggested the Parkway Inn on Ontario Street in St. Catharines.

Dennis Wood introduced an ingenious plan, worked out by Dave McCallum, which would keep the hearings in Smithville. He presented an estimate of the costs for renovations to the Old Farm Inn, the price for adding two portable structures, and a description of the accommodations available to members of the hearing panel, lawyers, consultants, and expert witnesses. Marilyn figured the Township had covered all the bases. There remained, it seemed, just one small problem: the members of the hearing panel, as well as the lawyers and consultants lived almost exclusively in Toronto. None wanted to spend the next two years living away from home or commuting back and forth to Smithville. The question of location remained unresolved.

After the morning recess, the parties tackled the question of transcripts—how many copies, to whom, and how quickly they should be transcribed. The representative for the Hearing Board stated that it was not customary to provide one-day turn around. Ian Blue contended that since seventy million dollars had been spent thus far, it seemed a shame to skimp on the transcripts.

Marilyn watched the hands on the clock crawl toward noon, twelve-thirty, quarter to one. The transcripts had nothing to do with her, and she found it hard to understand the intensity with which everyone debated the topic.

After the lunch break, the matter of the cutbacks in the requested funding again raised its ugly head. Dennis Wood, his dark good looks revealing the depth of his anger, snapped at one point, "Consultants, who do not bear the burden of the case, earn more than eighty-seven dollars an hour!"

Ian Blue bobbed energetically to his feet. "I sympathize with my learned colleagues. However, I suggest that if you want to save money—" He made a sweeping gesture toward the staff for the Region, the Township, and the Coalition, "—you should send some of your lawyers home."

Marilyn's mouth dropped open in fiery resentment, then snapped shut. "How dare he!" she muttered through clenched teeth as she surveyed the conference room crawling with OWMC personnel.

The tone had been set for the controversy to come.

* * *

Marie called Marilyn the next day. "Doug Draper's got a write-up in the *Standard*. For once, I like what he says—except for Stan Pettit's statement at the end, the basket!" Then, as always, she read the entire article to Marilyn.

As soon as she hung up the telephone, with supper turned down to simmer, Marilyn pulled on her boots and winter coat, and set off to buy the newspaper. As she stepped out into the early twilight, she felt a thrust of pleasure at the picture made by the neighbour's silos and barns, with their darkened shapes cast in sharp contrast against the frosted, rosy hue of the late afternoon sunset. Automatically, she hunched her shoulders against the winter's chill. Hurrying the short distance to the car, she unlocked the door and slid onto the icy seat.

Fifteen minutes later, as she waited for Clifford to come for supper, she sorted out a handful of markers and retired to her old orange recliner with the newspaper article. Draper said that Mary Munro had recommended provincial rates for the lawyers because experienced counsel would not only be of benefit to the interveners, but would also assist the joint board.

"Her recommendation,' Draper wrote, "would have cost the province only $600,000 more..."

When interviewed, Dave McCallum had said, "Then we could have had the lawyers we want." Marilyn paused, a prickle of anxiety around her heart. Did he mean, then, that the Township would definitely be depending on the very young, the very green Mr. Philip Morris?

Stan Pettit, chairman of the Region's Toxic Waste Steering Committee, had confirmed that Stanley Stein would be pulling out of the case as the Region's lawyer. "I can't blame Mr. Stein. But if you're going to do the job well, you need first-rate people...Unfortunately, these folk do not come at the legal aid rate." Then Marilyn came to Pettit's statement at the end of the article that had upset Marie. "If the province wants the OWMC facility in this location, they've got to assist us and themselves in making sure it is the best facility that can do the job."

Marilyn sighed deeply. Why did the politicians always discuss the fight in one breath, then speak of the facility as though it were a done deal in the next?

*   *   *

That same night, residents and Coalition members lined the visitor's seats for the conflict at the Regional council meeting over the funding. Stan Pettit, Mayor of Wainfleet and Chairman of the Region's Steering Committee, launched into a forty-minute address to Regional council. "If the province is attempting to save money, it's going about it in a strange way. The Ontario Waste Management Corporation has almost been given a blank cheque. Meanwhile, our lawyers told us yesterday at the preliminary hearing that they won't work for the legal aid rate, not when OWMC's lawyers are getting $200 an hour."

"Actually," Boggs interjected, "Market rate for a lawyer with the qualifications of Ian Blue is a minimum of $225 per hour."

Pettit continued. "Other cuts include a one-third reduction in the number of hours the Region estimated its technical consultants would need to study specific areas. And all travelling expenses for these technical consultants have been cut."

Wilbert Dick, Chairman of Regional Council, snapped, "The province...[has] put us in a lame duck position! We've prepared a better case than they had expected, and we don't think they like that kind of challenge."

Angrily, Niagara Falls Mayor Bill Smeaton echoed the threat of the Township. "I suggest that the Region consider pulling out of the hearings. It's time to tell Ontario...we're not going to play hockey without a stick. We should withdraw and see if they can hold a hearing in our absence."

The Region's solicitor, John Burns, advised the council members that the Environmental Assessment Hearings would go on with or without the Region's participation.

Mayor Smeaton retorted, "The province obviously wants the deck stacked in its favour."

"Unfortunately," Pettit told the regional councillors and the public, "the order-in-council which established the intervener funding panel specifically states interveners will be reimbursed for their legal expenses at the legal aid rate." He took a deep breath. "However, if we're doing a provincial job here, that money should come from the province."

Regional Administrator Mike Boggs, told the council, "The Region has three options in view of the funding cuts. It can withdraw from the battle, it can use regional tax dollars to make up the difference between legal aid and provincial rates, or it can try to get by with a cheaper lawyer. But at this stage, a new lawyer would be extremely prejudicial to our case."

* * *

Every day, Marie telephoned to apprise Marilyn of the rapidly unfolding developments detailed in daily newspapers. Meetings, meetings, meetings. Rhetoric, rhetoric, rhetoric. Yet, bound up in all the fist-shaking, information flowed in abundance.

In a rare display of openness, local politicians had invited the press to cover a two-and-a-half hour breakfast meeting held Saturday morning at the Parkway Inn in St. Catharines. According to the *Hamilton Spectator*, about twenty-five mayors and regional councillors had met with Ontario's environment minister, Jim Bradley, and elected Liberals from the Niagara area. Funding had topped the agenda.

By chopping up the numerous newspaper accounts and splicing the excerpts together, Marilyn gained a remarkably clear picture of what had taken place at the meeting.

The municipal politicians had met with Jim Bradley to express their anger over the amount of money they had been allocated. Stan Pettit told Bradley, "The province should be funding the opponents to the OWMC as fairly as it funds the OWMC."

Mr. Bradley had told him bluntly, "It's unlikely cabinet will give you more money."

Pettit had refused to be dissuaded. "The OWMC is...not limited in transportation and accommodation costs for its witnesses and experts. Nor is it limited to the number of lawyers it can hire or [restricted] to the same legal aid rates...It's difficult to find lawyers who are willing to work for the legal aid rate."

Bradley replied, "Perhaps the difficulty lies with these high-priced Bay Street lawyers, with no sense of public service. They have no regard for the difference between legal aid offered by the province and the fee they charge."

Peter Kormos, the New Democratic MPP for Welland-Thorold and a lawyer as well, snapped, "The intervener funding awarded is nothing but lip service by the province. The Region's involvement in the hearings shouldn't just be permitted, it should be sought!"

Irritated, Bradley had shot back, "Perhaps you've forgotten that before I became Environment Minister, there was no intervener funding. It was also my idea to hold an Environmental Assessment Hearing." Then Bradley had said accusingly, "You don't get credit for giving more money, you just get more requests."

Grudgingly, participants had admitted that yes, the province had provided more funding than in the past.

"It is possible," Bradley continued, "that the Environmental Assessment Board...could award a grant to cover costs of opponents once the hearings are over."

The councillors had promptly demanded that the province provide the money before the hearings began. "Otherwise, we'll either have to pull out of the fight or pay the difference—some $600,000—through increased taxes."

At that point, Bradley had issued a careful, non-committal assurance. "I think the Region can be relatively assured the assessment board will award funding—if the arguments are justified."

Stan Pettit suggested, "If awards can be made so easily at the conclusion, then why not just pay up front instead?"

The Liberal Minister of Natural Resources, Vince Kerrio, shot back, "We came here because we were invited to hear comments. We didn't come here to start running the government from this table!"

Thorold Mayor Bill Longo echoed the idea of dropping out. "It's useless to fight a giant with a slingshot."

"Please tell us now," demanded Bill Smeaton, Niagara Falls Mayor and Regional finance chairman, "if this funding decision is a fait accompli."

Mr. Bradley answered, "It is not."

With that, the breakfast meeting had adjourned.

Afterward, reporters had interviewed Regional councillors, seeking pithy comments to spice up their articles. Rob Dobrucki had stated bluntly, "It seems there is one set of rules for one side of the fight and another set for the other side... I did not hear any suggestions from Bradley that the OWMC's lawyers should be public-minded and work for the same rate."

Ivy Riddell had echoed Rob Dobrucki's sentiments. "I don't know of any public-minded lawyers who would totally commit themselves for two years to prepare and present the Region's case for fees substantially lower than what the OWMC's lawyers receive...We are fighting a battle for the whole province. The local taxpayers can't bear all this cost..."

More than all the political posturing that went on during public discussions, these informal remarks often laid bare a councillors' true position. Regional Councillor Eleanor Lancaster made a statement that sent a cold chill through Marilyn. "Even if the facility turns out to be a fait-accompli—as the Premier has indicated to me—we've got to force the OWMC to show the people of Niagara that it will be environmentally safe."

Marilyn's mind locked onto the finality of the statement. To know in her heart that Premier Peterson had long ago made up his mind in favour of the OWMC brought Marilyn anguish enough. To hear it confirmed from the lips of a regional councillor brought a suffocating sense of hopelessness. Marilyn dealt with it by hacking the article from the paper and shoving it into a file folder.

The road leading to the hearings seemed fraught with insurmountable obstacles—a road deeply rutted by shortfalls in funding and blocked by political manoeuvring that loomed like boulders. Marilyn found the continued furor highly distressing and sleep became more elusive than ever. Even John Jackson's assurances to the *Niagara Falls Review* couldn't dispel her worries. "Our group's lawyers will continue at the legal aid tariff rates. At the conclusion of the hearing, they will apply for a cost award to recover the difference."

For Marilyn, the question remained up in the air. Environment Minister Bradley had given no guarantee the money would be awarded. How long before the Coalition's lawyer re-thought his position and joined the exodus?

However, a week later, at the end of a day-long meeting between township and regional politicians, the Region's Steering Committee somehow managed to fill the shortfall-ruts for the Region's case and circumvent the political boulders, at least until the end of March. As a result, Stanley Stein agreed to stay on as the Region's legal counsel.

However, no solution had been reached for the Township's financial dilemma and like Eleanor Lancaster, Mayor Joan Packham spoke of the OWMC as inevitable.

David McCallum, West Lincoln's Toxic Waste Project Coordinator, revealed in an interview with the *West Lincoln Review*, "Mr. Wood has so far been paid $150 per hour for legal services to the Township. We cannot have Dennis Wood at the legal aid rate and we require Dennis Wood. Legal aid lawyers won't be able to carry the arguments at the hearings."

When reminded that Joe Castrilli, representing the Research Coalition, had said he would accept the legal aid rates, McCallum responded tartly, "The only reason [Mr. Castrilli] can go on is because he's got Stein and Wood as backup."

*   *   *

The following week's edition of the *West Lincoln Review* came up with a tongue-in-cheek solution to the Township's money problem. The OWMC had promised to pay the Township a windfall of $600,000 when it took out its building permit. Since the amount was almost exactly what the Township and the Region needed to make up the difference, the editor proposed that the Township should demand payment up front. Then, it could loan half of the $600,000 to the Region, at a modest rate of interest, and use the rest to pay its lawyers.

Marilyn might have found it amusing if it hadn't assumed that the OWMC would sooner or later require a building permit.

— 37 —

As the weeks rushed by, a day seldom passed that the newspapers didn't run an article regarding the multifaceted feud over funding. In the meantime, Marilyn sent a flurry of letters to the editor. One letter, written in John Dykstra's name on behalf of the Concerned Citizens, again challenged the validity of the OWMC's 500 to 1500 metre Impact Zones for "acceptable" impacts.

"Just a little dust, a little noise, a little ugliness," she reminded the reader.

> Yet, in June of last year, the OWMC announced it intended to establish two test plots...to monitor incinerator stack emissions...*ten kilometres* from the proposed site at points of approximately *zero* impact.

She emphasized the words ten kilometres and zero. "The OWMC's dinky little impact zones don't begin to cover the true areas of impact."

Another letter questioned, "When is a buffer zone not a buffer zone? When it is an OWMC 'technical' buffer zone." She then explained that although the 400-metre buffer zone was "technically" there to comply with Ontario regulations and to minimize risk to human health and the environment, in fact, most of the proposed facility spilled over into the buffer zone—the incineration plant, complete with the rotary kiln(and later a second kiln); the incinerator stack; the auxiliary boiler stack; the electrostatic precipitator; the administration building; the laboratory; the solidification plant and the associated conveyer and interim storage building; the landfill storm detention pond; the landfill surface water pond (potentially contaminated); and all landfill areas which surrounded the site.

"The buffer zone," she noted, "has been conveniently left out of the diagram on page eleven of the Environmental Assessment documents."

Letter number forty-one attacked more of the funding inequities, questioning how it was that at all the funding hearings, the OWMC did everything possible to block funding to the opponents. Or how was it that the OWMC had money enough to send Mary Lou Garr of the Smithville office to tour European waste facilities—but not John Jackson?

A week later, at the end of January, the OWMC broke its silence on the controversy surrounding the funding issue. Dr. Chant, praised in the *Standard* as "a long-time advocate of intervener funding," told reporter Doug Draper that by any comparative standard, the government had outdone itself. To date, the funds to opponents had been $5.4 million, which was totally out of sight with respect to any precedent anywhere in the world that he was aware of. "And certainly anywhere in Canada," he added. "That doesn't say it's enough or not enough, but it certainly says that the participants are better funded than anybody has ever been before."

As to legal fees, he refused to comment. However, he took issue with some of OWMC's opponents for saying they should get more funds because the corporation had spent about eighty million dollars since it began its work eight years ago.

> We are the ones that had to spend the money to develop a proposal. It's like an author writing a book. It might take a best-selling author four years to write a book and it takes a journalist or a critic half a day to write a book review. It doesn't take nearly as much money to review a highly technical piece of work that has been very expensive to do.

On the same day, Doug Draper, in a major opinion piece, wrote a scathing denunciation of the provincial cabinet's decision to keep the funding for the opponents' lawyers at the lowest end of the scale—the legal aid rate. Marie called Marilyn. Reading rapidly, she raced to the heart of the piece: "...It is only right that people who may be forced to bear such an unwanted burden...be granted a fair opportunity to challenge OWMC's plans at a public hearing..."

Marie stopped to editorialize. "This is the good part! It sounds as if we'd written it ourselves!" She read slowly, precisely, and with great emphasis:

> Is the facility really necessary, or are there better alternatives?...If there are flaws in the OWMC's plans, let's discover them now, not a decade from now...The public hearings...will only serve their purpose if all sides have equal access to lawyers and expert witnesses...

Then Draper pointed to the leaking PCB site in Smithville as an example where the failure to apply intense scrutiny to the PCB storage site had since shut down Smithville's water supply and had so far cost the taxpayers of Ontario more than $20 million to clean up.

> An additional $600,000...the amount of money at issue here, is...a bargain-basement price to pay if it helps assure that any potential for a mistake...is thoroughly reviewed before OWMC is allowed to proceed.

More excited than she had been in a long time, Marilyn hung up and raced to the corner store for a paper. If Doug Draper could finally see the light—that's the way she felt about it—then anybody could!

* * *

The appointment of three members to the Joint Hearing Board to preside over the Environmental Assessment Hearings caused a considerable stir. Lawyers, consultants, politicians, and residents alike tried to assess the standpoint of each appointee. How had each ruled in past hearings? Would any of the three be favourable to the opponents' case?

On Saturday, January 18, 1989, the Joint Board held a meeting in the Legion Hall. Dr. James Kingham, Vice-Chairman of the Environmental Assessment Board, had been appointed to chair the approaching hearings. Barbara Doherty, also a member of the Environmental Assessment Board, sat on the right; and Gordon Thompson, QC, a member of the Ontario Municipal Board, sat on the left.

Dr. Kingham, a distinguished-looking man with wings of grey hair, explained that the meeting had been called so that the newly selected Joint Board, which would conduct the hearings under the Consolidated Hearings Act, could identify the participants, decide on the location of future meetings, and determine the requirements for the hearing room. Admittedly, Ms. Doherty looked too young for the job. But like Dr. Kingham, and Mr. Thompson who peered over granny glasses perched on the end of his nose, Ms. Doherty had learned to school her face to reveal nothing.

The Concerned Citizens sat in a block of seats, close to the front of the room. As at most of these meetings, Marilyn found that so much of what was discussed made little sense to her because she had no frame of reference to fall back upon. Even though John Jackson tried to prepare Coalition members for each stage in the process, he simply couldn't know the vast gap between his knowledge and that of most residents. Why was it called a Joint Board? Marilyn wondered. What was the Consolidated Hearings Act? And why was it impossible for the parties to pare down their so-called hearing room requirements to fit the space available in Smithville?

*   *   *

Meanwhile, a showdown over funding between the Township and the Region seemed imminent. Closed-door meetings abounded. The Township voted for a tax increase of less than one percent during 1989 and 1990. Even so, the increase to the small community of 9,900 would generate only about $17,000.

The Township had then asked the Regional council to contribute $540,000 over two years to share in West Lincoln's legal costs, which would result in a tax increase of less than one percent to the Niagara Peninsula as a whole. The Region, facing a bill of nearly $2 million for its own battle, offered West Lincoln $91,000 as an interest free loan repayable within three years. Mayor

Packham said her community couldn't afford to repay it.

Rumours leaked out from the closed-door sessions. Bill Smeaton, Mayor of Niagara Falls and chairman of the Region's finance committee, didn't want to give West Lincoln a penny without a written guarantee that the Township's consultants would not point a finger to his community or anywhere else in the Region as an alternative site. Niagara Falls had been targeted as the second most-desirable location for the OWMC's proposed facility and had escaped by the slimmest of margins.

"Fair enough,' said an editorial in the *West Lincoln Review*. "Why ask the neighbours to pay for your fight and then give them a bloody nose?'

The melee continued. Mayor Packham said, "I have no problem with the stipulation. It's the Region, not the Township, which will examine site selection at the hearings."

However, questioned the *West Lincoln Review*, would the Region be able to argue fairly on behalf of the Township and still be able to cover its many backsides?

Although tales of the skirmishes between the Region and the Township continued to hit the newspapers almost daily, these stories remained unsubstantiated. Doug Draper lambasted the politicians for their closed-door policy, which effectively kept the public in the dark.

To Marilyn's consternation, the Township announced it had replaced its lead lawyer, Dennis Wood, with Phillip Morris of Toronto. Dave McCallum assured the *Standard*, "Mr. Morris has had experience in environmental cases, and Mr. Wood will remain in the background to assist him during the hearings."

Despite the bickering between the Region and the Township, Regional Chairman Wilbert Dick repeatedly assured reporters that the closed-door funding debates had been "positive and fruitful," and had been "marked by a spirit of co-operation."

"There's no way we want to see West Lincoln sitting out there on its own," he told Carol Alaimo of the *Standard*.

Marilyn felt certain she was developing an ulcer. Or maybe heart trouble. Or something. The misery in her midsection increased with every detail. She had discovered, in the course of the battle thus far, that she had a more rigorously logical cast of mind than she had supposed. But no one, logical or otherwise, could decipher the conflicting statements pouring out daily. The Township would withdraw from the fray. The Township pledged to keep fighting. The Township and the Region agreed to present a united front. The Township may break ties with the Region and go it alone. The Township may work in cooperation with the Coalition.

Poor pity the Coalition, thought Marilyn. John Jackson did his best to keep outside the eye of the hurricane. He gave few interviews, preferring to remain unavailable for comments.

On a different note, the *Niagara Falls Review* reported that Wainfleet's Stan Pettit and West Lincoln's Joan Packham were off to Swan Hills. Not only would the two mayors tour the toxic waste facility, but they intended also to talk to civic leaders in that western community and collect background information.

* * *

On February first, sparked by Mayor Packham's trip, the *West Lincoln Review* ran a feature article, based on a telephone interview with Swan Hills Mayor Margaret Hanson. From Marilyn's point of view, the write-up couldn't have been more damaging. Mayor Hanson's optimism ran like a racing tide throughout. In a nutshell, the town of Swan Hills, with a population of about 2,500, considered itself lucky to have the toxic waste treatment plant it had lobbied to get. The town, pleased with the facility and proud of it, was aggressively pursuing satellite industries to be located alongside it.

West Lincoln people, in their desire to get rid of the OWMC, "are missing out...on a great opportunity," Mayor Hanson told *Review* editor Judy McEwen. The facility, owned forty percent by the province of Alberta and sixty percent by Chem Security, a subsidiary of Bow Valley Resources, had, in her words, "put Swan Hills on the map."

When questioned about the Environmental Assessment Hearings, Mayor Hanson replied that she didn't know where the

OWMC—The "Wasted" Years: The Early Days

Environmental Assessment Hearings had been held or how long they had lasted. It didn't matter, she said, because "the residents knew so little of [the toxic waste treatment plant's] activities they wouldn't have known what to ask about."

Nor did she know the plant's capacity. But, thought Marilyn with chagrin, the mayor could certainly tick off on her fingers the pay-offs made to the community. "The facility, Mayor Hanson said, generously supports youth clubs, service clubs, the municipality and its schools and hospitals." The town was particularly proud of its indoor municipal pool established by volunteer groups and paid for by the community and corporate donations.

Marilyn felt a heaviness settle in the pit of her stomach. Far too many, with grasping fingers and a narrow, short-sighted view, would easily succumb to the seductive lure of so rich a shopping list. And the list didn't stop there.

Two hundred and fifty trees had been donated to spruce up the town. Radios had been given to the boy scouts to assist them when they hiked in the woods, and about $1,000 had been donated for promotional plastic garbage bags advertising the Grizzly Trail which was one of the attractions there.

And of course, jobs. Thirty-nine workers had been hired from Swan Hills for the blue-collar jobs. Another twenty, professionals, had been hired from outside the area. "The plant has brought professional educated people to a predominantly blue collar town," Margaret Hanson added. "They've provided a new look for us and brought new money into the town."

To fill the jobs vacated by the thirty-nine now employed by the Swan Hills toxic waste treatment plant, new people had moved into the area. As a result, about twenty new homes had been built. "A plebiscite in 1984 showed the people wanted the facility," Mayor Hanson told Judy McEwen, "and if another vote were taken this year people would still want it."

Suddenly Marilyn felt as though she were suffocating. I need to get out of here, she thought. I need to walk. She pulled on her snow boots and her ski jacket. With a woolly scarf wrapped across her nose and mouth and her fists crammed into heavy mittens, she clumped down the back steps and trudged out to the road. Facing west, she lowered her head and bucked into the cold wind.

If Clifford sees me, she thought, he'll be furious. She could almost hear his voice, shouting at her. "What's the matter with you? Are you *trying* to get sick?"

Right now, the idea appealed to her. Maybe she'd be lucky and develop pneumonia and die. That was the only honourable way out of this OWMC mess.

* * *

The following day, Marilyn called John Jackson and taking a pattern from Marie, she read the article aloud over the telephone. When she came to the part about the mayor not knowing where, when, or how long the Environmental Assessment Hearings had taken, John interrupted her. "They didn't hold any hearings," he told Marilyn.

"Are you sure?" she asked.

"I'm positive."

In mid February, columnist Bob Rooney wrote an article highly critical of those opposing the OWMC. Although Clifford and Marilyn had long ago adopted the attitude that everyone had a right to his or her opinion, Marilyn took offence at one statement in particular.

"Listen to this!" she stormed, when Clifford came in for supper. "'The anti-dump forces are not going to be informed, they are against it, and that is that.'" She tossed the newspaper onto the couch. "That sounds exactly like the sarcastic kind of thing Dr. Chant would say!"

"Write a letter," said Clifford mildly, exhibiting less pepper and mustard than Marilyn thought he should.

After supper, she sat down at the computer, which still resided on the dining room table. After quoting Bob Rooney's words exactly, she stated,

> In fact, the opponents are very well informed. Over the years of research and study, it has been our experience that

OWMC—The "Wasted" Years: The Early Days

those who are the most complacent about the OWMC facility are generally the ones who know the least about it.

She then invited Mr. Rooney to view the Coalition's video, *The Turnaround Decade*. "It shows the changes in attitude toward waste, the changes in processes... Some have achieved almost zero discharge. Others are moving in that direction." She concluded the letter with a rallying call. "Let us continue that which has already begun."

A week later, Bruce Piasecki, associate director of the Hazardous Waste Center at New York's Clarkson University, spoke at a meeting of NET Force. He condemned the Coalition's video, *The Turnaround Decade*. He called it very deceptive because, in his opinion, it gave the impression that industry could just snap its fingers and make waste reduction happen. Having been involved in the development of the waste reduction technology used by one of the companies, he knew it had taken several years to get the system up and operating. "There can be a lag time of 15 years to develop that technology. It does not happen overnight (and) you need an interim treatment facility."

He then gave his views on the OWMC's proposal:

> It's probably less risky to make improvements on proven treatment technologies. However, I have a few minor disagreements with the OWMC proposal. For example, above-ground storage rather than below-ground landfill.

The *Standard* carried John Jackson's response to Mr. Piasecki's criticisms:

> I agree that it takes time to develop waste reduction programmes. But one of our big beefs with the OWMC is that...the past eight years have been focused on the OWMC...[which] has delayed putting that time and money into getting waste reduction techniques implemented.

Marilyn heaved a heavy sigh. John's statement was no more than an add-on to the main article and, as such, carried little weight. She sighed again. For opponents, the attempt to gain any ground in this struggle was like trying to raise the water level of the Great Lakes by spitting into Lake Erie.

– 38 –

Three weeks after editor Judy McEwen's glowing report from Swan Hills, the *West Lincoln Review*, with equal vigour, soundly criticized the OWMC's work, based on a report produced by West Lincoln's consultants. This was the kind of hard information Marilyn craved to see—three columns of newsprint with virtually no wasted words, condemning the OWMC's conclusions. Marilyn circled, highlighted, and generally turned the article into a kaleidoscope of colours.

Incredibly, the Crown corporation had failed to provide something Marilyn would have thought crucial—a detailed profile of the wastes it expected to treat. To the Township's consultants, this suggested that the corporation didn't know what wastes would be arriving at the plant. Yet a detailed Wastes Profile, the report emphasized, "is critical to the review." As a result of the missing profile, the corporation had based its studies on pollution-control equipment at other facilities.

The report stated bluntly that the OWMC had failed to consider new technology. The rotary kiln system had been continually pushed because its design was an early policy decision, now entrenched. "Startup problems," the report noted, "should be expected because the incinerator technology chosen has a history of startup problems."

Another of the key issues, the landfill, also came under fire. The OWMC, in its early decisions, had made shallow entombed landfill a secondary criterion. Not until the site selection process had it became a number-one necessity. The consultants wanted to know why.

"I know why," Marilyn muttered. "They need a reason to explain why they picked West Lincoln."

The Crown corporation had failed to measure pollutants in combination with each other. It had used average conditions in worst-case scenarios. On and on went the consultants' list of flaws, right to the last line. Elated, Marilyn slapped down the newspaper and letting out a shout of exaltation that made her sound like a banshee, she thrust her fists toward the ceiling in a gesture of victory. Surely, now, people who had closed their eyes to the possible risks would take a second look. Surely the fence-sitters would jump off the fence and throw their support behind the beleaguered and battle-fatigued citizen groups. Surely OWMC's declarations of acceptable risk would raise the hue and cry of, "Acceptable to whom?"

* * *

In Marilyn's opinion, OWMC's logic had not improved one iota with the passing of time. "But, like old age," she declared to Clifford as she sat down at the computer, "I intend to fight it every step of the way." Clifford, his nose stuck in the latest issue of his weekly farm paper, *Ontario Farmer*, merely grunted.

Letter number forty-five, she noted as she waited for the computer to hum and whir its way through the boot-up stage. She resurrected early statements, showing that in the OWMC's view, "minimizing risk to human health" depended upon the number of people to be affected—the more people, the greater the risk; the fewer people, the smaller the risk. She then asked the reader to apply this same logic to a familiar situation—a dangerous school crossing. Using OWMC's criteria, authorities would judge the risk to be minimal if only two children used the crossing, rather than twenty. "But," she asked, "suppose those two children were *your* children. Suddenly, judged on an individual basis, risk is risk and it is the same for two or for twenty…"

She added several more examples of risk assessment from what she considered was the OWMC's warped point of view. She concluded by saying, "We must not let our attention become focused

OWMC—The "Wasted" Years: The Early Days

on the controversy of 'risk to many' versus 'risk to few.' The crucial word is *risk*. For risk to the individual means risk to all."

She read the letter to Clifford, gained his absent-minded approval, and immediately printed enough copies to send to the dozen papers that regularly printed her letters these days. At least with the computer, she thought, I no longer have to run to the library to make photocopies.

*   *   *

As the winter of 1989 melted into spring, the unrelenting round of meetings, meetings, and more meetings kept Clifford, Marilyn, and the others racing at a frantic pace. Hardly a day on the calendar didn't have a least one notation, and sometimes two.

On February thirteenth, the Coalition held a press conference at the Vineland Research Station to officially release the video, *Turnaround Decade*. Later that week, members of the group attended the regional council meeting for the video's first public showing.

On the way out, John Dykstra picked up an agenda for the next meeting of the OWMC Steering Committee. As he glanced over it, his face suddenly turned white, then fiery red. "We've been stabbed in the back!" he exclaimed. "The Township's finance committee has put forward a motion supporting a hearing location in the Toronto-Queen Elizabeth Corridor between Mississauga and Burlington and that 'where feasible—' His voice oozed scorn. "'Where feasible,'" he repeated, "'some sessions should be held locally.'"

"Well guys," he said with a small laugh that held no humour, "I guess we have to go to West Lincoln council on Monday night." Mary said nothing. But her eyes took on a weary look of resignation.

Marilyn felt the strength drain out of her. Don Austin loudly declared his displeasure, making it plain he'd had more than his fill of running to meetings. Marie, on the other hand, her eyes flashing and her voice tart, prepared to do battle.

On the following Monday evening, the Concerned Citizens filed into the West Lincoln council chambers. John kept his tone conciliatory, yet passionate. After considerable discussion, West Lincoln Council narrowly agreed to rescind the motion. Victorious, the

Concerned Citizens filed out of the council chambers and stood outside on the walk in the crisp night air. Grinning from ear to ear, John declared, "Now when we meet with the Regional Steering Committee in the morning, we at least have the backing of our own council!"

*   *   *

In the past, the Concerned Citizens had come away from Tuesday morning meetings with the Regional Steering Committee feeling bloodied and discouraged. Now, they encountered a warmer reception. Most on the committee had begun to treat both the Concerned Citizens and the Coalition as allies. True, some still looked upon the group as a nuisance and a threat of some sort. But enough committee members welcomed them with smiles to make it easier to make their plea for a hearing location in Smithville or as close by as possible.

Not that it was an easy sell. Marilyn began to understand some of the complications as Eleanor Lancaster laid out one of the strategies for a successful hearing. "The hearing board and the lawyers need to be able to get home each night. I've seen it happen. In a long hearing, they get so sick of motel rooms and restaurant food that all they want to do is to get the hearing over with. Although they don't mean to, they stop giving it their best shot. If we want to win in a long battle, we have to keep the board and the lawyers comfortable and content. That means, they have to go home at night and see their families, eat their own food, and sleep in their own beds."

The discussion continued. Every decision was weighed with an eye to winning. John revealed that, as of the night before, the Township now favoured a hearing location within Regional Niagara, with portions to be held at the Old Farm Inn. Half in jest, Eleanor Lancaster said, "Holding the meetings in Niagara would certainly make the members of the Hearing Board appreciate the local transportation routes and the distances involved in hauling dangerous goods from the industrial areas in the province to Smithville."

When the vote came, despite the nays, the Regional Steering Committee voted to hold out for the hearings to be held somewhere in the Niagara Peninsula.

OWMC—The "Wasted" Years: The Early Days

* * *

A week later, in a secret closed-door meeting held on February 28, 1989, West Lincoln council pushed through a version of the old defeated resolution, naming the Toronto-Q.E.W. corridor as a suitable location for the upcoming Environmental Assessment Hearings.

* * *

For the first council meeting in March, the Concerned Citizens once again filed into the Township's council chambers, their numbers swelled by angry residents. Again John spoke on behalf of the group, protesting the location decision: "We feel very strongly that a location near Toronto would be unjust. The citizens have done what they were supposed to do. We became involved. We became informed. We have made our concerns known to various levels of government. We have given of our money and our time and our strength. We have *earned* the right to have the hearings held in a place accessible to those who are most concerned."

Marilyn searched the faces around the council table for some slight sign of yielding, some flicker of sympathy for the residents' plea. She searched in vain.

With closed faces and expressionless eyes, the council thanked the citizens and agreed to consider the request. As always, the Concerned Citizens filed quietly from the council chambers and stood outside on the walkway, shivering in the chill night air.

"Well, *that* was a waste of time!" declared Marie. "Now what do we do?"

John shrugged. "Make our plea to Dr. Kingham, I guess."

* * *

While the Concerned Citizens and other residents fought to have the hearings accessible to them, Doug Draper featured former West Lincoln mayor, Allard Colyn, in a half-page article in the *Spectrum* section of the *Standard*:

> Mr. Colyn argues it would be a mistake for the Township to devote all its energy simply trying to keep the OWMC

facility out of West Lincoln. "You can't always have what you want," [Mr. Colyn] said, "and my prediction is that the (OWMC) technology...will be successful at the environmental hearings."

- 39 -

The newspapers carried the notice of the second Preliminary Hearing, scheduled to last three days. By quarter to ten on Tuesday, March 28, the Legion Hall in Smithville swarmed with residents and Coalition members, as well as politicians from West Lincoln, the Region of Niagara, and other municipalities. Philip Morris now represented the Township and Doug Hodgson had replaced Stanley Stein for the Region. The court reporter sat near the front corner of the tables arranged in the familiar rectangle, with her back to the window. At ten o'clock precisely, the Hearings Officer had the assemblage rise. The hearing panel entered and took their places at the head of the table.

Once the formalities had been dispensed with, Ian Blue submitted the OWMC's phasing proposal. For once, copies were available to the public and Marilyn made notations as she attempted to follow the proceedings. "At least it's getting easier," she whispered to Clifford. "I understand more of the terminology now."

The hearing panel, otherwise known as the Board, meant Dr. Kingham, Ms. Doherty, and Mr. Thompson. A witness panel meant a group of expert witnesses—consultants or OWMC personnel—who would sit together at one table to give joint evidence on one particular topic. A Phase referred to a group of witness panels, composed of related topics. Thus far, there were six Phases and fifty-nine witness panels.

Immediately, the Region, the Township, and the Coalition submitted their revised version of the OWMC's phasing proposal.

Hodgson rose slowly to his feet. "With respect, Mr. Chairman,

Panel 4, "Benefits of OWMC's Undertaking," cannot be examined before all other evidence has been introduced and examined. I suggest Panel 4 be placed at the end of the hearing in Phase VI."

Marilyn could see the reasoning. The OWMC wanted to establish early in the proceedings and early in the Board's mind that, yes, the OWMC's facility would be beneficial. The interveners wanted to prevent that. As she glanced over the list of fifty-nine panels, the names alone overwhelmed her. "Facility Capacity and Profiles; Alternative Treatment Processes: Physical/Chemical; Design, Construction, and Commissioning..."

The debate moved into the mechanics of the hearing days. Mary Munro had predicted the hearing would last between 254 and 380 sitting days. It would amount to approximately three years. Ian Blue, in a plea to aid the Board, the province, and to reduce the expense to the taxpayers, suggested the participants shorten the time by increasing the hearing days from four days a week to five, and the hours per week from twenty to thirty.

Joe Castrilli came swiftly to his feet. "Mr. Chairman, five days is an unworkable number. It assumes that all preparation for the hearings ended at the beginning of the case. In fact, preparation will be a continuing process throughout the hearings." He glanced down at the notes he had made during Blue's presentation. "In my experience, if a hearing sits for more than five and a half hours, information fatigue sets in."

The instant Joe Castrilli sat down, Philip Morris jumped to his feet. "Mr. Blue is giving the Board a wrong impression. By his suggestion, the interveners are being made to look uncooperative. Mr. Blue, as counsel for the OWMC, would have numerous staff working behind the scenes on a full time basis. The interveners will not have a similar support system. It is recognized that at the level of twenty-four sitting hours in a week, the witnesses become tired and argumentative, the lawyers become tired and argumentative, and the efficiency of the hearing has not improved."

Ernie Rovet, representing Tricil, rose next. Of medium height, heavy-set and perpetually jovial, he had a sense of humour befitting the class cut-up. Yet his round face exhibited a good-natured shrewdness. "May I suggest that by lengthening the hearing days, we run the risk that the hearing panel itself would become too tired to *hear*."

"Thank you for your concern," said Dr. Kingham as a ripple of amusement ran over the room. Dr. Kingham deferred an immediate decision. "The Board needs time to distill what it has heard."

Not pleased to have his motion put on hold, Blue bobbed to his feet, his face drawn into a dark scowl. Making a sweeping gesture with his arm toward the opposing lawyers, he said, "With respect..." His tone and manner made a lie of the word. "Counsel are all highly-paid professionals. They should be able to hold up their end of the hearing."

Obviously in the habit of command, Dr. Kingham passed over Mr. Blue's objection and smoothly moved the focus to other issues—the cost of duplicating documents and how to decide which documents were relevant and which ones were not. Ian Blue half stood and said tersely, "A document won't be provided if I decide it's not relevant."

When the issue of scoping came up again for debate, Mr. Watters of the MOE rose and began to speak. Dr. Kingham interrupted him. "The Board thinks the MOE should be a disinterested party. Therefore, you should have the last presentation."

Just as Marilyn had assumed that adequate funding for opponents was a right, so she had assumed that the role of Ministry of the Environment would be that of a neutral party, present to provide information if asked. So when she wrote Dr. Kingham's statement down, she gave it no thought. But the lawyers argued the issue heatedly. Hodgson, Morris, Rovet, and Castrilli all took exception to the ministry's supposed neutrality. "It's difficult for Mr. Watters to be considered neutral when he referred to the OWMC's proposed facility as *our* facility," Hodgson stated.

Blue retorted, "The province needs this facility. The ministry has funded us...However, the OWMC should argue last. As the Attorney General's representative at these hearings, I have the right to argue last. Mr. Watters needs to be educated by the evidence and then cross examine. But Mr. Watters does not have to make the final judgment."

Philip Morris objected. "Mr. Blue is not here on behalf of the Attorney General but on behalf of the Ministry. The OWMC is a creature of the provincial government. This hearing is not the same as other hearings."

When all had been said, Mr. Watters rose heavily to his feet. Much the build of Ernie Rovet, he wore wire-rimmed spectacles and a benign expression. With his middle finger, he pushed his glasses farther up onto the bridge of his nose. "I'm troubled that the objectivity of the ministry is the subject of debate." He sounded aggrieved that the other lawyers had turned on him so fiercely. To Marilyn, it seemed like a tempest in a teapot and a tremendous waste of time.

Morning recess came, then noon. Marilyn took voluminous notes, afraid to leave anything out because she didn't always seem to grasp what was important and what was not.

As the wrangling filled the hours, Blue said irritably at one point, "I'll call the case as I see fit. The opponents can respond as they see fit."

Marilyn was astounded at the lack of respect he showed for the Hearing Panel. Would not Dr. Kingham, Ms. Doherty, and Mr. Thompson make those decisions? Yet time and again Mr. Blue all but thumbed his nose at the Board. Marilyn watched him make his proclamations, his head held high as he pivoted periodically on one foot to encompass all hearers. He's like a bad kid in the classroom, she thought. He's pushing to see how far he dare go. She stared at Dr. Kingham. Why didn't he put Ian Blue in his place, now, and settle the matter once and for all?

Just before the afternoon recess, Joe Castrilli reminded the Board, "I have received no documents on waste quantities. Yet waste quantities is in Phase I."

After the recess, Philip Morris noted that eleven documents had not been made available to him. "One won't be available until at least May 15." When Morris told the Board that he had requested a list of all of the documents, back to 1981, Dr. Kingham placed a honeyed smile on his face. "Mr. Blue should produce as much as he is willing without the Board ordering it."

The hearing was still in progress when Clifford and Marilyn

reluctantly rose to leave at twenty after four. Every Tuesday night after supper, no matter what else came up, Clifford and his brother, Ernest, loaded pigs to ship out the next morning.

* * *

The second day of the Preliminary Hearing followed the same pattern of wrangling over every tiny detail. To Marilyn, the lawyers seemed overly picky. They treated each bit of minutia like a hair-line-crack in a dam, which had the potential to widen to disastrous proportions.

They cited Rule 47 and Rule 11. They argued that the Environmental Assessment Hearing could not start before November if the OWMC's Witness Statements would not be ready before May 31, two months away.

They spent hours arguing over timing—when the OWMC's documents should be released. Hodgson, for the Region, insisted, "OWMC has to put it all on the table. It has to have the courage to do that."

Blue argued, "Twenty-seven documents pertaining to Panel 5 only came in last month. We need time to study them." Marilyn looked up Panel 5 on the schedule. Waste Quantities—the missing data on what toxic materials would be coming to OWMC's proposed facility and how much.

Joe Castrilli referred to Regulation 205. Blue responded that the witness statement for Panel 54, on monitoring the facility, would not be written until the end of the hearing.

Young Philip Morris complained about his incomplete document list and missing documents. "Mr. Blue says we're not entitled to know his case to the end." He paused for emphasis. "I thought trial by ambush had disappeared from the system."

Ernie Rovet made one of his brief, pithy comments. "The OWMC should have its case together by now. I suggest Mr. Blue should disclose all he can."

Blue jumped to his feet. "Mr. Chairman, Mr. Castrilli and Mr. Morris have 289 reports. What difference does it make if eleven are missing?"

But, Marilyn thought, one of the missing reports contains the crucial waste quantities data.

Philip Morris half rose. "Actually, I'm short nineteen reports, not eleven."

The Chairman called a brief recess. Afterward, the parties ploughed through more issues, from funding, funding, funding, to Blue's contention that the Hearing Panel needed to hire its own expert staff to assist them in understanding the evidence.

Marilyn saw the Chairman's lips tighten ever so slightly. He conferred for a moment with his colleagues, his hand cupped to hide his mouth. Then, with a trace of a smile and a tilt to his head, he stated, "The Board will not engage expert staff. The parties will put forth the evidence in a manner understandable to the Board—and to the public."

On the subject of transcripts, the Board stated that twenty hard copies and eight disks would be available. The Board would need three. The rest of the participants could divide up the remainder among themselves. Marilyn still didn't see why everyone made such a big deal over the transcripts. They were like the minutes of a meeting—old business. Why on earth did so many people require a copy of them?

The residents drove to the Old Farm Inn for lunch. The OWMC trouped en masse to the near-by restaurant in the Village Square Mall. Neither side wanted to be anywhere near the other.

When the hearing reconvened, the Chairman said, "The Board requests clarifications as to who the Coalition is?"

Puzzled to find that the Coalition remained a question mark in the Board's mind, Marilyn swung round in her chair to look questioningly at Edith Hallas. Edith looked equally puzzled and raised her shoulders in a prolonged shrug. Joe Castrilli, after consulting with John Jackson, replied, "The Coalition consists of seven member-groups, totaling over 2,000 individual members, coming from all across the Niagara Peninsula."

Blue immediately attempted to undermine the Coalition's sphere of influence. "Most of the groups have only a handful of members. The number is derived from larger agricultural organiza-

OWMC—The "Wasted" Years: The Early Days

tions who, although they have given tacit support, are not active in the Coalition."

And so it went, with the parties hammering away at every issue. Marilyn, ignorant of the content of evidence panels and witness statements, found it difficult to follow much of the debate. Finally, the proceedings arrived at the thorny problem of where to hold the hearings. The Chairman polled the parties. The Coalition preferred a location as near to West Lincoln as the available facilities would allow. The Region opted for the Parkway Inn in St. Catharines, with local issues heard at the Old Farm Inn.

Philip Morris, for the Township, kept his gaze away from the residents. "The Township of West Lincoln favours the Ennisclare Centre along the Queen Elizabeth Corridor near Oakville."

An angry uproar arose from the audience. "Traitor!" exclaimed Lynda Bradley from two rows behind Marilyn.

"Backstabber!" hissed another voice.

Marilyn felt a wave of fury engulf her. "Council doesn't want us there!" she snapped in a stage whisper. "We're too knowledgeable!"

Hodgson offered a letter from Harry Pelissero, MPP, supporting the residents' wish to have the hearings held locally.

Ian Blue stood, his head tilted back and rocking on his heels. "If convenience to the lawyers and cutting costs were the only considerations, then the OWMC would favour the Ennisclare Centre. However, the OWMC strongly believes the citizens should be able to come to the sessions. Therefore, the OWMC supports a location in or near Smithville."

Marilyn's mouth dropped open and a frown formed between her eyes. Why had the OWMC supported the residents?

Marilyn knew she was terribly naïve about most things. She knew that past performances did not add up to the benevolent Crown corporation she had expected. Yet part of her believed Blue's pronouncement contained a genuine concern. After all, had not Dr. Chant always been a champion of public hearings and public participation? Then the spirit of scepticism, formerly foreign to her nature, kicked in. Or did Dr. Chant and Mr. Blue know that the hearings would be held in Oakville and it was safe to play the role of the good guys?

"Oh Lord," she whispered. "I hate politics."

For the first time, Philip Morris dared glance at the residents. Marilyn could almost hear the pleading in his voice as he sought their understanding. "It would cost between $120,000 and $130,000 to locate in Smithville. We wrote to both Environment Minister James Bradley and Dr. Chant, apprising them of our difficulty. We received no response from either one. Therefore, believing our funds better spent on consultants than accommodations, we have opted for the Q.E. Corridor."

Marilyn stared at her note pad, her feelings in turmoil. If it was true that the lawyers and consultants charged not just mileages but their hourly rate when driving to Smithville, as she had read somewhere, what choices would she have made in council's place? She didn't know. She ran her hand across the back of her neck. The ordeal of attending the hearings in Oakville loomed like a mountain she must scale with no peak in view.

The Chairman called for a fifteen-minute recess. Not wanting to talk to anyone, Marilyn headed for the restroom. In the privacy of the small cubicle, she let her tears drip silently into a wad of toilet paper. I can't make the trip to Oakville every day! she thought. I just can't do it.

The sound of the concerted movement of many feet upon the old wooden floors told her that the Board had returned. Swiftly she splashed her face with cold water and tried to hide her blotchy cheeks with makeup. Then she hurried up the stairs and tiptoed to her seat.

Ian Blue was all excited. "I think we've come up with the perfect solution. What if we set up a viewing room with a large monitor where people could watch the hearings in progress? Maybe it could be done with cable TV. Or a satellite dish into the homes."

Hodgson suggested videotapes that people could borrow.

Lynda Bradley leaped to her feet. "No! We want the hearings in Smithville. We want to *be* there, able to respond and take part. Watching the hearing on a monitor is unacceptable. It's the difference between actually being at a ball game or watching it on TV!"

The Chairman's glance encompassed the entire audience. "The Board recognizes the good intentions of the residents. However, it is our experience that the continuing, daily, and ongoing intervener involvement in the hearing is likely to come from counsel representing the citizen's local governments, rather than the individual residents."

Marilyn had no time to get nervous. She jumped up and flung her words directly at the Chairman. "You don't know what I intend to do. When I say I'll do thing, I do it. When I say I intend to come to the hearings, I mean it." As she glared at him, she could feel her knees begin to tremble.

"And you are?" the Chairman inquired politely.

"Marilyn Gracey."

"Thank you for your comments, Mrs. Gracey."

Shaking all over, Marilyn stepped back to her chair and LaReine Foden took her place. A real estate agent and a member of the Coalition, LaReine offered additional information on several possible hearing locations in St. Catharines.

Lynda stood up again and directed a hard stare at Mr. Watters. "If the MOE wants to be fair, you should urge the government to grant West Lincoln the money it needs."

After some additional haggling, the hearing adjourned until the next day. The members of the various Coalition groups gathered in the parking lot to vent their anger. "They're going to hold the hearings way off in Oakville, I just *know* they are!" stormed Lynda.

The group moved aside to let the dark-suited lawyers pass. Hodgson, with his unruly grey hair blowing in the cool breeze, said to Philip Morris, "I'll be with you in an minute." Then he switched his briefcase to his left hand as he walked purposefully toward the citizens. "I want you people to listen to me." His expression intense, he moved his hand in a chopping motion as he talked. "I was the proponent for a proposed landfill site in a place so small—maybe sixty people in the hamlet—that there wasn't even a restaurant. Every day, for five months, the local ladies provided lunch, homemade cookies, coffee and tea." His gaze swung over the group, engaging every eye. "I'm convinced I lost that case because the Board couldn't look at those ladies every day and then stick them with a landfill." He bore down with emphasis on each word. "Keep doing what you're doing. Show up at the hearings."

Wearing only a suit coat, he suddenly hunched his shoulders against the chill. He broke into a smile and waved as he strode toward the waiting car. "See you tomorrow!" he called over his shoulder.

Marilyn watched the car pull from the parking lot and head for Toronto. It was the first time any of the lawyers had spoken to the residents, or made them feel their presence was something other than a nuisance. She had stopped speaking to Dennis Wood, the Township's lawyer, when she passed him on the stairs. He never answered. He just looked past her as though she didn't exist.

Now Hodgson had told them their presence at the hearings was important. They weren't wearing themselves out for nothing.

As Clifford took her arm and steered her toward the car, the old three-part litany ran through her mind. Read the reports. Write letters. Build the team. Now she added a fourth: Be present at the hearings.

*   *   *

Day three of the second Preliminary Hearing varied little from day one and day two. Mainly, the parties cleared up details that had been left hanging. Blue again defended his right to call his case as he saw fit.

The Chairman inquired, his head tipped inquisitively, "Would you have a right under Phase VI to introduce new evidence, as you say, 'as long as I can pay for it'?"

Blue snapped to his feet. "Why wouldn't you want the best, most up-to-date evidence?"

Edith Hallas raised her hand. "Would we get the same chance to introduce new data, for example, on the food chain or MISA?"

The Chairman made no ruling but thanked Edith for her question. Mainly, Marilyn found the discussion humdrum. How and when to file documents, whether they should be numbered or lettered, alphabetically or by subject—it all seemed irrelevant to her. The lawyers hashed over the matter of transfer stations. Should other areas be notified, so that they could participate in the hearings?

Not until Philip Morris objected because the OWMC had provided a copy of the E.A. documents to the Board did she perk up.

"They should not yet have been filed with the Board," he declared.

Aggrieved at the suggestion that the Crown corporation had acted inappropriately, Blue rose to respond. "The Ministry of the Environment instructed us to do so. The Board should read it. The case is based on the EA."

Rovet offered his opinion. "The Board should not read it yet."

Watters pressed his palms on the table before him and pushed himself to his feet. Peering over his glasses, perched on the end of his nose, he assured the Board, "There is no law preventing the Board from reading the EA. And I offer a word of caution to Tricil's counsel." He turned to glare at Rovet. "Your opinion is a badly disguised effort to delay the work of this hearing." Then he lowered himself to his chair and stared at the page of notes before him.

The Chairman glanced at his colleagues, then gave a quick rundown of the items that would appear on the agenda for the third round of preliminary hearings. He spoke to Mr. Blue in particular. "We will meet here again at ten a.m. on July 4, 5, and 6. And just to be safe, Mr. Blue, please reserve this room also for the seventh."

* * *

On April 7, Regional Council voted 15–13 against a two-million-dollar budget to fight the OWMC. The *Hamilton Spectator* called Marilyn for her comments. Candidly, she told Janet Lees,

> I am constantly astounded at the level of ignorance on the part of the regional councillors concerning the OWMC. However, I must say, if that vote had been taken a year ago, I would not have expected it to be so close.

She sighed heavily as she considered her next words. "You know, I can have all my facts, and be as right as I can be, but if I have no one to fight for me at the hearings, it doesn't matter a particle."

* * *

If Marilyn and the others needed proof that the Township was secretly angling for the OWMC facility, it came with the revelation of a letter written by Joan Packham, mayor of West Lincoln, to the

Minister of the Environment. Basically, West Lincoln council wanted to know how to re-designate farmlands for industrial purposes, "particularly in the immediate area of the proposed toxic waste facility."

Meanwhile, Marilyn continued her letter campaign to educate the public. She had great faith in the ordinary person's common sense. Once in possession of the facts, logically presented, she fully believed public opinion would sway toward what was best and what was right. Letter number forty-nine extolled the need for the costly environmental hearings as a forum for testing:

> Any new product, whether it's a car or a new drug, needs to be properly tested before it comes onto the market. No one wants to buy a car with a braking problem or see another thalidomide disaster.
> Yet over and over we hear the foolhardy statement: "Stop wasting time and money. Build the OWMC facility and get on with it."
> In other words: Don't test! Just build!
> This kind of reasoning would be unacceptable with any other product seeking approval in the market place. Opponents do not believe that the OWMC proposal will stand the test of a full and fair Environmental Assessment Hearing. But a full and fair hearing *must* be held if the flaws are to be revealed.
> Do we *really* want to skimp today, only to find that we have left a legacy of disaster for the next generation? In our parents' day, that was known as being penny-wise and pound-foolish.

* * *

On the first Sunday in May, Clifford and Marilyn woke up to an inch of snow. "Thank goodness the wedding for your brother Bill and Betty was yesterday," Marilyn declared over breakfast. She smiled at the remembrance of the joyful celebration. Both had lost a mate and had been lonely for a long time.

That same day, John Jackson was elected president of Great Lakes United. He was the first Canadian president of this coalition of two hundred environmental, citizen, sports, and labour groups from throughout eight Great Lake states, Ontario, and Quebec.

May seemed to be a month for fun and respite. The Niagara Residents for Safe Toxic Waste Disposal entered their float in Fenwick's Lion's Parade. The Concerned Citizens voted to attend the launching of the *NIMBI* from Port Maitland near Dunnville. Chuck and Pat Potter, the husband and wife team that first blew the whistle on Smithville's PCB catastrophe, planned to use the tug to monitor pollution and educate school children and others on the importance of the Great Lakes. Excursions aboard the *NIMBI* would be limited to the Grand River and Lake Erie.

The name, *NIMBI*, was a take-off on the slur often cast on the opponents of the OWMC. Pat had explained to Marilyn, "NIMBY, with a Y, stands for Not-In-My-Back-Yard. NIMBI, with an I, stands for Now-I-Must-Become-Involved."

Marilyn read the invitation at the Concerned Citizens' meeting on Monday night. Already they had begun to plan for their fundraising barbeque in August. "We could do with a little fun!" declared John.

Marilyn couldn't agree more. "Please, God," she said, "Let it be warm and sunny." A picnic in the rain held no appeal for anyone.

Reggie said, "I think we should wear our Concerned Citizens hats." She broke into a conspiratorial smile. "Maybe we can win over some more support."

Marilyn had to admire Reggie. She kept up her cheery, unfailing smile, despite the fact that she had lost her husband to cancer only five months before. All through the months after his surgery, until his death, Reggie had read documents, attended meetings, and carried on as though nothing was wrong in her private life.

On May 10, the *West Lincoln Review* published a letter to the editor written by Willa Leach. At the top of a list of cheers and jeers, she wrote, "Cheers and more cheers for Clifford and Marilyn Gracey and the terrier-like tenacity of their letters-to-the-editor...based on pure unassailable logic."

Marilyn, seated in her old orange recliner with the foot up and the paper spread across her lap, pressed her fingertips to her lips as she read and re-read the last five words: "based on pure unassailable logic." Willa Leach could have searched the world over and not found five words that brought more joy to Marilyn's heart. Logic. People responded to logic. And they lost confidence when logic lapsed.

Her joy shattered two weeks later when the Board chose the Ennisclare Centre along the Q.E. Corridor as the site for the formal Environmental Assessment Hearings.

- 40 -

Once the Board issued its decision on the location of the hearings, Marilyn's telephone began to ring with reporters wanting her reaction. She made no effort to hold back her bitterness when she spoke to Janet Lees of the *Hamilton Spectator*. "The good old boys will get together and decide our fate for us. We feel...abandoned." What Marilyn didn't say was that she had learned by chance that the Good Old Boys from all sides had already gotten together for a meeting in Toronto, a meeting not made known to the public. Personally, she believed it had played a part in the Board's decision, but she had no proof.

"Will you still be attending the hearings in Oakville?" asked Janet Lees.

Marilyn sighed. The words of Lillian Southwick one night recently at prayer meeting flashed into her mind. "My goodness, you sigh a lot," Lillian had remarked.

Marilyn sighed again. It was a sigh weighty with the knowledge that she must face whatever had to be faced. "We and the other residents have effectively been cut off. It was going to be a hardship to attend them, even if they were close by. Now, with our work and the hour's drive there and back, there just isn't any way that we can attend with regularity." Her voice turned harsh with resentment. "They don't want us involved. That was all just words. We've been shafted again."

Marilyn heard Janet Lees thumbing through pages. Obviously referring to previous notes, she told Marilyn, "Rob Mens, administrator for the Township, says that public information sessions will be held in West Lincoln to keep the residents abreast of the developments."

Marilyn snorted in derision. "That's just not good enough. If this is the type of treatment we get on something as simple as the location, then how can I have much faith in the process itself? Is the hearing just an expensive charade?" Her voice, suddenly low and sorrowful, betrayed the hurt of disillusionment. "Our hope has always been that whether we won or lost, it would be fair. Now we'll never know for sure whether or not it was a fair hearing."

Marilyn could hear Janet Lees flipping through more pages. "John Jackson says that choosing Oakville for the hearings makes a mockery of the province's promise to encourage community input. He says his case could be hampered by the lack of grass-roots attendance."

Marilyn didn't respond.

"Mrs. Gracey?" queried Ms. Lees.

Marilyn swallowed and pressed her fingers to her trembling lips. "I'm sorry," she apologized. "I can't talk any more."

Later in the day, Marie called to read from Paul Forsyth's account in the *Standard*. "Dick says the Oakville site will undoubtedly work to the OWMC's advantage, but he doesn't expect the Region will challenge the decision." She took a moment to skim the story. "He told the reporter, "We can't afford to waste time on lawyers arguing where the hearings are going to be. We will concentrate on trying to prepare our case by the expected fall start of the hearings."

The sobering reminder of the possible October date, a mere four months away, brought the telephone conversation to an end.

* * *

Two days later, the Region announced it would fight the hearing location in Oakville. Meanwhile, the Township had appealed once again to the Region for funds to fight the OWMC.

Marilyn tried to keep her spirits up despite the rumours of in-fighting and bad relations between the Region and the

Township over money, the hearing location, and just about everything else. By day, she put on an optimistic front. Only in the wee hours of the night did she admit to the fear that crawled all through her innards.

Although the onset of the growing season on the farm daily increased Marilyn's workload, she loved the warm sunny days of June. With the huge maple and ash trees in full leaf in the front yard, the sun made bright lacy patterns on the grass as she weeded flower beds and mowed the lawn. The vegetable garden, newly planted, showed the first signs of emerging rows as the beans, beets, and sweet corn pushed up through the soil. Only the stand of old-fashioned strawberry rhubarb looked decimated, evidence of the on-going feast of pies and rhubarb crisps that had adorned the Gracey supper table for the past six weeks.

As Marilyn mowed around the clothesline pole and down the slope toward the barn, her mind assembled the next letter to the editor. The topic? The OWMC's continual and infuriating insistence that all risks associated with the proposed facility were "acceptable". She stepped on the clutch and put the lawnmower out of gear. Straining to get her hand into her pocket, she retrieved a stub of a pencil and a folded piece of paper.

"Acceptable to whom?' she jotted down in a cramped hand. "No matter how you cut it, risk inevitably comes down to risk to human health." She put the mower in gear and started up again, her mind occupied with the quote she would take from Water and Pollution Control, the publication put out by AIS Communications Ltd., and the article titled *Water Quality—The Public's Perception*.

I need to do this right now while it's clear in my mind, she decided. She stopped the mower and hurried into the house. Snatching the publication from the table where she had dropped it, she highlighted with a yellow marker the part she would use.

- ...In Canada, the environmental ethic runs very deep...especially when there is a likely effect on future generations.

Writing along the edge of the crumpled paper she had pulled

from her pocket, Marilyn then tied the quotes to the issue of the proposed toxic waste facility. "OWMC's concept of acceptable levels of risk is out of step with the rest of Canadian society. We who stand in opposition to it do not stand alone."

* * *

Marilyn rushed to have an early supper and be ready to pick up Reggie for yet another trip to regional council. This time they were going in support of the Township's third request for money from the Region to cover the shortfall in funding to fight the OWMC.

When Marie called shortly before supper, she sounded grave. "You're not going to like the write-up in the *Standard* tonight:"

> After being turned down twice in recent months, the Township yesterday convinced the Region's finance committee to lend it $116,500 to fight the OWMC. Regional council still has to approve the loan.

Marie stopped reading to clear her throat.

> The decision sparked some frayed tempers among councillors fed up with the frequency of funding requests—at least a dozen times in the last four months—from both West Lincoln and the Region's own OWMC Steering Committee.

"Here's where it sounds bad," Marie remarked before continuing. "'It's getting sickening,' said Niagara-on-the-Lake Mayor Stan Ignatczyk." Niagara Falls Mayor Bill Smeaton had agreed. "I've got better things to do than sit here and see this thing back at the table every two weeks."

Marilyn interrupted. "Who *hasn't* got better things to do?" Marie hesitated, then said, "It gets worse."

> Fort Erie Mayor John Teal wonders if the OWMC might triumph from the "wearing down" of the Region over funding struggles. "It gets to the point where any right-thinking person would say, 'put the —— thing in and forget it.' This

thing's going to go on forever. It's going to bankrupt the whole —— Region," Mayor Teal lamented.

In a rare display of despondency, Marilyn exclaimed, "I'm every bit as sick of all this as they are! I half-wish they'd put it in and get it over with!"

"You don't mean that!" Marie declared.

"Well—no."

"Anyway," Marie continued, "they okayed the request. They decided that since the Township and the Coalition only have part-time lawyers because of the funding squeeze, the Region would have the only full-time lawyer. They figure two heads are better than one, so they've agreed to loan the Township $116,500."

"If they've already agreed to the loan, why are we going to Regional Council tonight?"

"Because Council still has to approve it."

"Oh, shoot. I thought I might get the night off."

Marie chuckled merrily. "No chance. See you tonight."

* * *

Marilyn looked at Paul and Marija as they stood with their backs to the deserted reception desk at regional headquarters. "How many times have we come to Regional Council in the last year?" she asked. It was a rhetorical question and she expected no answer. Paul threw up his hand and raised his shoulders in a prolonged shrug.

It was the third time in five months that West Lincoln's Mayor Joan Packham had asked the Region for funding to help cover the Township's shortfall in its fight with the OWMC and according to the newspapers, Regional tempers were running short.

John Dykstra's gaze darted over the Concerned Citizens group. "I guess we're all here," he announced. With a gesture like that of a captain in the cavalry urging the soldiers into battle, he waved them into the Regional council chambers one more time.

Ray Konkle, Mayor of Lincoln, jumped up from his chair and rushed to shake hands with the various members of the group.

"Loved your last letter," he told Marilyn. "Keep up the good work."

Stella Ziff and several other members of Regional Council also manoeuvred around chairs and fellow councillors to shake hands with the citizens. Finally, as the group took their seats, Marilyn whispered to Reggie, "Sure is different from when we first started coming here to meetings."

Reggie chuckled quietly. "Now, if they'll just give the Township some money..."

An hour later, the residents filed from the council chambers and out onto the broad walkway. Regional Council had voted unanimously to lend the Township $116,500.

"It's not exactly a win-win situation for the Township," remarked Reggie, "not if they have to pay the money back in two years—with interest."

John Dykstra, always on the alert when it came to money, reminded the group, "At least the Region reduced the interest to prime less two and a half percent."

Mary said, "Anyway, now the Township can hire a lawyer full time for the hearings."

"Most important," Steve Dinga reminded them in his quiet voice, "the Region and the Township have finally agreed to stop fighting and join forces against the OWMC."

* * *

Clifford and Marilyn's elder daughter, Judy, had been dating Glen Brough exclusively for a year and a half. However, when Glen telephoned the Gracey home on Monday morning from Burlington, Marilyn couldn't hide her surprise until he asked, "Could I come and visit you and Mr. Gracey this evening?"

Marilyn felt a lump form in her throat. "Why—yes, we'll be here."

When Clifford came for lunch, Marilyn said, "Judy's Glen is coming tonight. He didn't say why—but I suspect he's thinking marriage."

* * *

Glen looked nervous as he sat down on one section of the chesterfield in the Gracey home. He leaned forward, his elbows on his knees as he rubbed his hands together. "I guess you know why I'm here."

Marilyn smiled encouragingly. "Why don't you tell us?"

"I'd like your permission to ask Judy to marry me."

Although Marilyn had been prepared for it, she found she was not. She wanted to laugh and to cry. She wanted to tell him to go away and not make the first break in their family. She wanted to embrace him and welcome him to the family. She did none of these. "Have you asked Judy yet?"

"No. I thought I should ask you first."

"We appreciate that," Marilyn said. Clifford remained silent. It was times like this when he most reminded her of his father—silent, non-committal, his face registering none of his thoughts. To lighten the moment, she asked, "What would you do if we said no?"

"I'd ask her anyway," he replied nervously. "I just thought it was proper to ask you first."

Marilyn knew that he loved their daughter. One weekend, not long before, Judy had volunteered to help her dad clean pigpens. Returning to the house, hot and sweaty, wearing no makeup and her hair as straight as a string, she had clumped to the house in huge rubber boots. In a pair of her father's combinations, extra large tall, with the cuffs rolled up and the crotch down to her knees, she had looked like a bedraggled waif.

Glen, leaning on the kitchen counter and watching her through the window, had murmured, "A-ww-w, isn't she cute?" Marilyn had seen his face and eyes, soft and glowing with the sight of her. Marilyn had known then that he genuinely loved their daughter.

Now, here he was, formally requesting Judy's hand in marriage. So far, Clifford had said nothing. She knew he would need time, because behind the expressionless features, lay the loving heart of a father who also wanted to tell Glen to go home and not make changes to their family.

Marilyn broached Glen's views on marriage, divorce, and

numerous other topics. At one point, she said, "I want you to promise me something."

Glen looked startled. "If I can," he replied hesitantly.

"I've seen families get into smozzles over something and end up with hard feelings that don't end, even at the grave. I won't mean to speak out of place, but I have to be truthful. I have been Judy's mom for a lot of years. And I know me. At some time or another, I'm going to do the mother thing. I'm going to give unwanted advice or step on somebody's toes."

Her voice took on an almost demanding tone. "I want you to promise me that, no matter how angry you may be with me at some point, you will agree to mend the rift. You will accept my apology. You won't stay eternally angry. Promise me that our family will never be split up over some word spoken in haste."

Glen rubbed his hands together thoughtfully. "Yes, I can agree to that—as long as you don't do the mother-thing more than twice a year!"

It took a second for his slow grin to register. Then Marilyn burst into laughter. This young man had a dry sense of humour, and she had better get used to it.

The following Friday, Judy, Kathy, and Glen came home for the weekend. Saturday was June 10, 1989, Judy's birthday. With Kathy in her own bed where Glen usually slept, he had been relegated to the sectional chesterfield. The next morning, while Judy was still upstairs, Glen whispered to Marilyn, his eyes dancing from excitement, "I've got the ring. I'm going to give it to Judy for her birthday."

"Oh no!" Marilyn exclaimed. "You mustn't do that!" Glen looked as though she had struck him. "You don't give an engagement ring as a birthday present."

"Why not? You mean I can't ask her today?"

"Oh yes! You can give her an engagement ring *on* her birthday, but not *for* her birthday!" Marilyn suddenly looked aghast. "Oh dear! I've already done the mother thing and you're not even married yet!"

"Well, then," said Glen with his slow grin, "I guess this

one doesn't count."

Judy came downstairs just then and Glen took her hand. "Let's go outside and sit at the picnic table."

Marilyn looked at the trees whipping and thrashing in the wind as Judy tried to refuse. Glen insistently coaxed her toward the door. Outside in the yard, they sat at the picnic table, with their backs to the house. Seconds later, Marilyn eased open the front door and snapped three photos—one with their heads bowed over the tiny ring box in his hand, one as they faced each other, and one with his arm around her and her head snuggled on his shoulder.

* * *

Because it was Judy's birthday, the family ate in the dining room. "I've always wanted my wedding reception to be held in our yard!" Judy exclaimed excitedly. "So we're going to get married next year on the last Saturday in May. That way, I can have lily-of-the-valley for my wedding bouquet."

Marilyn felt a ton of weight fall on her shoulders. "Oh Judy! We can't have it here."

"Sure we can. We've got a whole year to get ready."

Marilyn tried to smile in all the right places. Yes, she would help her pick her wedding dress. Yes, she and Dad would help with the expenses. But all the while, Marilyn felt herself being swallowed by a great dark hole. A garden reception required flower gardens, trim and neat and bountiful with colour, not the neglected weed patches that graced the Gracey yard.

"It won't cost much," Judy declared, her eyes alight with the fruition of glorious wedding plans formed in her teen years and lovingly nurtured until now. "And we can cater it ourselves."

Marilyn's despairing protests fell on deaf ears. Weddings were never easy, she knew. But to cater it themselves? Especially now, when the Environmental Assessment Hearings were about to begin way off in Oakville? Yet from sad experience, she knew she dared not mention anything connected to the OWMC within the hearing of either daughter.

She listened as though from a distance to the excited voices making elaborate plans. "Kathy's going to be my maid-of-honour, and Melody Burnham will be my bridesmaid, and…"

- 41 -

Wedding or no wedding, the battle against the OWMC raged on. On June 14, 1989, four days after Glen proposed to Judy, Marilyn slapped the latest issue of the OWMC's publication, *Update,* down on the table, then snatched it up again. "Enough is enough!" she stormed and marched upstairs to the computer. She and Clifford had finally bought a second-hand oak desk and crowded it into Judy's bedroom. Now, she spread copies of *Update* from other months over the bed as she gathered her ammunition. It was high time the OWMC's newsletter, paid for with taxpayers' money, stopped promoting the impression that the OWMC's toxic waste facility was a foregone conclusion. It was time the writers stopped saying "will' and started inserting the word *proposed*.

She barely waited for the *let-ed-2* file to open before she began picking out the words:

> Public information put out at the public's expense is one thing. However, the manipulation of public opinion…raises a serious question of ethics…We think it's time that those in charge of public information for the Crown corporation stop…"filtering the presentation"…and take an oath to tell the truth, the whole truth, and nothing but the truth.

The letter practically wrote itself. Two days later, she shipped copies off to the twelve newspapers across the Niagara Peninsula that regularly published her round-robin letters.

In no time flat, Leslie Daniels from the Smithville office wrote a letter to the editor refuting Marilyn's claims. Marilyn, astounded that Ms. Daniels had chosen the March/April 1988 issue as her example of vindication, promptly shredded the lead article on the front page in a second letter to the editor. Thirty times in that article alone, the word "will" had been used.

> *Monitoring Programs Will Help Protect The Environment* is a prize example of OWMC's continuing practice of leading the public to believe that the location and technology of the proposed facility have already been settled upon and only the details are left to be worked out.

Marilyn then listed a number of examples.

- "...will be carried out at every stage of facility development..."
- "...will be used before and after the facility is built..."
- "...will be extensive during the facility's first year of operation..."

Only *twice*, she pointed out, had the word proposing been used. But in each case it referred to a monitoring program that the OWMC *proposed* to set up after the fact.

"If this is not manipulation of public opinion," Marilyn concluded, "we don't know what is."

To Clifford, she grumbled, "According to the way the *Update* reads, the facility's a sure thing, but monitoring it is only a maybe."

* * *

As always, the minute Marilyn heard the mailman turn the squeaking box on Wednesday, she dashed out the front door for the *West Lincoln Review*. The front-page headline stopped her dead in her tracks. "Township avoids debt by hiring new lawyers."

"Again?" she exclaimed. How could any lawyers get a handle on a case this complex when they kept changing at every little whipstitch?

But the more she read, the better sense it made. Roger Cotton had been with the same legal firm as Dennis Wood and Philip Morris, and had appeared at the preliminary hearings. Now, he had moved to another firm. While he would still require big bucks, he would only work two days a week, whereas his assistant lawyer, Mr. Glen Bell, would work full time for the legal aid rate of $83.75 an hour.

In an interview, David McCallum, the Township's Toxic Waste Coordinator, said gleefully, "We pursued Mr. Cotton and won him back. We wanted Roger. Glen Bell is a bonus!"

With relief, Marilyn read that Mr. Bell was no novice. Nor did he demand Mr. Morris's $108 dollar an hour fee. Glen Bell had seventeen years experience in environmental hearings. He had successfully defended Canadian Natives against the MacKenzie Valley Pipeline, winning them a moratorium on the project.

This time, Marilyn called Marie.

* * *

The third block of preliminary meetings had been scheduled from July 4–6 at the Royal Canadian Legion Hall in Smithville. Canada's July first celebrations, marking the one hundred and twenty-second year of Confederation, had passed. July fourth, the American holiday, was just another working day in Canada. Even before the meeting began, Marilyn lifted her wilted blouse away from her chest and longed for a swim in Lake Erie.

She was startled out of her daydreams when she was handed a copy of a letter written to Ian Blue by Doug Hodgson, the lawyer for the Region. In it, Hodgson told Blue that the OWMC must fully disclose its case and that since the hearings were slated to commence sometime in October, it should release the remaining fifty-three or so witness statements. "Partial disclosure," Hodgson wrote, "is not the same as non-disclosure, but it comes very close."

On the last page of the letter, he accused the Crown corporation of obstruction.

> …You have limited us by arbitrarily deciding to provide us with an inadequate number of copies of OWMC reports. You have limited us by refusing to produce reports which you know are relevant but which you choose to label "irrelevant" until we force the issue…

Marilyn stopped reading, her memory jogged by the word irrelevant. During the sessions in March, the interveners' lawyers had argued over the cost of duplicating documents and how to decide which documents were relevant. Ian Blue had partly risen from his chair and proclaimed, "A document won't be provided if I decide it's not relevant."

At the time, Marilyn had deemed his answer high-handed and arrogant-sounding, but perfectly reasonable. Now, she realized how easily a crucial matter could slip past a person not schooled in legal technicalities.

But Hodgson, for one, refused to accept the limitations imposed by Mr. Blue's assessment of what documents the interveners needed to see. A tiny stirring of hope flowered into excitement. At last! Someone who would stand up to the Crown corporation's lawyer and face him nose to nose.

Marilyn took her seat for the third preliminary hearing. The Township's newly-hired Roger Cotton, younger than Dennis Wood, but older than Philip Morris, rose to speak. Despite the intolerable heat and closeness of the room, he wore the accepted uniform—dark suit, white shirt, and a discretely-striped tie. "I propose five avenues of participation for those of the public who wish to present evidence or to observe," he announced.

He proposed that the first week of the hearings be held in Smithville, that there be public days and/or or evening sessions for interested parties to present submissions, and third, that there should be satellite hearings. Marilyn wrote down the term without knowing what it meant. He wanted the Township's evidence-in-chief to be part of a satellite hearing. And lastly, he proposed that the closing be held in Smithville.

As the lawyers debated the Township's suggestions, Marilyn wrote like crazy, her handwriting fast deteriorating into an illegible scrawl.

At the end of his submission, Joe Castrilli took a swift survey of the audience. "All of the people in this audience are Coalition members...If the residents have to travel, they should be reimbursed."

Then Ian Blue rose. His tall, spare frame, fine-boned and fit, captured every eye as he pivoted on one heel to address the audience as well as those seated around the rectangle of tables. "If any portion is held in Smithville," he stated, "it can't be much.

Otherwise, an expensive facility in Oakville would stand empty. Phase 3 and Phase 5, if held locally, would need all of the facilities...It would double the expenses. It would be like trying to move the legislature around. So we say, give us Oakville or give us Smithville—but not both."

Reading from a sheet of paper, he suggested using videos and large screen viewing rooms. "We could have a live inter-active mode. A resident could press a button to get the chairman's attention in order to participate." However, when he revealed that it would cost a hundred thousand dollars to install the cable equipment, etcetera, the parties unanimously rejected it, declaring the exorbitant cost unjustifiable. The debate reverted to the idea of bringing tapes, or edited tapes, to Smithville.

Edith Hallas tentatively raised her hand. Then, without waiting to be acknowledged, she stood. "We are not interested in reviewing video tapes. The session would be over and done. There would be no chance for input."

Dr. Kingham glanced pointedly at the clock on the wall, then pressed the parties for some sense of a decision. "Wherever the hearing is held—or what parts of it—there is the need for at least eight weeks to make renovations, receive tenders, and so on. The parties also need time to move in," he told them.

To Marilyn, nothing in these meetings ever seemed to be resolved. Having hashed over the problem of location and added to the complications, the lawyers and the Board moved on to other topics. But the question of public notice of Transfer Stations, filing dates for reports and witness statements, and whether or not the Board should be reading the Environmental Assessment documents prior to the public hearings all held a familiar ring for Marilyn. Dr. Kingham, in a polite but definite statement, informed the parties that, yes, he would be reading the Environmental Assessment Reports. Period.

Dr. Kingham then asked for the list of names of those represented by the Coalition. Joe Castrilli apologized. "I'm sorry, but we need agreement from one of the groups before we can release the names."

Coffee breaks came and went. At noon, the hearing was adjourned until three p.m. The day ended with more bickering over what Marilyn considered minor details. Did it really matter that much who filed what and when?

The hearings days set aside for the rest of the week were cancelled. The date for the interrogatories was extended to September 15, 1989. Ian Blue assured the Board that henceforth, the OWMC would try to answer interrogatories within twenty days. Dr. Kingham stated that the parties would be notified when a date had been set for the fourth preliminary hearing. With that, the hearing adjourned.

As Marilyn closed her notebook and stretched, she decided that she should make an effort to find out what interrogatories were and why the lawyers argued over them so heatedly.

- 42 -

In between picking beans and helping Clifford in the field, Marilyn began the task of weeding and cultivating long-neglected flower beds. With perspiration dripping from her nose and her chin, she fell back onto the cool shaded grass. Just then, she heard the telephone. Lurching to her feet, she half-ran, half-staggered to the back door. At the top of the stairs, she snatched the receiver from the hook on the fifth ring.

Between gasps, she explained to Marie where she had been and what she had been doing. "I can't leave it until next spring," Marilyn explained despairingly.

"I'm sorry to bring you in, but I thought you'd want to get the *Standard*." Marie began to read about a feud between the Region and the Ontario Environmental Assessment Board. The Region complained that it had waited six months to be reimbursed for bills tallied up in the OWMC battle. The Board informed Michael Boggs, the regional administrator, that the Region couldn't be reimbursed for cash spent on consultants, technical staff, meals, travel, and other expenses unless it stuck to the rules in submitting invoices.

"Mr. Boggs blames the problems on the assessment board. He said the Board's forms are much too complex…The Region can not attempt to pre-judge expenses as eligible or ineligible for a refund."

Marie's chipper voice took on a merrier tone than ever. "Here's the good part!" she exclaimed with a chuckle.

…A citizen's coalition, a small-scale environmental

group,...has successfully completed the refund forms and received its money. Said Board spokesman Nada Davidovic, "The Coalition doesn't seem to have any problem complying with the board's invoice policy and if they are able to do it...I don't know what it's going to take for the Region to comply."

"Good show!" Marilyn declared. "I wonder if John Jackson's seen this yet. It's his work, no doubt, that Ms. Davidovic is praising."

* * *

With Judy determined to have a garden-type reception at the farm, Marilyn launched the month of August by eyeing the huge rolling yard with growing dismay. Nowhere could she see a level place large enough for a dining tent.

Her gaze came back to the old horse corral, west of the house. Pock-marked with years-old hoofprints, and grown waist-high with weeds, it was the only spot they could use.

Marilyn knew that Clifford would have every reason under the sun to oppose the idea. In truth, he was simply too harried at this time of the year to take on another task, let alone purposely make a rod for his own back by substantially increasing the size of their yard.

Knowing her husband well, she did the only thing she could. She resorted to subterfuge. She took the lawn tractor into the corral and with the mower raised as high as possible, commenced to chop down the weeds. It took her several hours just to get a good-sized start on it. Then she hauled out the rotor tiller and let it bounce and jar over the rough ground, making little impact.

"This isn't going to work," she said with a sigh. Then, viewing a section of harrows leaning against the barn, she dragged it to the rail fence enclosing the coral and man-handled it over the lowest rail. With a chain wound around it, she hooked it to the lawn tractor. Back and forth, around and around she went in the hot sun, making little dint in the hard-packed soil. She stopped and brought out an ice-cold Coke. Sprawled in the wooden lawn chair, she surveyed the bush across the road and listened to the birds. Already she could sense the fall migration entering into their noisy chattering.

Suddenly, the familiar sound of Clifford's tractor penetrated her reveries. She tossed her empty Coke can into the recycle box and dashed to the lawn tractor. With a single turn of the key, it jumped into action and she began the patient but wily circling round and round in the old pasture.

As Clifford pulled into the driveway, he stopped the tractor and sat there watching her. Then, as if she had only just seen him, she waved and continued her circling. In truth, if Clifford didn't come to her aid with large equipment, she was fully prepared to circle and circle until she accomplished her goal—a flat piece of lawn.

"What do you think you're doing!" he demanded over the roar of the mower as he lifted his peaked cap, pushed back his hair, and resettled his cap.

"Working up lawn for the wedding."

With a disbelieving shake of his head, he ran his gaze over her pitiful efforts. "And just how long do you think it will take you?"

"As long as it takes," she replied with stubborn stoicism.

"And you couldn't wait until I have time—"

"No. There's no time to be patient. We have to have a full-blown yard before the end of May." She could see the struggle going on inside him. He needed to get on with his own work, but he couldn't bring himself to leave her to hours of fruitless labour when a few turns of the farm equipment would make short work of the job. And she knew when to cater to him. She jumped off the lawnmower. "I'll get you a cold drink," she said and dashed to the house.

Minutes later, she watched as he unhooked the equipment he had been using in the field and hooked onto the disk. An hour later, the old corral looked like a worked field, ready for planting. But not ready for grass, she knew.

"Now, just leave it!" he ordered. "I'll finish working it tomorrow."

"I can live with that," she grinned and gave him a conciliatory peck on the cheek. Minutes later, he pulled out of the driveway and she jumped into the truck, dirty face and all. She dashed up to the hardware store and bought twice as much grass seed as she ordinarily would have needed. "We need a thick lawn—fast!" she told Jimmy as he put the purchase on Clifford's bill.

Friday, Clifford worked the pasture again. On Saturday, he worked it one last time. Then with a plank fastened behind the tractor to help level the ground, he scraped and dragged until it looked reasonably flat. With rakes, they gave it a final grooming. Then Clifford filled up the seeder he used to walk the clover seed onto the fields in the spring, and began the task of walking on the grass seed. Up and down, back and forth, he crossed and crisscrossed, turning the handle that spit the seed in a spray. Marilyn lightly raked the seed into the soil, then they strung hose from the drilled well at the barn and started to water the newly seeded lawn.

When Judy, Glen, and Kathy arrived for the long weekend of Civic Holiday, they exclaimed in astonishment at the transformation. On Monday, they all donned the most ratty-looking clothes Marilyn could scare up and raided the loft in the barn for straw. With pitchforks, they broke the bales apart and lavishly strewed the straw over the huge addition to the lawn.

Marilyn rushed to the house for her camera. The weekend before, she had taken pictures of Judy as she tried on wedding dresses. Now she took pictures of her looking like a scarecrow. But nothing could erase the bridal glow from her face.

* * *

On Saturday, August twelfth, the Concerned Citizens held their second annual fundraising barbeque. As Marilyn gazed out over the crowd, with more waiting at the door, she could feel her heart begin to sing. Three years ago, almost to the day, she had written her first letter to the editor. In it, she had quoted a statement from Hansard, made by a Mr. McGuigan. He had told the parliamentary committee of a group of ladies, fighting the first site in Cayuga.

> They held cookie sales, auction sales, and had singsongs in church basements...(But) in addition to raising...money, every person who attended one of those meetings and

bought cookies or whatever had became part of their team. They just kept building and building that team so that it was overwhelming…"

The statement had never left the back corners of her mind. Everything she did, everything she urged the group to do, had but one goal in mind—to keep building and building the team.

As the official hostess, she made her way through the press to welcome Chuck and Pat Potter. "This is all your fault!" Marilyn reminded Pat. "You're the one who sat in this very room and said, 'Read the documents, read the documents.' That's what got me into this fight in the first place!"

"Good!" declared Pat. "I'm happy to take the credit."

- 43 -

August rushed past in an exhausting haze of canning, pickling, and freezing. Even so, Marilyn managed to send out two more letters to the editor.

On September thirteenth, the Ministry of the Environment announced it had completed its review of the OWMC's final Environmental Assessment documents. The hearings had just taken a giant leap closer.

* * *

To Marilyn, the events controlling her life kept escalating so fast, she could scarcely keep track of them. Regardless of the OWMC or Judy and Glen's wedding, two daughters meant two young women with men in their lives. Kathy had been dating Donald Almas for some months, and it had come time to invite the Almas family to the farm for a barbeque. While Marilyn cleaned the house in preparation, mowed the lawn, and everlastingly moved the hose to water the newly seeded lawn, Clifford worked ground and planted winter wheat.

Meanwhile, the conflict surrounding the OWMC's proposed facility seldom died down long enough for her to catch her breath. At best, it simmered and bubbled on the back burner like her chili sauce. Early in September, Regional Chairman Wilbert Dick hit the front-page headlines of the *Standard*. "The deal has been cooked," he told environment reporter Doug Draper. "The Board might as well write the decision now...Holding the hearings in Oakville is a deliberate move to take the decision-making away from the [people] most affected."

Dick was incensed because for the past two years, Premier Peterson had refused to meet with the Region. "This is a deliberately orchestrated charade," he said. "It is distressing that they have only been willing to talk through lawyers, consultants, and unlimited money."

For Marilyn and the others, the days grew blacker as the hearings loomed like a dark cloud over the coming months. Dick's revelations only made the cloud darker and more ominous.

* * *

The news that the ministry had completed its review of the OWMC's proposal stirred renewed interest in the toxic waste issue. The men's club from Lundy's Lane United Church invited Clifford and Marilyn to come and explain the proposal and the problems it posed. At the same time, the Niagara Residents for Safe Toxic Waste Disposal manned a booth at the Welland Fair. Or, as it was now called, the Niagara Regional Exhibition.

"We should go and lend them our support," Marilyn told Clifford at breakfast. "We're all on the same side, even if we go about things differently."

That afternoon, Clifford and Marilyn bumped shoulders companionably as they roamed the fairgrounds in the warm autumn sunshine, munching caramel popcorn and sipping cold drinks. Marilyn hitched her camera higher on her shoulder as they stepped into the stuffy confines of the arena. Booths lined the walls and stood back to back in the centre space, forming long aisles.

She saw the buzz of excitement before she actually saw the Niagara Residents' booth. Adults and children stood in a line, waiting to outline their hands on a three-feet wide, two hundred and sixteen metre long "hands on" petition, then sign their names in protest to the OWMC's proposed facility. Along the back and sides of the booth, the Niagara Residents had pinned up children's posters on the environment, entered in a recent contest.

"Okay if I take a picture?" Marilyn inquired of Mike Jones and Ron Hansen. The two men stopped to pose and Marilyn pretended to snap their photos. Then, when they weren't looking, she took pictures of them actually busy in the booth.

That night, both the *Tribune* and the *Guardian Express* carried pictures of Welland-Thorold MPP Peter Kormos adding his name to the nearly six hundred signatures collected during the five-day fair. During the photo session, Kormos declared, "You people are the victims of yet another broken promise." The statement took clear aim at the inadequate intervener funding provided by the provincial government. "Keep on fighting!" he urged.

Marilyn stared at the photos in the newspaper. The Concerned Citizens had determined to remain free from political ties. And she was sure that for their group, it was the right decision. But she had to admit that having a Member of Provincial Parliament come out on the residents' side certainly seemed like a plus.

* * *

Before the month ended, the Region called for a delay to the OWMC hearings. Lawyer Doug Hodgson told Doug Draper of the *Standard* that the ministry had concluded that the OWMC's E. A. documents—dubbed the "Blue" review by opposing lawyers and consultants—had not satisfied the requirements of the Environmental Assessment Act. "To put it in a nutshell, the technical reviews of the ministry say that this landfill is going to leak into the groundwater." As Marilyn read the article, she could picture Hodgson's intense expression, his handsome craggy features beneath his shock of unruly white hair, and his hands making impassioned chopping motions as he hammered home his points. "Mr. Bradley has the power…to require the OWMC to address the chloride problem…before the hearings."

"Chlorides!" Marilyn exclaimed aloud. "So there *is* a problem with the chlorides, just like we figured!" Eagerly she scanned the article for more information. Maybe, just maybe, this would bring the nightmare to an end. If the OWMC couldn't fix the problem of chlorides—salts—in the landfill, there was no point going any further with the hearings.

But as always, Draper couldn't write a column without letting the OWMC put in its two cents' worth. And the further she read, the more her hopes turned into ashes.

OWMC spokesman Michael Scott had told Draper that the corporation's technical staff was confident it could resolve the problems. Hodgson had revealed that information had reached him "in an informal way" that the Ministry of the Environment and OWMC were holding behind-the-scenes discussions over when the concerns should be addressed. OWMC wanted to deal with them during the hearings so there would be no delays getting the hearings under way.

"I think it's disturbing (that the discussions) are not occurring in public," Mr. Hodgson said, adding that he had requested but not received copies of the correspondence between the two parties.

"Disturbing!" Marilyn exploded. "I think it stinks!" The utterance was freighted with fury as she hurled the words across the empty kitchen. She slapped the newspaper onto the table and marched out to the garden. Never had she yanked out dead cornstalks and cucumber vines with such violence or with such speed.

"What are you *doing*?"

Marilyn nearly jumped out of her skin. "I didn't hear you come up behind me," she snapped at Clifford as she slammed more rubbish into the wheelbarrow.

"I asked you what you're doing."

"I'm bashing OWMC and ministry personnel—one by one, name by name." She smashed a particularly huge over-ripe cucumber. "And that's for sneaking around behind our backs!" Her strength suddenly spent, she started to laugh.

"What's so funny?" Clifford asked with a perplexed expression.

"Nothing. But if I don't laugh, I'll cry."

*   *   *

Scarcely had the fourth Preliminary Hearing been brought to order before Dr. Kingham chastised OWMC's lead lawyer. Ian Blue had sent a letter to the Board regarding the provisions of the Intervener Funding Project Act 1989. In the letter, he had challenged the Joint Board's view that the provisions of the Act "may apply to this hearing" and accused the Board of basing this view on something other than information received during previous preliminary hearings.

In a voice that could have frozen ice cubes, Dr. Kingham told Mr. Blue, "I'd like to make it clear that that was not the Board's intent and it should never be transcribed as the Board's intent. When we arrive at a decision, we do so on the basis of the facts before us...I hope that is understood by all parties and that the words will be very carefully chosen in the future."

The reference to the funding issue quickly plunged all the parties into a free-for-all over the terms of the Intervener Funding Project Act. Once again, Marilyn took copious notes, without really understanding what she wrote.

After the noon break, the Board ruled against Mr. Blue's interpretation of the Act and the adverse ruling made him feistier than ever. Finally, the Chairman issued a reprimand. "Mr. Blue, that's the third time to day...you've questioned the Board's motives in regard to this matter of the Intervener Funding Act...It's hardly becoming to your client, I think, Mr. Blue."

Blue stood and demanded, "Mr. Chairman,...could the Board answer one straight question?"

Marilyn, who dreaded confrontation, shrank low into her seat. After several heated exchanges, Blue snapped at Kingham's answer, "It's not responsive to my question, Mr. Chairman. I'll withdraw it."

Dr. Kingham glanced at the clock on the wall. "It's 4:20...It's been a long day...I'm wondering...whether we should call it a day."

Mr. Blue responded, "Mr. Chairman, I would suggest Item 6, we can deal with that one."

Item 6 dealt with the question of reimbursement to residents for travel expenses to and from Oakville.

Blue rose, immaculate in his navy suit and white shirt even at the end of a long tiring day. "Mr. Chairman, until this morning, I was going to have to take the position that we couldn't pay the transportation. I'm happy to say that we've negotiated an arrangement with the ministry whereby they've authorized OWMC to pay transportation costs..." As he talked, he referred to notes he held in his hand. Obviously concentrating on each word, he swayed slightly as he talked.

OWMC—The "Wasted" Years: The Early Days

"The second point I wish to make about it is that now the OWMC has this responsibility, it is going to establish criteria for eligibility and accounting that meet the provincial auditor's guidelines and those will be non-controversial, I'm sure. They will be ones that people can easily comply with.

"The third point I want to make is that reimbursement for transportation costs is not going to be available to Mr. Castrilli's clients..." He pivoted on his heel and made a sweeping gesture with his arm that plainly demonstrated his contempt for those it encompassed. "People represented by the Ontario Toxic Waste Research Coalition—Mrs. Bradley, Mrs. Gracey, and entourage—should get their transportation costs covered under funding that's already been granted to the Ontario Toxic Waste Research Coalition..."

As cries of outrage swept across the hearing room, Mr. Blue kept right on talking. "...So certainly if there are people not represented by the Coalition who want transportation and they meet the option criteria, we will reimburse them for their expense."

Marilyn declared aloud, "So we're to be punished for taking an active role in this issue!" Her eyes stony with anger shifted from Ian Blue to Dr. Kingham.

Joe Castrilli looked as dumbfounded as his clients did. It took a moment or two before he voiced his objections. "Mr. Chairman, very simply, the funding that my clients received for intervention in this hearing had nothing to do with provision of assistance for purposes of transportation. The budget is tight enough as it is. It's directed to legal fees. It's directed to expert witness fees and it's directed to the disbursements in connection with preparation for the hearing. It is not directed in any way, shape, or form to provision of transportation...I would ask the Board to direct Mr. Blue to include my clients who are in good measure the only ones likely to attend the hearing in Oakville...Otherwise...his grandiose gesture is quite meaningless..."

The Chairman remarked with the hint of a smile, "I wasn't convinced that it was made in a grandiose style, Mr. Castrilli." The smile disappeared. "The Board remains of the view that a way has to be found to ensure that those citizens from this area, for those parts of the hearing that will be held outside the region, have a way of getting there, which is not to their disadvantage."

The Chairman gazed out over the audience. "Now, Mr. Blue has suggested a way for some of those citizens, namely those non-members of the Coalition. [Mr. Castrilli], your submission is that Coalition members are essentially most of the people who would likely be attending. Therefore we have a mutually exclusive set of hypothetical persons...This Board is determined that such costs are not assigned individually to people who want to participate. Some way has to be found to do that."

Ian Blue insisted, "Well, Mr. Chairman, Mr. Castrilli has got $847,000. It's a simple matter of him taking some of that money away from legal fees and consulting fees and put it on getting the people who live here to the hearing...We're saying we'll make money available for somebody else. We've had to reallocate our priorities..."

Indignant at hearing herself and the others called by name, yet discussed as though they were not present, Marilyn set down points for a response to Mr. Blue's outrageous proposal.

- Residents of the Region and the Township are eligible.
- Only Coalition members don't qualify.
- Have we forfeited our citizenship?
- I have not resigned as a resident of the Township of West Lincoln.
- I have not resigned as a resident of the Region of Niagara.
- Such treatment sends a clear statement to any who would dare to oppose a Crown corporation. You will be discriminated against! You will be punished!
- Only the Coalition members are required to wear themselves out fundraising. The OWMC operates on *my* money!

Suddenly spent, Marilyn found she could no longer concentrate. Although she continued to take notes, they made less and less sense. At twenty minutes to five, day one of the fourth preliminary hearing adjourned.

OWMC—The "Wasted" Years: The Early Days

*   *   *

The next morning, Marilyn approached the court reporter. "Excuse me. How can I go about getting some pages from yesterday's transcript?"

"We don't generally provide individual pages," she answered.

"But I can't afford to buy the whole day's transcript."

The woman wrote out a telephone number and a name. "You'll have to call the office of Farr & Associates and speak to the supervisor. And you need to know which pages you want."

Marilyn started looking for someone who would let her skim through his copy of the transcripts. Letter to the editor, number fifty-eight, she vowed, would contain, word for word, Mr. Ian Blue's scandalous statement from the day before.

The second day of the fourth preliminary hearing had barely gotten under way before Marilyn suddenly sat bolt upright in her chair. The Environmental Assessment Hearing had just been postponed from October fifteenth until the first week in December.

As the legal jousting continued, Hodgson complained that he was still waiting for fifty-two witness statements. Mr. Blue argued that since the hearing would be delayed, he proposed to hold back the witness statements, promised to the interveners by October first. "I'll serve them on November seventh," he told the Board. "It's only fair..."

Marilyn made no effort to hold back her derision. "Fair? He doesn't know the meaning of the word fair!"

"...It'll allow us to finish and polish the witness statements," Blue continued. "I have to tell you they're all—they're virtually all done at the present time. And if the hearings were going to start on October fifteenth, I'd release them. But if the hearing is not going to start as planned, I would like a chance for one final polish."

Mr. Blue then accused the interveners of showing no desire and no intent to scope the issues. Furious at the accusation, Hodgson spelled out the Region's efforts to comply and the agreement they had reached with the OWMC. "An agreement," he told the Board, "which they just tear up when it's not convenient." Then he let fly with a scathing denunciation of the Crown corporation. "Mr. Blue's client has taught my client one lesson, Mr. Chairman, and that is, that an agreement with the OWMC isn't worth the paper it's written on...What Mrs. Gracey and the other members of the public...have learned today is that OWMC...isn't to be trusted and that the Ministry of the Environment won't make OWMC keep its promises...Mr. Chairman, today we've had here a metaphor about a glacier. Mr. Blue [says he] is steering a glacier." Hodgson indicated the members of the public. "Out there in the audience we have the people who are to be crushed by the glacier...Is this Crown corporation to be trusted to build a facility of this nature if they can't make a simple promise and keep it?"

Suddenly overwhelmed by the image Hodgson had portrayed of the iceberg bearing down on the residents and crushing them, Marilyn bolted for the rest room.

Ten minutes later, she returned to hear both sides arguing whether or not a limit should be placed on the cross-examination of witnesses. Blue said yes. Rovet said no. Marilyn didn't know what the rest had said. Brynell, for the ministry, said it should be at the discretion of the Board. The matter was left there.

The minutes dragged by like hours. Every time she let her mind wander, it conjured up the picture of a vast iceberg, reaching high into the sky and bearing down inexorably upon her and Clifford.

– 44 –

For the Coalition's October 1989 meeting, John Burton lugged a television set and a VCR up the stairs to the meeting room at the Vineland Experimental Station and set them up, ready to show two new videos on handling hazardous waste. The group was hosting an Open House to bring members and others up to date on the latest developments and to prepare them for the upcoming hearings.

Amid applause, the Concerned Citizens turned in a cheque for $1,000. In July, John Jackson had informed the members that the Coalition's lawyer needed an additional eight to ten thousand dollars a year. The Concerned Citizens had raised $305.60 at their Canada Day baked goods and yard sale. By adding most of the proceeds from the barbeque, they had been the first group to meet their commitment of $1,000 a year. But $1,000 certainly wasn't $10,000.

Excitement rose to its highest peak in months when John Jackson distributed two documents. One was a letter from Mott's Canada to Michael Scott, OWMC's Director of Communications. The second was an internal memo from Welch's. Both had been written on behalf of the parent company, Cadbury Beverages Canada Inc. Both documents found the idea of locating such a facility in the Golden Horseshoe area of Ontario "confounding and freighted with unforeseen risk."

The Welch's memo, written mainly in point form, listed concerns based on a review of OWMC's site selection synopsis.

Synopsis: Wastes to be processed...will include PCBs.
Thoughts:...A few years ago a road surface was applied which used

recycled transformer oil (the prime source of PCBs). We would not accept grapes from the adjoining vineyard...

Synopsis: Prevailing winds are from the West, which would move emissions toward the grape growing areas.
Thoughts: The first time there is evidence or even a suggestion that toxic emissions are present, the media will initiate a level of public concern that will [reduce] the vineyards [to]...low value real estate...

Synopsis: Most of the data generated is based on theoretical models...
Thoughts: Now *there's* a confidence builder. Ontario will be a test case. I'm not real comfortable living only two hours away.

Synopsis: The OWMC were also interested to know whether we (Cadbury) should consider U.S....grapes if contaminants became an issue.
Thoughts: A rare possibility. At present sales rates, we need [Niagara's] grapes.

Synopsis:...On page 14, worst case scenarios were submitted.
Thoughts: The likeliness of a tank farm fire leading to the release of methylene chloride, estimated to be not more than once in 100 years, is not reassuring. That could be one day after the facility opens...I do not believe the existence of monitoring programs will do much to alleviate public perception.

The writer concluded: "Even if one assumes all the risks are well in control, it is not a good location for this type of facility..."

The second document was a formal business letter addressed to Michael Scott from Mott's Canada. "There are a number of...points in the [site selection synopsis] causing us concern. Firstly, the evaluation of the facility seems to be largely theoretical as there isn't a similar plant of equal size in existence..."

The company expressed extreme concern over the possibility of toxins getting into their grapes. Niagara grapes, used in jams, jellies, juices, and grape drinks, had been marketed under the Welch's name for over 40 years, and were also sold in raw or processed form to Welch US and other Canadian manufacturers. OWMC's summary had identified grapes as sensitive to both visible and bioaccumulation.

Because Welch's concentrated its grape juice, any level of contaminant would increase in the product. For the public, the issue would not be the level of contaminant, but whether a contaminant was present at all.

As for importing grapes from the U.S., the idea was fraught with difficulties. The North American grape market had little available excess tonnage. Because of differences in acidity, sweetness, and flavour from some areas of the U.S. (e.g. California), Welch's would have to blend grape varieties to match the current product. Citing added transportation costs, duties charged at the Canada/U.S. border, and the need to compete for a tight market supply, the letter concluded that these factors would make prices unfavourable for the company to buy from the U.S.

The Coalition meeting concluded with a social time of coffee, fruit juice, and home baked goodies. The room rang with more laughter and spirited conversation than Marilyn had heard for a good long while. John Burton, with a wide grin and his usual pithy humour, waved his copies of the Mott/Welch/Cadbury letters in the air and declared, "Couldn't have said it better myself."

* * *

On the weekend, Coalition members held a garage and bake sale in front of the Vineland United Church to raise more money. Throughout the hearings, only the Coalition would have to carry out such fund raising activities.

* * *

Marie Austin meticulously scoured the *Tribune* and the *Standard* every night. Marilyn kept track of the *West Lincoln Review* and the *Pelham Herald*. Between them, everything they read indicated that the fifth round of Preliminary Hearings had been set to begin six weeks away, on November 23, 1989. The actual hearings would commence on December first, provided there were no delays. Delay, however, seemed imminent. The Region insisted that the ministry should force the OWMC to address the chlorides problem before the hearings began.

Nevertheless, on Tuesday, October 11, 1989, the Office of Consolidated Hearings gave notice that the public hearing would start on November 23, 1989 as scheduled. OWMC's Leslie Daniels told the *West Lincoln Review* that "the review board and the Ministry of the Environment...have agreed that the outstanding issues can be resolved at the hearings."

"Still to be ironed out is the scoping of issues: which parties will be in charge of which areas of concern at the hearings."

On the same day, Mary Lou Garr from the Smithville office manned an OWMC information table set up in the Smithville Library.

A week later, the *West Lincoln Review* carried the two-page official Notice of Public Hearing, complete with maps, from the Office of Consolidated Hearings.

Douglas Hodgson, the lawyer representing the Region, reported to the Regional Steering Committee that he had not receive OWMC's witness statements until two weeks after the October first deadline originally agreed upon at the third Preliminary meetings back in July. Marilyn knew that Ian Blue had tried to get permission from the board to withhold them until November seventh. Had the Board agreed? She didn't know. So, to all intents and purposes, he had delivered them earlier than he wanted to.

Hodgson's opinion was very clear. "They never intended to file them by October 1...because we would read them...We could find out whether (Ian Blue) has a case or a leaky boat."

However, when Doug Draper of the *Standard* contacted the OWMC office in Toronto, spokesman Michael Scott said, "...The corporation notified the Region and other opponents last month it would ask the province for an extension on filing its witness statements because it was having difficulty completing them by the October first deadline...The corporation wanted all parties to know we were having difficulty and was not trying to deliberately deprive them of the material."

Marilyn felt her eyes blaze with fire. Michael Scott's excuse did not jibe with Ian Blue's assertions at the last preliminary hearings regarding the witness statements. "And I possess the pages from the transcript to prove it!" she declared triumphantly as she took the stairs to Judy's room two at a time.

Seated at the computer, she wrote, "Since neither Doug Draper nor Michael Scott were present at the preliminary hearings on October third and fourth, we would like to quote from the transcript..." She then painstakingly repeated word for word what Ian Blue had said about the witness statements. "'...They're virtually all done at the present time. And if the hearings were going to start on October fifteenth, I'd release them...' (Transcript pp. 1079, 1080)'

"Was the corporation truly 'having difficulty' meeting the deadline?" she asked. "Or were the witness statements 'virtually all done' by the October third and fourth preliminary hearing days as stated by OWMC's lawyer? Could the documents have been released at that time? Or were they purposely held back?"

Then, using an expression she had recently picked up, she accused the OWMC of "filtering the presentation."

* * *

Late October brought a welcome chill to the night air. Yet the days were warm and lazy, still filled with summertime. The sun shone with muted brilliance as it gilded the maple leaves clinging to the trees. On quiet afternoons, as Marilyn lay on the picnic table, she shaded her eyes with her hands as she watched wavering ribbons of Canada geese pass over the house, noisily squawking and honking as they trailed south against a backdrop of deep blue sky tufted with white clouds.

Despite running from city to city with Judy, looking for the perfect wedding dress, and making sausage for their customers, Clifford and Marilyn managed a few outings. They drove to Elmira with John and Mary, and then to Kitchener Market. In between, they attended

Cecil and Violet Gracey's fiftieth wedding anniversary. Marilyn raked up huge mounds of leaves and with the abandon of a child, she buried herself and their young beagle in them.

One particularly glorious afternoon, Marilyn convinced Clifford to take the canoe down from the rafters in the machine shed, load it into the box of the half-ton, and drive to Wellandport Park. Raised in Welland along the bank of the Welland River—her father had always called it the Chippewa—she never felt more at peace than when she was paddling past old trees leaning out over the water and hanging with wild grape vines. Turtles basked on logs and a family of ducks kept their anxious distance.

The halcyon days of Indian Summer ended abruptly during the night of October thirty-first. Along with dull skies and cold, nasty winds, November plunged Clifford and Marilyn into another hectic month.

– 45 –

On November 8, 1989, area newspapers exploded with news of a letter from Donald Chant to Gary Posen, deputy minister of the environment. It had been kept secret from the opponents since January 10, 1989.

The headline in the *West Lincoln Review* shouted, "OWMC seeks tax cash but no market shown." Dr. Chant's letter to Posen, dated ten months earlier almost to the day, contained two requests. One, for a law to ensure that OWMC's facility would be utilized. Such a law would require some waste generators to use the facility, regardless of cost, for the treatment of some hazardous wastes. The second request was for a government subsidy to ensure the proposed facility would be competitive. Direct grants or annual subsidies—either would be fine. The corporation just needed something in writing that it could use at the upcoming hearings.

Mr. Bell, the lawyer for the Township, had told Blair Burgess of the Review, "It's clear OWMC wants to monopolize the market by using funding to undercut their competition. If they were not a Crown corporation, they would be prosecuted under the Competitions Act."

The article concluded with Mr. Bell's complaint about OWMC's plan to wait until witness panel 51 to present the expert testimony supporting the actual need for the waste facility. "We may have run out of resources by then," he told Burgess. "The cost of witness panel 18 alone is going to stagger the imagination."

Just as Marilyn reached for a fistful of highlighters, Marie called to alert her to similar articles in both the *Welland Tribune* and the

St. Catharines Standard. "It hit page one in the *Standard*!" Marie exclaimed with bubbling excitement. "OWMC seeks tax subsidies."

Michael Boggs, the Region's chief administrator, told reporter Paul Forsyth that OWMC's request was contrary to what the public had been told. "All along, (OWMC) have said publicly...that this facility will be self-supporting."

But perhaps the most damaging revelation came from Welland Councillor Rob Dobrucki. He said that the Region was told by the OWMC, *after* the January 10 date of the letter, that the facility would not be subsidized.

Marie chuckled with glee. "I love the way this guy talks. "At best there has been a misrepresentation; at worst, we've been lied to," Dobrucki said."

"How did they get hold of a letter from Chant to the ministry?" Marilyn wanted to know.

"John Jackson found it." Marilyn could hear the rustling of pages. "The *Tribune* says, 'The Ministry of the Environment has been withholding information about the OWMC from the Region for nearly a year...'"

"A year!"

"That's what it says. 'The Region's lawyer did not get a copy of a January 10, 1989 letter by OWMC chairman Donald Chant until the middle of September. He received a copy only because John Jackson, coordinator of the Ontario Toxic Waste Research Coalition, found the original by chance in some OWMC files he was researching in Toronto.'"

"Trust John to ferret out something like that!" Marilyn declared.

"I'll save these clippings for you," Marie told Marilyn.

"Great! Thanks for calling."

As Marilyn hung up the receiver, she decided she couldn't wait the two days until Marie came on Friday afternoon for the meeting of the Concerned Citizens with John Jackson. Grabbing the car keys, she dashed to the corner store and bought the papers for herself.

Minutes later, ensconced in her dilapidated recliner, Marilyn unfolded the *Welland Tribune*. Under the heading, "Will toxic waste site be a 'white elephant'?" reporter Joop Gerritsma emphasized that Ontario was a net importer of hazardous waste and had imported 75,000 tons in 1988 for treatment in private facilities across the province. Said Hodgson, lawyer for the Region "Any government subsidy to the OWMC will amount to 'the spectacle of Ontario taxpayers subsidizing generators of U.S. waste.'"

With a happy sigh, Marilyn put up her feet and commenced to color-code the pertinent quotes.

* * *

The next day, Leslie Daniels, communications officer from OWMC's Smithville office, told reporters that opponents of the project were "using the letter for an argument which doesn't stand up under scrutiny."

She then tried to explain away the request for subsidy as common business practice. "The corporation will need financing in the beginning of operations as start-up capital..." she told Ken Avey of the *Welland Tribune*. "We'd be poor corporate citizens if we built a facility and then couldn't afford to run it."

The *Guardian Express* carried a quote from an unnamed spokesperson for OWMC. "If opponents of the Crown corporation's proposal think the information will lead to its defeat, they are pinning their hopes on a non-issue."

Marilyn read the statement, and then read it again. "They are pinning their hopes on a non-issue." Niggling doubt forced a crack in the reservoir holding her high hopes and she felt her elation of the past twenty-four hours begin to trickle away. With the workings of corporate financing a mystery to her, combined with the overwhelming might and influence wielded by the Crown corporation, she realized she and the other residents had been thinking like wishful fools. The revelation of one lone letter from the OWMC to the ministry would have no more effect on the issue than a rifle shell bouncing off the side of a destroyer.

As she leaned back in her recliner and stared sightlessly at the shadows on the ceiling, her lips moved soundlessly. A non-issue... It's just a non-issue. Relentless in its pursuit of truth, her uncompromising logic dredged up similar situations. Hadn't her father run his sporting goods business on the bank's money? Didn't every farmer have an operating loan or a line of credit? She sighed half aloud. It's just a non-issue, she thought. John Jackson only latched onto the letter because the OWMC had assured the public repeatedly that its facility would be self-sufficient.

From a habitual need to tidy, she shook out the newspapers on her lap and began folding them one by one. When she came to the *West Lincoln Review*, a notice caught her eye, reminding her that the Township had scheduled a public meeting for that night at Wellandport Hall. For the past three years, the Township's committee of aldermen and hired legal guns had been meeting behind closed doors. Now, for some reason, they had decided to hold a public meeting, including a question and answer period.

Marilyn felt more discouraged than she would ever dare let on, even to Clifford. Just as John Jackson's unfailing optimism kept the Coalition members from faltering, so it was that many in the community looked to Marilyn for hope.

To stoke her own tiny, wavering flame, she picked up her Bible and turned to the tattered pages in II Chronicles. "O Lord God…art not thou God in heaven? and rulest not thou over all? and in thine hand is there not power and might so that none is able to withstand thee?"

* * *

That evening, Clifford and Marilyn smilingly entered the Wellandport Hall amid the greetings and handshakes of friends and neighbours. "You're doing a great job. Keep up the letters," they were told over and over. From a table just inside the door, Marilyn picked up two copies of the Posen letter.

An air of euphoria seemed to permeate the meeting, summed up in six words uttered by Roger Cotton, lead lawyer for the Township, "We believe we can win this."

Funding, however, remained of major concern, or more accurately, the projected shortfall. Anger flamed afresh at Cotton's reminder that, as a result of the shortfall, the Township had been forced to choose Oakville as the location for the hearings. An hour's drive for most residents, combined with the heavy workload born by the farming community, those present argued that they had callously been cut off from the proceedings. Added to that was the most recent injustice, the Crown corporation's magnanimous offer to cover travel costs for residents of the Region and the Township—but not if they were members of the Coalition.

Marilyn remained incensed. "Belonging to the Coalition does not mean I've renounced my citizenship!" she declared amid the furor.

As always, she took pages and pages of notes, her handwriting deteriorating into a scrawl in places. Roger Cotton reaffirmed the Township's position of opposition. Project coordinator Dave McCallum gave a general run-down of the work done by the consultants. Stan Pettit, Chairman of the Regional Steering Committee, after stating the Region's opposition, tacked on his assurances that, in the event they lost in the hearing, the Region would seek mitigation.

"Who can fight all out to win," Marilyn grumbled aloud, "and make plans for losing at the same time?"

Yet the reminder that the Region had kicked in $700,000 out of its own coffers put teeth into the claimed opposition.

Glen Bell, the alternate lawyer for the Township, hammered away on the issues raised by the Gary Posen letter. Then, holding up one of OWMC's thick reports, he said, "This is OWMC's 4B report. Our consultants tell us the four Bs stand for bash, burn, boil, and bury!"

The hall exploded into laughter and applause.

"This hearing is turning into a war of attrition," Mr. Bell continued, once the commotion died down. "OWMC plans to call eighteen witnesses for Panel 18. They'll be paid by the hour to sit there. We'll cross-examine six at the most. The rest will just sit there. Because the OWMC has money to burn, it can afford to indulge in this form of intimidation."

Mr. Bell ended by asking the area residents to read the witness statements." Make your comments to Dave McCallum. Tell us," he urged, "when OWMC hasn't got it right."

By this time, Marilyn had heard a lot about witness statements. Witness statements being late. Witness statements being held back. Witness statements being released. What she didn't know was, what was a witness statement and why should she read it?

The meeting was then opened to questions from the public.

"Will the Township support the Coalition at the next Procedural meeting?" asked one resident. "Will they take a strong stand that Coalition members should be given travel expenses, like all other residents?"

Mayor Packham responded, "We have already sent a letter to OWMC, protesting this discriminatory action."

Roger Cotton assured the residents, "We will support the Coalition at the next preliminary session on November 23."

To cheers and applause, Ziggy read portions of the letters from Motts/Welch/Cadbury,

Following him, former Mayor Allard Colyn, while not actually saying it, suggested OWMC had a good chance of winning. "Donald Chant is smart. The Crown corporation has spent $100 million dollars designing the plant and planning its defence. No technical arguments have been given by opponents to show they can defeat it..."

At that point, fourteen residents got up and walked out.

Marilyn, who had begun to believe the "we can win this" theme, felt her hopes dim. The opponents might be able to chip off a few pieces around the edges. But did they actually have anything that would utterly shatter the OWMC's case?

* * *

The next morning, Clifford drove to the bus stop in Grimsby to pick up John Jackson. The Concerned Citizens were meeting at the Gracey home that afternoon to discuss their information for the upcoming hearings. While Clifford was gone, Marilyn put the finishing touches to the lunch of cold cuts, salads, and homemade batter rolls. Then she assembled the tray of cups and saucers, and the fixings for tea and cookies for the Concerned Citizens following the meeting later. With a harried gesture, she raised her hand to push her hair back from her face and realized she had not taken the brush rollers from her hair. What a hectic week this has been, she thought, noting the dark smudges around her eyes as her fingers tugged at the picks and rollers. And the week isn't over. Tomorrow was Saturday, and Saturday's were chronically chaotic. And Sunday meant Sunday school...She felt herself tense in sudden recollection. She had not yet dug out the flanalgraph figures she intended to use with her lesson.

John Jackson and Clifford came in the back door just as she finished setting the table. Not until they had eaten lunch and Marilyn had cleaned up the kitchen did she broach the topic that had been on her mind for the past two days.

"John," she asked, as she collapsed into her easy chair, "how did you happen to find the Posen letter?"

"Purely by chance!" His eyes flared with animation as an exultant grin threatened to stretch from ear to ear. "I remembered that the OWMC was required to send the ministry material describing the project. And that these documents, mainly correspondence, were kept in the library in the offices of the Environmental Assessment Board in Toronto. So, with the hearings getting so close, I got the idea that maybe I should check out the ministry's file on the OWMC."

"How'd you get permission?"

"I didn't need permission. It's in the public record. Actually, I almost gave up. I'd been there for five hours and I hadn't seen anything new. But I made myself keep looking. Then I came across this letter. So I had it photocopied and when I got home, I called Joe. I told him, I think I've got something hot. The lawyers took one look at it and they agreed."

"So it's important?" She sounded doubtful.

"Oh yes! It's hard evidence that even the OWMC knows the facility will be a white elephant. I mean, you've got statements, straight from the president of the Crown corporation, showing that the OWMC would need subsidizing and that the government would have to force companies to use the facility to ensure a supply of waste. So I let Tricil know that the OWMC would need to run them out of business." He made no effort to hide the wicked glint of pleasure he had taken in stirring the pot. "There's already bad feelings between the two corporations. A few years ago, the OWMC attended hearings in Sarnia and had restrictions placed on Tricil's landfill."

Marilyn felt her spirits cautiously rise. "When did you find the letter?"

"Back in September."

Marilyn was silent. Two months ago. Why hadn't John told them before this?

As if reading her mind, John said, "I couldn't tell anyone because the lawyers wanted to keep the letter quiet until they decided how best to use it."

"So they released the letter to the media two days ago."

"Actually, that's not quite what happened. During these ongoing preliminary hearings, anyone who wants to take part in the actual Environmental Assessment Hearings has to apply for party status. To do that, you have to prove that you have a good reason for wanting party status."

Marilyn's gaze drifted to the wood paneling behind John's head as she absorbed this new information. "So . . .?"

"The Board didn't want oral submissions, only written ones. So we had to submit a written affidavit to the Board, outlining our reasons for wanting party status. In defining the issues, we used the letter to show why we wanted to investigate need." He broke into a low rolling chuckle. "When the OWMC read in our affidavit that we had found the Posen letter, they went crazy. Just this past Monday, I had to go to some lawyer's office in Toronto. I had to sign a deposition and undergo questioning." Indignation made his face redden. "It was outrageous! The hearings haven't even begun and I had to undergo two hours of cross-examination."

"Did—Did you have to go alone?"

"No, Joe Castrilli and Doug Hodgson were there to protect my rights. Ian Blue, Jim Micak, and some others were there for OWMC."

"How did you know what to do? Weren't you scared?"

His grin flickered and faded as his voice turned earnest and solemn. "Yes," he admitted, "I was nervous. This was a *big* deal. But the lawyers spent time beforehand preparing me. They told me things like, 'Answer the questions, but don't elaborate. Take a second before you answer. Give us time to object.'" His face lit in amusement. "It was nothing more than a fishing expedition, with the intention of intimidating me." Laughter rumbled in his throat. "It worked! I was intimidated! But I wouldn't want the OWMC to know it."

Marilyn cringed inwardly. "I'm glad it was you and not me."

Just then, cars began pulling into the driveway as the Concerned Citizens gathered to discuss information for the hearings, which now loomed less than three months away.

– 46 –

With the hearings approaching ever closer, the Coalition's November meeting was jam-packed with business. Most of the twenty-four members present were regulars and had been busy, busy, busy over the past weeks. Fundraising, lobbying, more fundraising, and outreach. Holy Name School in Welland had requested information from the Coalition on toxic wastes in the Great Lakes. Gracia Janes from PALS emphasized the importance of attending the upcoming public meetings being held by the Region. John Jackson was scheduled to show the Coalition's video, *The Turnaround Decade*, at the House of Commons on November ninth. To alert the members of parliament, Doris Migus, Doug Emslie, Joyce McEwan, and Ruth Burton had mailed out two hundred and ninety-five letters, urging the MPPs to attend the viewing. Doris had also taken on the task of contacting selected members of each political party by telephone.

John Jackson informed the members that he would be flying to Manitoba for three days, to meet with citizens and discuss the plans of the Manitoba Waste Management Corporation. He also reported that Alexander and McCormick had sold ninety copies of the Coalition's video thus far, one of which had been purchased by the staff of the Ministry of the Environment.

Edith Hallas thought the Coalition should hold a protest parade on the day the hearings opened in Smithville. Marilyn let her eyes rove from face to face as members of the group threw themselves into making plans. After all this time, five years for some, eight or ten years for others, still they possessed the spirit to stand against the impossible odds and shout, "No!"

OWMC—The "Wasted" Years: The Early Days

Although no hard and fast date for the opening session had been named, Marilyn, like everyone else, was allotted a page of names from the telephone book to call, urging support for the rally. She dreaded the job more than a toothache.

From then on, every day she had to argue with herself, forcing herself to make at least five phone calls. Usually by then, her stomach was in such a state of turmoil that she couldn't dial another number.

But once she had suffered through her ordeal for the day, she raced upstairs and devoted herself to what she did best—compiling a series of letters to the editor. Letter number sixty refuted OWMC's claims that it would use the best available state-of-the-art technology and she took her ammunition straight out of an article in OWMC's *Update*.

The article stated that although the levels of nitrogen oxide emitted from the incinerator stack would be very low, the Crown corporation intended to build in provisions "to allow a nitrogen oxide removal unit to be installed if the government brings in tighter controls." She wrote the word *if* in italics.

"So in other words," she wrote, "*if* the OWMC is forced to it, it can do better, but only *if* it is forced to it. No wonder the residents of Niagara refuse to put their trust in OWMC's 'best available technology.'"

She ended the letter with a passionate plea to residents to give strong support to the upcoming Protest Parade.

* * *

On October 12, 1989, to ensure that intervener funding would continue for the Coalition after the end of 1990, John Jackson had written a letter to the Honourable James Bradley, minister of the environment.

It took more than six weeks for John to receive an answer. But when it came, it confirmed that since the Joint Board had determined that the Intervener Funding Project Act applied to the upcoming hearing, it would be appropriate for the Coalition to apply for funding. Signed by Jim Bradley, the letter stated: "…If for any reason additional funding is not available to a party under this legislation, I would ensure that a request for necessary assistance by Order-in council be appropriately considered in future."

Jubilant at the minister's response, John explained that to ensure the funding would be available, the Coalition must submit an application for a nominal amount—one dollar. This would give the Coalition party status and make them eligible to apply later for additional funds.

The day before the fifth and final series of preliminary hearings began in the Old Farm Inn, Doug Draper of the *Standard* interviewed Helen Kzan. "I guess we're glad they're getting under way," she said of the hearings. "But we feel resentful of the whole process. OWMC's proposal has put our lives in limbo for the past four years and it will take at least two more years before there is an answer. We still don't know what is going to happen to us," she said. "You just try to go on with your lives as best you can."

The Globe and Mail ran a full-page article about the start of the hearings, including a four by five picture of Helen. She looked particularly drawn and weary. John Jackson, when interviewed for the same article, raised the question of long-term safety. "The incineration process, the high temperatures, actually create new chemicals that we don't even know exist. We don't even know what to measure for."

When reporter Craig McInnes interviewed Marilyn, he raised the over-worked argument that the OWMC had spent eight years and $80-million dollars, as though that alone should satisfy the opponents. Marilyn retorted, "If you have a bad plan, then you have a bad plan, and all the money in the world won't turn it into a good one!"

Near the end of the article, McInnes quoted Dr. Chant. Marilyn could almost hear Dr. Chant's refined, well-modulated voice as he lightly brushed aside the concerns of the opponents. "...The OWMC has consulted with the public extensively in an effort to defuse opposition. To a certain extent, it has been a success. Local residents have become less active in their opposition to the plan."

Marilyn flung the paper aside. She had long been convinced that the Open Houses and the infamous questionnaire had been nothing more than an expensive pat on the head, meant to pacify and distract, like the shaking of a rattle to quiet a cranky baby. If she had needed steel added to her oft-times wavering resolve, his claim that the OWMC's much-touted public relations programme had defused public opposition did the trick. Her eyes narrowed menacingly as she muttered through clenched teeth, "Dr. Chant, just you watch us!"

But though she had been fortified by renewed determination, she found it no less agonizing to make her daily quota of telephone calls.

* * *

Officially, the Environmental Assessment Hearings into the Ontario Waste Management Corporation's proposed treatment facility for the province's hazardous waste would get under way at ten a.m. on Thursday, November 23, 1989. In truth, no evidence would be heard. The two-day session, called to allow those seeking the right to speak at the hearings to apply for party status and for funding, would take the form of just another in the long drawn-out string of preliminary hearings.

However, in response to the official two-page notice placed in the newspapers by the Office of Consolidated Hearings, the media showed up in numbers.

As Marilyn stepped into the banquet hall of the Old Farm Inn, she hesitated, struck by the startling impression of a room coloured in black and red against neutral walls. Like a restless black oil slick, men and women in dark suits filled the room as they congregated in loosely shifting groups against a backdrop of crimson table-cloths. Some faces were familiar. Many were not. Yet in the midst of the oppressive sea of black, like a beacon of hope, stood John Jackson in a long-sleeved blue and white striped shirt, open at the throat, his hands shoved deeply into the pockets of light grey trousers. Laughing and relaxed, he chatted with Dave McCallum and lawyers Doug Hodgson, Joe Castrilli, and Glen Bell.

By contrast, Ian Blue made a poignant and solitary figure. He stood at the end of the OWMC's table, his hands resting on the back of a chair, watching John and the lawyers.

John and Mary Dykstra and Don and Marie Austin had saved seats near the front for Clifford and Marilyn. Precisely at ten a.m., Nada Davidovic, Deputy Hearings Registrar for the Office of Consolidated Hearings, called the session to order. Twenty-three applicants had come seeking party status.

"We'll start with the municipalities," stated Dr. Kingham. "Mr. Hodgson?"

Ian Blue promptly jumped to his feet. "The OWMC requests the right to question the applicants." The Board conferred briefly, then refused permission.

Hodgson outlined the areas the Region wished to investigate. Roger Cotton declared the Township of West Lincoln unalterably opposed to the OWMC's proposed project. Lawyers representing the City of Toronto and the Town of Milton had shown up at the eleventh hour, concerned lest they find themselves the unhappy hosts to transfer stations. Marilyn suddenly found herself on high alert, her brain humming and her pen skimming over the pages as a representative from the Niagara Peninsula Conservation Authority listed the possible impacts connected to the OWMC's plan to withdraw and discharge water into the Welland River.

Then came representatives from the private sector—the Mott/Welch/ Cadbury Company, Tricil, the Canadian Association of Metal Finishers, the Liquid Waste Carriers Association, Med-Track Waste Management Inc, the Canadian Portland Cement Association, and so on. Most feared the OWMC would impact negatively upon their businesses.

Shortly before lunch, Joe Castrilli applied for full-time party status for the Coalition. At that point, he submitted a list of the seven member-groups.

Blue rose hastily. "Mr. Chairman, you should not mark that list as an exhibit until we see if the Coalition is granted party status."

Despite Blue's objection, the Board gave the document the number C-3. C for Coalition and three, meaning it was the third document submitted into evidence. Joe Castrilli then gave a detailed rundown of the Coalition's objectives. "To examine the OWMC's proposal, to research alternate solutions, to educate the public, and to propose methods for the handling and disposing of hazardous waste. The Coalition is also concerned with the possible negative impact the OWMC's facility would have on the development of other methods for dealing with Ontario's hazardous waste, the impact on members of the Coalition, and—"

"Isn't the Coalition simply six people and Mr. Jackson?" interrupted Ian Blue, his voice snappish with sarcasm.

Exclamations of anger swept over the audience. Joe looked the picture of offended dignity. "Mr. Chairman, that's outrageous! It's intimidating and bullying! I refuse to answer that!"

Marilyn could have sworn that she saw just a hint of amusement pass over Dr. Kingham's stern countenance, giving rise to the suspicion that he enjoyed the occasional strike of flint against stone.

West Lincoln's Town Crier, Mr. Green, applied for the right to participate at various times and he asked for intervener funding on a part time basis. Mr. Blue pushed forward the opinion that Mr. Green should be represented by the Coalition, despite the question of incompatible objectives.

As the hours wore on, Marilyn felt her mind go numb. At long last the procedural preliminary hearing adjourned for the day, with the notice that the Friday session would begin at 9:30 a.m.

*   *   *

On Friday morning, Dr. Kingham handed down the Board's rulings on party status. Of the twenty-three applicants, the Region, the Township, the Coalition, Tricil, Mr. Green, and a handful of others had gained the right to take part in the hearings and receive intervener funding. The OWMC entered into evidence the cross-examination of John Jackson's deposition regarding the Gary Posen letter. The hearing then relapsed into the same time-worn catalogue of complaints—OWMC's failure to file witness statements on the agreed date, funding, and most important, what the interveners deemed OWMC's abuse of the phasing process for the hearings.

"James Micak is scheduled to make thirteen appearances," Castrilli noted in his list of multiple appearances by numerous witnesses. "The OWMC wants to be able to call evidence as long and as often as they like, as long as they can pay for it. Or until the witnesses get it right. This will put pressure on the other parties to call repeat witnesses to counter the OWMC's latest evidence. Quite frankly," he told the Board, "the Coalition can't afford it. Our resources will be exhausted, putting us at a severe disadvantage. Mr. Chairman, a phased hearing doesn't work unless the witness panels close at the end of the phase."

After the morning recess, Hodgson reminded the Board, "This is the ninth day of procedural discussions and we're still dealing with matters I thought finished during the July fourth meeting." He, too, faulted the structuring of the OWMC's case. "It was impossible to know in March, from the first seven witness panels, of the OWMC's abuse of the process. We had a list of the panels, but not the witnesses. The OWMC shouldn't have to call a witness thirteen times. That's an abuse of the public purse." He spieled off a bunch of figures. "Over the next two years, the OWMC itself could spend more than $7.5 million for lawyers and consultants during the hearings."

Marilyn recalled reading a report in the *Tribune* of a heated tirade by Hodgson to the Region's OWMC steering committee. He had been incensed at the time because a list released by the OWMC showed that the Crown corporation intended to call 74 witnesses, some as many as 12 or 13 times. One panel alone would have 18 witnesses. Hodgson had told the committee that the Region and the other opponents would run out of money just by cross examining OWMC's witnesses, before they could call their own expert witnesses to testify. According to Joop Gerritsma of the *Tribune*, the equivalent of twenty-five percent of the interveners' budget could be used up just on that panel of eighteen witnesses.

Marilyn dragged her attention back to the present as Hodgson proceeded to give his view of how the case should be structured. "Witnesses should appear one at a time and only once-in-chief." He swiftly ran down a list of the witnesses who would be called a number of times. "Mr. Blue should be required to justify the calling of witnesses more than once. Multiple appearances will only

OWMC—The "Wasted" Years: The Early Days

lengthen the Environmental Assessment Hearing." He moved for phasing to be done away with.

He insisted that the whole case hinged on need. "If it is proven there is no need [for the facility], there is no need to continue with the hearing." At that point he introduced Dr. Chant's letter to Deputy Minister Gary Posen. "The request for a commitment to some form of government subsidy in order to keep prices competitive suggests that even the OWMC has doubts about the need."

Mr. Bell continued the attack on the structure of the case. "Early on, I saw opportunities for the OWMC to split its case. My suspicions have been confirmed. Opponents won't know the OWMC's case until the last witness leaves the witness box." He noted that other panels were nothing short of patch-up panels. "This case has been structured to require constant vigilance. The opponents can't afford constant vigilance. OWMC has the great capacity to generate new evidence and the Board will be loathe to reject that new evidence. So the OWMC will have constant opportunities to patch-up the evidence. For example, the OWMC says it will set prices. Where are the studies on pricing? We won't know the case until Mr. Blue sits down and says, "This is my case"."

The issue of how the case should be structured resumed after the noon break. Finally, Ian Blue rose and proclaimed, "My friends' complaints are 'castles of conclusion built on the sands of speculation.' That's a quote from a book on life in outer space. I will not resuscitate witnesses." He adopted a martyred tone as he sat down. "I'm entitled to *some* trust."

The Board called a fifteen-minute break. After the break, the Board agreed with Hodgson. "Panels 1 to 7 had few witnesses and little duplication." Dr. Kingham tipped his head to one side in his familiar speculative fashion as he lined up his reasons. "If the opponents had had Panels 8 though 14, they would have seen the multiples of witnesses and the considerable duplication."

With ill-concealed anger, Mr. Blue flung out his arm in a scornful gesture of dismissal. "*They* want the order of the witness panels changed. *They* want witness to appear once and never again. Well, OWMC doesn't want to call its case that way. I have a right to call my case as I see fit." He locked stares with Dr. Kingham. "You have no right to tell me how to call my case."

All around her Marilyn heard gasps of shock at the rudeness of his words and his tone. Except for a slight tightening of his lips, Dr. Kingham revealed no change of expression.

Mutiny oozed from every furious syllable as Ian Blue unleashed his defence. "It would cost thousands of dollars to prepare witness statements again. I have a right to cycle witnesses. As for the interveners complaining that they'll run out of funds, they won't go short. The government promised them they would have what they need."

At this, Mr. Watters, the legal counsel representing the Ministry of the Environment, vigorously shook his head no. The debate deteriorated once again into a hassle over funding. Finally, as the hands on the clock approached five, Dr. Kingham adjourned the proceedings until the following Friday. As Marilyn wrote down the date, she felt a wave of panic. December 1, 1989, and she had barely started to prepare for Christmas.

* * *

In the days leading up to the next hearing date, rhetoric filled the newspapers. Regional Chairman, Wilbert Dick, wrote to the premier of Ontario, complaining that, "the odds are stacked heavily against a fair hearing...Not only are you and your cabinet the proponents, but in an appeal, you are also the judges."

The *West Lincoln Review* reported the proceedings under the headline, "New foes of OWMC seek $1 million." The editorial on page four, titled "Trough reopens," charged that the companies who were recent newcomers to the hearings weren't reading anything, had no concrete arguments, and cared little about what the three major interveners were doing.

* * *

On the day of the last procedural hearing, Marilyn sat on the edge of her chair. What the Board decided this day would have a crucial impact on the fate of the hearings. For the very first thing, Dr. Kingham stated that all motions by all parties had been denied.

Marilyn, not sure what the motions had been, felt her hopes slide into despair. Ian Blue nodded his head vigorously up and down in agreement as each motion dropped into oblivion.

Dr. Kingham then addressed his remarks to the audience as well as to the lawyers. "The Board has the OWMC's assurance that it will not split the case and we believe that phasing will assist the Board. Also, it would be unfair to the OWMC to change the rules now."

Marilyn glanced at Marie and Don, and John and Mary. Their faces reflected the grim, grey look of defeat she felt inside. Behind her, Marilyn could hear Marija's angry mutterings to Paul in her native language.

Dr. Kingham inclined his head to one side as he spelled out the Board's decision. "It seemed to us that the interveners' complaints relate to how the OWMC has organized its case within the phases. Had the Board anticipated that any party would use the rules on phasing to construct witness panels of more than five witnesses, with several witnesses appearing more than twice within a given phase of the hearing, we would have more explicitly limited the witness panel privilege."

Dr. Kingham faced Mr. Blue. "In the Board's view, the proponent does not have unfettered discretion to organize its case as it sees fit." He continued to outline the Board's decision in considerable detail. He noted that the evidence, in this form, would cause hardship and expense to the interveners. "Not only would it be manifestly unfair, but we are convinced it would not facilitate and expedite the hearing. And we don't think we would understand evidence brought in this fashion."

Again his gaze shifted to the audience. "We are requiring the proponent to put the evidence in a way that we can understand it. And let no one be mistaken on this point. The purpose of this hearing is for the Board to be clearly informed on the matter, in a public forum, so that we may make a reasonable decision."

He asked Ms. Davidovic to pass out to the parties a revised outline. "We may have erred on some details...but our intent is sufficiently clear. We are not asking Mr. Blue to change the content of his witness statements—just to put them in a more meaningful order."

Ian Blue shoved his chair back roughly as he came to his feet. "I demand an adjournment—right now!"

"Why do you want an adjournment now?" Dr. Kingham inquired, his tone excessively reasonable. Then as if he suddenly realized he ought not to deny the request, he glanced at the clock. "It's ten twenty-five. We'll reconvene at eleven."

Thirty-five minutes later, Blue returned, armed with examples of other cases and prepared to do battle. When no argument seemed to prevail, he snapped, "You don't have the right to meddle in my case! You don't have the background!" A hush fell over the hearing room.

Mr. Thompson, his face an angry red, stated, "Some on this Board *do* have the background, so I would appreciate it if you didn't make those kinds of statements." Mr. Thompson, so Marilyn had been told, had been a criminal lawyer. Whether it was true or not, she didn't know. But she knew him to be a QC—Queen's Counsel.

But Queen's Counsel or no, Ian Blue made no pretense of backing down.

"OWMC is not inclined to revise its witness statements one wit! You can disallow a panel. But that does not give you the right to restructure my case!"

Surprisingly, the lawyer for the City of Toronto entered the fray. "According to the Consolidated Hearings Act, the Board may impose or give such directions as it sees fit."

Hodgson added his opinion as to the Board's authority. "A Board wanted a poll of fifteen lawyers in a hearing before the Ontario Energy Board involving deregulation in the natural gas distribution industry. The lawyers were asked how long they thought the hearing would last. The answer came back: one month. The chairman then told them, 'The Board doesn't have a month. The hearing will start on Monday and end at Wednesday noon.'"

Dr. Kingham waited for the ripple of subdued chuckles to die down, then called a recess. Upon the Board's return, Dr. Kingham stated, "The submissions we have heard this morning have, if anything, reinforced our determination to proceed as we outlined earlier. We will give you, Mr. Blue, to the end of January to restructure the case in accordance with the principals we enunciated earlier. Failing that, we will do it ourselves."

"I object to that ruling, Mr. Chairman."

"Thank you, Mr. Blue. We have now to deal with the matter of timing and the possible abridgment..."

The hearing lasted another hour. Afterward, Marilyn spoke to reporters with cautious optimism. "The Board's decision is definitely favourable. If the OWMC had not been instructed to cut its witness list, it would have put a terrible strain on the interveners who don't have the resources of the OWMC."

* * *

On the second day of December, Marilyn sent out a letter refuting Mary Lou Garr's statement that "the West Lincoln site was attractive because of its Golden Horseshoe location" and therefore it would "minimize transportation costs."

"It's difficult to see," Marilyn wrote, "how the West Lincoln site would minimize transportation costs when 93% of Ontario's waste is generated outside the agricultural area chosen by OWMC and *all* additives and reagents would come from outside the Niagara Peninsula."

The following week, elderly Mrs. Ella Krick called Marilyn. "I read your letter," she told Marilyn in her quavering voice. "What I want to know is this: Why would it cost too much to ship the lawyers and consultants down here for a couple of years but it wouldn't cost too much to ship toxic waste to West Lincoln from all over Ontario for the next fifty years?"

Marilyn burst out laughing. Mrs. Krick might be in her eighties and have serious health problems, but her mind was as sharp as ever.

"They talk about their state-of-the-art technology," she told Marilyn. "I call it state of the ark—Noah's ark!"

* * *

That same week, at its annual convention, the Niagara Fruit and Vegetable Growers took an official stand against the OWMC.

* * *

On December thirteenth, Clifford and Marilyn received a copy of a letter sent out by John Jackson to Coalition supporters. It gave an upbeat assessment of the proceedings thus far, noting that the Coalition had already sold more than a hundred copies of its video, *The Turnaround Decade*. The letter ended with a plea for financial support. "The Ontario Toxic Waste Research Coalition is making substantial progress in showing that there is a better way and that the OWMC is not the only possible approach to Ontario's hazardous waste problems."

* * *

As if the OWMC hearings weren't enough for one small community to deal with, the first preliminary hearing into the use of a mobile-PCB-destruction unit to clean up the old D and D storage site in Smithville was held at the Legion Hall on Wednesday, December fourteenth. One of the OWMC's flock of young lawyers, John McGowan, had been dispatched to seek party status on behalf of the OWMC. "To show that our work is complementary," he said. Lawyers for the Township, the citizens committee, the Ministry of the Environment, and Ensco, the firm hired to burn the PCBs, all argued against allowing the OWMC to participate. "OWMC is still more of a concept than an operating reality," they stated.

In the end, as the *West Lincoln Review* put it, "the Ontario Waste Management Corporation was asked to butt out, and it did. None of those who gained party status was opposed to the $13-million dollar proposal."

- 47 -

The hearing of evidence into the Ontario Waste Management Corporation's proposed toxic waste facility would begin on February 5, 1990. With that announcement, preparations for the Protest Parade went into high gear. Marilyn flooded the newspapers around the peninsula with letters to the editor, attacking everything from the "it has to go somewhere" mind-set to the dinky little landfill depicted in all of the OWMC's news releases. She contended that any darned fool could see that a facility intended to run for at least fifty years would need a landfill that would last longer than 20 years. And that even at twenty years, the landfill pit would require every square inch of the proposed site and then some. She described a vast continuous landfill pit fifty feet deep, gobbling up acre upon acre, and the probable impacts on the groundwater.

She bombarded the public with facts on the number of trucks hauling clay out, trucks hauling additives in, construction, and later, employee traffic to and from the proposed site. She revealed that the proposed facility would be classed Heavy Industrial, making it, according to OWMC's documents, "unsuitable to be sited in commercial areas because it would not be compatible with commercial activities."

"What makes it compatible with agriculture?' she demanded to know.

She quoted OWMC's documents, which gave two reasons why the facility would be incompatible with commercial areas. First, "private investment may have been committed for commercial development on the basis of existing official plan designations."

Second, "the potential closure of small business is of concern."

With unswerving logic, she flung down her arguments. "Didn't the farmers of Niagara commit their private investments on the basis of existing official plan designations?" she questioned. "And what about our small businesses?" Then she called upon all residents to join the protest parade on February fifth and signed off each letter with the familiar slogan, *There is a better way.*

* * *

Early in January of 1990, the Dunnville Chronicle sent Angus Scott to the Gracey home to interview Marilyn. First, he took photos of Clifford and Marilyn in the backyard, with the barn in the background, and their coat collars hunched up around their ears.

Indoors, Marilyn showed off the posters she had designed to use in the protest parade. "It's taken me weeks to do these!" she exclaimed with pride. One showed a huge dump truck bearing the word "taxes" as it dumped a load of fluttering dollars marked "subsidy" into a deep hole in the ground labelled "OWMC's bottomless pit." Another turned the letters OWMC into the form of a waste facility, with smoke curling from the twin smoke stacks made by the arms on the "W." But her favourite was the highly colourful *Bash, Burn, Boil,* and *Bury* poster. Each word depicted its meaning. The letters in *Bash* appeared crumpled. *Burn* spouted flames. *Boil* had a fire lit beneath it and steam rising above. And the word *Bury* showed a huge pit with a dinosaur peering curiously over the edge as a statement refuting OWMC's state-of-the-art technology.

Then, curled up in her old orange recliner, facing the young reporter, Marilyn rolled her eyes and said, "I used to think people who did this were oddballs. We've never been public people."

She paused in thought from time to time as she went back to the earliest days, relating how she and Clifford had gradually been drawn into a battle they had never intended to fight. "We expected a Crown corporation to be fair to the local residents and to present their case truthfully. But here's the latest example." The sense of betrayal made her voice harsh as she handed him a photocopy of Dr. Chant's letter to Gary Posen. "The OWMC never said publicly that the facility would have to be subsidized by the province of Ontario."

Angus Scott shook his head slowly, his eyes reflecting a shadowy sense of disbelief. He leaned forward and rested his elbows on his knees. "You seem to be the most unlikely couple imaginable to be outspoken environmental activists. What gives you the courage to keep going?"

She started to say, "Our faith in God." But her courage deserted her. She shifted her gaze to the window beyond the young reporter and exhaled slowly as she hastily reviewed a slew of safer, more acceptable answers. She could speak of her father's influence on her life and it would be true. She could point to their belief that they were fighting against wrongdoing. And that also would be true. She could stress the heartbreak of packing up and moving away from family and friends because they couldn't face living the rest of their lives with the OWMC on the next road.

She silently drew in her breath. She would tell him those things, but later. Now, with her hands clenched in her lap and her gaze wavering nervously from the waiting young face to the carpet under his feet, she said, "Clifford and I have a deep belief in God. It's our faith that gives us the strength to carry on when we want to quit." She glanced up, half-expecting to be met with an expression of scorn. Instead, with his expression intent, he wrote rapidly as she spoke and she fell silent, waiting for him to finish. When he looked up, she hesitated, then spoke directly and honestly. "I know this is not a popular thing to say these days, but we're born-again Christians. We're always praying for wisdom to do the right thing." Again she paused, watching his hand until it slowed. "With our letters to the editor, we try to be completely truthful. We may make mistakes," she admitted, "but we're never deliberately dishonest."

"When do you think the battle will be over?" he asked.

"They say it'll be two years, at least. And then there will be the appeal." Half laughing, half bitter, she added, "I always thought I'd live to a ripe old age. But this is taking years off my life."

"What will you do with your time when this is all over?" he asked.

"Believe it or not, I had a life before the OWMC came along. I had things I was doing, things I wanted to do. I'll go back living a normal life..."

She felt tears form suddenly behind her eyes and she pressed her fingers to her lips. Would she and Clifford ever again have a life they could call normal? Oh Lord, she thought, what do You have in store for us?

* * *

On January 24, 1990, less than two weeks before the hearings were scheduled to begin, the Dunnville Chronicle ran a full page spread titled, "The Fighters." It featured Chuck and Pat Potter, and Clifford and Marilyn Gracey. The eye-catching sidebar had been written in large print.

> They do what many of us would like to do. They take on governments, big business, the polluters, the politicians...They are called activists. They make waves. They don't back down. They are the conscience of a community. They wager their homes, their energy, and most of their time on their crusades. The Potters are from Port Maitland; the Graceys from Wellandport. Here are their stories.

Marilyn felt a wave of relief and gratitude as she skimmed over the words. Young Angus Scott had treated his interview with Marilyn with integrity and respect. "[The Gracey's are] tall, open-faced farmers who have a deep belief in God," he wrote, "but [they] believe they're right in their fight and stand a good chance of winning."

* * *

The funding issue had been settled at last. The Township, the Region, and the Coalition had each been awarded one dollar.

"One dollar is not a large amount by anyone's standards," ran the story in the *Tribune*, "but Regional Niagara considers getting it a major victory."

"It is not the $1 that is important", said Regional Steering Committee Chairman Mayor Stan Pettit of Wainfleet. "What's important is that the province, by approving the award, has acknowledged that the Region is entitled to funding under the new Ontario Intervener Funding Project Act to compensate it for its hearing costs."

From day one, Marilyn recalled, the three groups had maintained that the $3,218,379 awarded to them collectively was not enough to mount an effective case against the OWMC's proposal. Now, they had the promise of additional funds to see them through to the end.

* * *

With the Protest Parade scheduled for the opening day of the hearings just three weeks away, John Dykstra and Clifford took to the roads, distributing flyers to businesses and newspapers. At the office of the *Pelham Herald*, Clifford gave an on-the-spot interview.

"We know that the weather may not be ideal," the *Herald* printed a week later, "and we know that 9 a.m. on a Monday morning is not the best of times, but we urge the people to come out anyway."

Meanwhile Marilyn answered questions over the telephone from the *West Lincoln Review*. "The proposed march has been met with enthusiasm. Dr. Chant said that the opposition has waned. We want to say that the opposition hasn't died and we want to have a good time doing it."

Dr. Chant responded by telling Blair Burgess of the *West Lincoln Review*, "We live in a democracy, thank God, and it is certainly people's right to launch a parade if they want to. I certainly wouldn't stand in the way of that." Dr. Chant then said he was not worried that a large parade turnout and the resulting national media coverage would rally more residents against the facility. "I think (the march) will let people blow off some steam, but I doubt it will change the situation at all."

Marilyn's jaw jutted out as she clamped her teeth shut. Blow off some steam! she fumed. He makes it sound like we're children having a temper tantrum!

In the same article, Leslie Daniels attempted to defuse public indignation over Dr. Chant's statement that opposition had waned.

"He really meant that it is difficult to keep people involved over a long period of time."

John Jackson told the Review that the response to the rallying cry had been extremely positive. "The hearings aren't just something going on between lawyers and consultants. It's a people issue."

Three days before the hearings opened, Dr. Chant blasted critics for accusing the Crown corporation of seeking government subsidies and a monopoly in the toxic waste disposal business. He told Doug Draper of the *Standard* that the Region, the Township, and the Coalition were being "paranoiac and overly picky" about the letter he had written to Gary Posen. "I was only asking for bridging funds to offset operating costs until the plant reaches full capacity and can pay its own way. If the wording was clumsy, so be it," he said.

Michael Boggs, the Region's chief administrator, said, "The Region reviewed the letter before attacking it last fall. I'm not going to debate in the news media the way in which Mr. Chant now rationalizes the wording of the letter."

John Jackson insisted that the letter was very clear in its requests for subsidies. "The Coalition opposes subsidies because it would smother incentive for industries to reduce their waste volumes."

In a separate article, Doug Draper waxed eloquent as he described the OWMC's proposal and the upcoming hearings. Marilyn noted a distinct change in tone. No longer just an incinerator stack, it had become a *towering* incinerator stack. It would be upwind from more than 130,000 people...and thousands of acres of *prime* foodland. Niagara would become the final destination for thousands of tonnes of *poisons* generated throughout Ontario and, possibly, from other provinces and the United States.

> Dozens of tanker trucks weighed down with a witch's brew of solid and liquid industrial garbage, would roll down the highways and roads of Niagara every day. They would eventually pull into a plant on what was once a sprawling pasture on part of Stan and Helen Kszans' dairy farm. The Kszans, of course, would be gone.

He quoted the Coalition's belief that the proposed plant would impose an unfair and unnecessary burden on one community. He then outlined the Coalition's preference for the development of safer and more progressive approaches, including portable treatment systems, and new technologies for reducing and recycling wastes.

In yet another write-up, the *Standard* gave a brief history of the fight in the late 1970s over plans to use the Walker Brothers Quarry to dump toxic waste. "Ironically," the paper noted, "the leader of the citizens' group challenging the Walker Bothers plan was Leslie Daniels, a Thorold resident who is now coordinator of regional communications for OWMC. A large part of her job with the corporation involves preaching its gospel to the citizens of West Lincoln."

Mrs. Daniels dismissed any suggestion that she sold out when she joined the OWMC in 1985. "Walker Brothers' Quarry, on the brink of the Niagara Escarpment, was an awful site for a toxic waste plant. If there was an issue today like Walkers', I'd still be standing on the other side of the fence."

It seemed the area papers couldn't get enough news about the OWMC onto their pages. Old quotes were resurrected, such as Dr. Chant's assertion that he was "not running a popularity contest." He had stressed that he had "no intention of letting emotions and fears get in the way of finding an 'environmentally secure' site." Nor had he hesitated to use words like "silly," "crazy," and "asinine" to describe some criticism. He once called politicians in Niagara Falls "idiots" for their concerns that the tourist trade would suffer if a waste plant were built there. And he had referred to critics near a candidate site in Milton as "fat cat hobby farmers." He then repeated one of the statements that scared Marilyn the most: "It's got to go somewhere." Too many people, unacquainted with the issue, would be swayed by that simple declaration.

"We've got to have these treatment facilities and I'm adamant on that, and some community is going to be unhappy. The whole concept of modern society is that individuals and communities may have to suffer in the broader interests of everybody. That's the way we run our affairs."

Marilyn could feel her sense of injustice boil over. Neither she nor any of her neighbours had volunteered as sacrificial lambs, not in the beginning, and not now.

John Jackson defended the right of the residents and the people of Niagara to object. "Yes, they're the ones doing the criticizing. And therefore you can call it NIMBY. The reality is that it is those who have to live with the threat, who are most aware of the negative impacts..., who are looking for alternatives and real, long-term solutions. There is nothing like being under pressure to make you creative. It certainly doesn't mean it's selfish."

He reminded the interviewer that the members of the Coalition were not arguing the plant should be located somewhere else. "They are saying, no one in this province should have to live next door to this type of facility. It is the wrong way to go."

On Saturday, February third, the *Standard* ran another two pages of coverage written by Doug Draper, including a feature article in *Spectrum* titled "A Glimmer of Hope." The article portrayed a very optimistic John Jackson. "The reason I'm feeling more confident," he said, "is that the world has changed and OWMC hasn't."

Draper used statements from a wide array of participants and experts, summing up with quotes from both Dr. Chant and John Jackson.

"Dr. Chant, an internationally respected scientist and environmental activist...insists, as he has for the last decade, that "it would be a disaster" for the environment if OWMC's proposal is rejected."

For once, John Jackson had the last word. "All the experts and scientists are telling us we have to stop (pollution) at the source. You don't do that by building incinerators and landfills."

John Jackson then condemned the provincial government for being inconsistent. "Here we have [Environment Minister] Jim Bradley...telling municipalities they have to reduce their [garbage going to landfills] by fifty percent by the end of the century. He gives them clear instructions and tells them they have to do it. Why do we not do the same for hazardous industrial waste?"

On the same day, in another article in the *Standard*, Doug Draper described his trip with the politicians to Biebesheim, Germany in September of 1987. "'There it is,' said one of the Canadian visitors as eyes peered out at the 20-storey stack towering above fields of sunflowers and vegetables. It was OWMC's version of The Emerald City,...' he wrote, referring to the fabled and magical city in *The Wizard of Oz*.

Dr. Chant was quoted as saying that there was no reason to be concerned. "I've seen these plants in Western Europe, the counterpart of what we're proposing to build...I have seen that nobody has died. There are no two-headed cows. There are no crops that can't be sold. There has never been a serious accident...What's the big deal?"

The big deal, Marilyn thought, is that according to the mayor of Biebesheim, crops from those fields were sold to a large baby food company.

– 48 –

On February 5, 1990, the morning of the protest parade, Marilyn woke to the first grey light of dawn. The sky looked cold and overcast, but at least it wasn't snowing. She winced as her bare feet hit the icy floor of the hundred-year-old farmhouse. Quickly she pulled on heavy socks and the thermal long johns she would wear under her blue jeans. She glanced at the clock. Clifford had gone to the barn more than an hour ago and would soon be in for breakfast. Shivering, she slipped into her fuzzy housecoat and zipped it up to the neck. As two of the volunteers, she and Clifford had to be on the spot as soon after eight as possible. John Dykstra in his tractor would lead the parade. Clifford in his tractor would bring up the rear.

Marilyn barely made it to the kitchen before Clifford came up the back steps. He showered and dressed before wolfing down bacon and eggs, toast and coffee. Within half an hour, they were on the road. Marilyn followed behind the tractor in the car, the slow-moving caution lights flashing.

On the final curve leading into Smithville, Clifford slowed and pulled to the side of highway #20. There, he parked the tractor, caution lights still flashing, and helped to direct traffic to the assembly area in the parking lot of the Village Square Mall. Already, one minor fender-bender had taken place in all the excitement. Marilyn tied the posters she had made to the tractor and to the car, and then headed uptown to position herself on one of the street corners. There, with six rolls of film in her pocket, she would make a record of the attendance. The Coalition had decided not to allow

OWMC—The "Wasted" Years: The Early Days

anyone—media, politicians, or the OWMC—to downplay the attendance. On another corner, another member of the Coalition would take a video of the parade. Still others, like Edith Hallas, would take a body count for immediate release to the media.

At five minutes after nine, by car, by truck, by tractor, and on foot, the protesters wound their way through Smithville, bearing signs voicing their opposition. Carol Haynes, a school bus driver, showed the picture of a school bus and a toxic waste tanker under the question, *Side By Side On Rural Roads?* The Township's fire truck and one of its dump trucks followed John's tractor. Women pushed their children in strollers or grocery carts. Don Austin and Marija Balint carried the Concerned Citizens' banner. Four men shouldered a small black coffin, labelled *Niagara*. Others wore protective suits and makeshift gas masks, urging *Ban the Burn* on hand-lettered cardboard. As the vehicles and marchers streamed past snow banks in a seemingly endless line, Marilyn waved at friends and neighbours, then snapped their photos. Tractors towed manure spreaders, covered in signs and slogans. Lloyd Comfort drove the family tractor and spreader, which bore a huge sign on the side that read *OWMC Toxic Waste Spreader*. In the spreader, supporting a straw figure representing Dr. Chant, Lloyd and Mary's son Bruce shouted through a megaphone, "I know all the answers! Trust me! Trust me!"

Wiley Brothers' delivery truck displayed the picture of a bottle of their juice and the wording, *100% Grape Juice*. Behind it came a brown K-car bearing the message, *The Brant County Environment Group Supports You!*

Still the vehicles and the marchers flowed past and Marilyn could hardly believe the huge response. She stamped her feet from the cold and blew on her fingers to try to keep some semblance of feeling in them. Finally, with the end in sight, she raced to fetch the car. She parked it at the arena where the Smithville Branch of the Canadian Legion was serving coffee and hot chocolate. Then she raced across the street to the rally already in progress in the parking lot of the Old Farm Inn.

The protesters marched twice around the Old Farm Inn, chanting, "There Is a Better Way!" and "Save Our Environment for

Our Children's Sake." Meanwhile, inside, the three-member hearing panel, made up of Dr. Kingham and Ms. Doherty representing the Environmental Assessment Board, and Mr. Thomson, Q.C. representing the Ontario Municipal Board, were attempting to hear the parties' opening statements amid the uproar.

Outside the inn, the straw figure of Dr. Chant was burned in effigy. One of the first speakers was Ziggy Sojka. Members of the Niagara Residents fired up the crowd to renew the chant, "There Is A Better Way!" for the television cameras as they unveiled a huge roadside billboard, designed and painted by Mike Kicul. The sign said, *No O.W.M.C. Hazardous Toxic Waste Facility*. The final stroke of the giant "N" had been shaped like a smokestack. Over the "N," a huge hand giving the thumbs-down-gesture had plugged the stack with its thumb.

Regional Chairman Wilbert Dick, after walking with the marchers, had gone inside to observe the proceedings. Summoned outside to speak to the protesters, he reached for the hand-held microphone. "They're in there in the warmth, in their pin-striped suits and we're out here in the cold. They want to *leave* the residents out in the cold." Cheers punctuated each statement. "That's why the hearing is only beginning here before it moves to Oakville, where none of you can attend. The government doesn't want you there." Louder and louder came the cheers from the crowd. "If OWMC builds this facility in the bread basket of Ontario, the rest of Ontario or Canada will not want the produce that's grown here!" he warned as whistles and shouts cut the frosty air.

John Jackson spoke last, his hands stuffed into the pockets of his parka. "The OWMC has accomplished virtually nothing in the ten years it has existed. What they've done is spend $100-million of our tax dollars...They've produced tons of reports that have no substance to them. They've sidetracked the province of Ontario from going after the real solution to hazardous waste—which is not to produce it in the first place."

His head thrown back to project his voice, he told the crowd, "In the ten years the OWMC has existed, Ontario industries have changed. They now realize eliminating the toxic waste problem means to stop producing the materials. OWMC, on the other hand has remained in the dark ages...Opponents will have to prove *'There is a Better Way.'* You are the people who have found and fought for that solution. It isn't the lawyers. It isn't the technical experts. You're told you have no power. We are *not* powerless!" he declared as the crowd cheered and shook their signs in agreement. "...But continued opposition is essential if we are going to convince the provincial government to change its way of thinking."

As Marilyn gazed over the assemblage, she saw Niagara regional councillors and West Lincoln aldermen, their noses red from the cold. She saw people of all ages. Incredibly, even elderly Ella Krick had walked in the parade.

With the rally over, reporters swarmed around the speakers, seeking quotes to add to their notes. John Jackson reminded Shaun Smith of the *Guardian Express* that the hearing board, at the request of the intervener groups, had instructed the OWMC to downsize its witness panels from 250 to 160 individual [appearances]. "This decision comes as a major victory for us...Never before has a hearing panel instructed a proponent to restructure their case to the extent they did...The reduction," he explained, "will prevent [the OWMC] from dragging the hearings on in an attempt to exhaust the opponents' funding."

Close by, Paul Hallihan of the *Hamilton Spectator* interviewed Edith Hallas. "OWMC threw down the gauntlet, and we've answered," she declared. "I think this was a fantastic turnout for a cold Monday morning."

With the rally over and the cheering done, Clifford and Marilyn joined the throng of protesters crowding into the welcome warmth of the Old Farm Inn. Still carrying her sign, Marilyn squeezed past other spectators and sat in a chair behind the railing of a small raised area near the windows. Ian Blue, the lawyer for the OWMC, had already launched into his presentation. "The third point I want to make is that..."

Day One of the dreaded hearings had begun.

– Sources of Information –

This account is as historically and chronologically accurate as I could make it, with several notable exceptions:
- In chapter 2, the meeting at Wellandport Hall was held on June 4, 1984, not in the fall of 1985.
- Also in chapter 2, the meeting in Vineland Public School was held on October 31, 1986, not in October of 1985.
- In places, when an exact chronology would have confused the reader, related events are clumped slightly out of sequence.
- All stated dates are exact.

As well as the sources of information listed below, I also relied upon the meticulous minutes of the Concerned Citizens meetings taken by Marie Austin, the minutes and notes from Coalition meetings, old calendars, and other documents too numerous to list.

Chapter 1
"Residents' Fears are Confirmed as Schram Site is Selected," *West Lincoln Review,* 27 September 1985, p. 1.

"Knows it won't be popular Chant outlines reasons for OWMC decision," *West Lincoln Review,* 27 September 1985, p. 2.

"Task Force promises to battle proposal," *West Lincoln Review,* 27 September 1985, p. 1.

"Quick fact' summary on OWMC proposal," *West Lincoln Review,* 27 September 1985, p. 4.

"Quick fact' summary on OWMC proposal, Major Characteristics of Site LF-9C, "*West Lincoln Review,* 27 September 1985, p. 4.

Chapter 2
Chant, D. A., "Chant responds to delegation ," *Pelham Herald,* 26 October 1986.
"OWMC Communications Director fields questions," *Niagara Farmers Monthly,* 4 June, 1984.
"Chant speaks at Brock," *Guardian Express,* 19 November 1986, p. 13.
Draper, Doug, "West Lincoln promised 'crown jewel'," *The Standard,* 18 November 1986.
OWMC, "Facilities Development Process, Phase 3 Report," March 1984.
Tait, Eleanor, "In our back yard but on your dinner table," *Spectator,* 1 November 1986.
"Andrewes, Chant discuss concerns over OWMC plan, "*West Lincoln Review,* 31 December 1985, p.1.

Chapter 3
"New Mayor Claims Difference in Philosophy—Disbands Previous Committee As Too Radical."
Haynes, Mr. and Mrs., "The OWMC battle: What's going on?," *West Lincoln Review,* 24 December 1985.
"Revised committee with battle OWMC," *West Lincoln Review,* 24 December 1985.
(photo) "OWMC opponents urge attendance at anti-waste rally Saturday," *West Lincoln Review,* 9 July 1986, p. 1.
Draper, Doug, "Site for hazardous waste facility passes soil test," *The Standard,* 8 August 1986, p. 1.

Chapter 4
Jackson, J., Weller, P., & the Waterloo Public Interest Research Group, "Chemical Nightmare," Toronto, Between The Lines, 1982.

OWMC—The "Wasted" Years: The Early Days

Jackson, John A., Resume
Draper, Doug, "Questionnaire riles resident," *The Standard*, 11 July 1986.
"Residents buck survey, say questions 'obscure'," *West Lincoln Review*, 17 September 1986.

Chapter 5
"Speakers lash OWMC plan at "People Power' event," *West Lincoln Review*, 23 July 1986, p. 1.
"People Power Rally, Dump the dump," *Pelham Herald*, 23 July 1986, p.1.
(Photo) "CHANT: Schram from God's Farm Country—We don't need Chemical Burgers, Smoke Stack Stew, or Fall-out Hot Cakes, *West Lincoln Review*, 23 July 1986, p. 12.
"Speakers lash..." *Review*, 1986.
"People power lining up for a long battle," *Guardian Express*, 23 July 1986, p.5.
"If You Wait Till After The Dump Is Here it Will be Too Late," *Niagara Farmers' Monthly*, August 1986.
"If You Wait," *Niagara Farmers' Monthly*, 1986.
"OWMC open houses, OWMC wants to hear your views," *West Lincoln Review*, 23 July 1986.
"West Lincoln, not Waste Lincoln," *Pelham Herald*, 6 August 1986.
"Hansard, Official Report of Debates, Legislative Assembly of Ontario, Standing Committee on Resources Development, Estimates, Ministry of the Environment, Second Session, 33rd Parliament, Speaker: Honourable H.A. Edighoffer, Clerk of the House: R.G. Lewis, QC," 26 June, 1986, p. R-94.
Di Ramio, Al, Migus, John, & Jackson, John, "Position Paper' on toxic waste," *Pelham Herald*, 30 July 1986.
Tilden, Deryk, "Love Canal analogy unfair, Chant states after rally," *West Lincoln Review*, 30 July 1986, p. 1.
Gracey, Clifford B., Gracey, Marilyn J., based on taped conversations of Open House at Wellandport, July 31, 1986.
Hansard, p. R-93.

Chapter 6

"OWMC open houses," 23 July 1986.

Nelson, Norman, "OWMC projects "coming down to the wire'," *Pelham Herald*, 6 August 1986.

Gracey, Clifford B., Gracey, Marilyn J., "Wastes only slightly toxic?" *West Lincoln Review*, 17 September 1986.

Hansard, R-90.

Hansard, R-95.

Tait, "In our back yard,"1 November1986.

Gracey, Clifford B., Gracey, Marilyn J., "Still riled over OWMC questionnaire," *Pelham Herald*, 7 January 1987.

Chapter 7

Nelson, Norman, "Taking on the OWMC," *Pelham Herald*, 10 September 1986

Ramsay, Laura, "W. Lincoln residents vow to fight dump," *The Hamilton Spectator*, 5 September 1986.

Dunlop, Greg, "Not the only choice' claims the task force," *The Tribune*, 13 September 1986.

Giovannone, Frank, "We the people have the last say," *Pelham Herald*, 1 October 1987, p. 20.

McEwen, Judy, "It's time for us to survey OWMC," *West Lincoln Review*," 17 September 1986

Chant, D. A., (Speech to public health inspectors)
Part one: "Demanding action before it's too late," *Pelham Herald*, 1 October 1986.
Part two: "Ontario badly needs waste disposal facility," *Pelham Herald*, 8 October 1986.
Part three: "Nothing in our world is failsafe," *Pelham Herald*, 15 October 1986.

"Scuttling toxic waste dump would be very serious," *Pelham Herald*, 24 September 1986, p. 1.

"Balloon release," *Niagara Farmers' Monthly*, October 1986.

Chant, D. A., "Experts say rotary kiln "state-of-the-art'—Chant," *West Lincoln Review*, 10 September 1986.

Chant, "Chant responds to delegation."

Gracey, Clifford B., Gracey, Marilyn J., "OWMC misleading," *Guardian Express,* 22 September 19986.

Hansard, p.R-94.

"More OWMC answers to residents' questions," *West Lincoln Review,* 29 October 1986.

Scott, Michael, "Capacity design realistic," *West Lincoln Review,* 8 October 1986.

Chant, "Chant responds to delegation."

Gracey, Clifford B., Gracey, Marilyn J., "Wastes stockpiled while facility under construction," *West Lincoln Review,* 15 October 1986, p. 5.

Hansard, p. R-104, R-105.

"Task Force wins first place," *Pelham Herald,* 8 October 1986.

Chapter 8

Chant, Part three: "Nothing in our world is failsafe."

Gracey, Clifford B., Gracey, Marilyn J., "Sticking to the facts," *Pelham Herald,* 5 November 1986, p. 4.

Hansard, R-86; R-96.

Chant, D. A., "Weigh facts, ask questions," *West Lincoln Review,* 17 September 1986.

Draper, Doug, "Toxic Waste Fight Has Changed Their Lives," *The Standard,* 24 October 1986

"Politicians to tour 'sites'," *West Lincoln Review,* 1 October 1986.

Chapter 9

Chant, "Chant responds to delegation."

Forsyth, Paul, "Waste facility could aid area: Chant," *The Tribune,* 6 November 1986.

OWMC, "Final Report Site Selection Process Phase 4A Selection of a Preferred Site(s) Engineering," May 1986, table 2.2, p. 47.

Committee, West Lincoln task Force Communication, "Task Force formed to oppose OWMC," *Pelham Herald,* 29 October 1986, p. 2.

"Waste meeting tonight," *West Lincoln Review,* 5 November 1986, p. 1.

"Pollution Probe," *The Standard*, 8 November 1986.

"OWMC here Thursday," *Pelham Herald*, 5 November 1986.

Notice: "OWMC Drop-in Centres," *Pelham Herald*, 19 November 1986, p.6.

"Answer to toxic waste problem is no mystery: OWMC chairman," *The Tribune*, 7 November 1986, p. 6.

O'Brien, Paul, "Chant dangles OWMC carrot," *Guardian Express*, 8 November 1986, p.3.

"Construction, operation of site will result in economic benefits," *The Tribune*, 6 November 1986, p. 6.

"Room at the trough," *West Lincoln Review*, 19 November 1986.

O'Brien, Karen, "Grass roots waste battle is growing," *West Lincoln Review*, 12 November 1986, p. 7.

Draper, Doug, "Toxic waste facility opponents still waiting for provincial cash," *The Standard*, 31 December 1986, p. 20.

Nelson, Norman, "If we get the dump, give us the goodies," *Pelham Herald*, 12 November 1986.

O'Brien, Paul, "No more Mr. Nice Guy as Dick vows to get tough with OWMC's chairman," *Guardian Express*, 21 November 1986, p. 3.

Chant, D. A., "Chant Outlines Toxic Waste Problems In Speech To Chamber," *Niagara Farmers' Monthly*, December 1986, p. 8.

O'Brien, "Chant dangles OWMC carrot."

Sadleir, Dick, "Unity urged in OWMC battle," *The Standard*, 25 November 1986.

Chapter 10

O'Brien, Karen, "Dick slams Chant over funding issue," *West Lincoln Review*, 26 November 1986

Herod, Doug, "Region takes on OWMC—Dick maintains Niagara won't foot $355,000 bill," *The Standard*, 21 November 1986, p. 1.

Forsyth, Paul, "No leachate for minimum of 800 years," The Tribune, 27 November 1986, p. 5.

"We already have free trade, unhappily, it's in pollution," The

Tribune, 28 November 1986, p.1.

Nelson, Norman, "Environmentalist says risks underestimated," *Pelham Herald*, 3 December 1986, p. 1.

O'Brien, Karen, "Crowd is sparse at people Power Convention," *West Lincoln Review*, 3 December 1986, p. 1.

Gracey, Clifford B., Gracey, Marilyn J., "Public opinion opposed to dump," *Pelham Herald*, 17 December 1986.

Herod, Doug, "OWMC fight coordinated," *The Standard*, 5 December 1986, p. 8.

"Another balloon is found," *West Lincoln Review*, 10 December 1986.

Nelson, Norman, "Government should get off high horses," *Pelham Herald*, 17 December 1986, p. 4.

"Andrewes: September election, waste is the issue," *West Lincoln Review*, 7 January 1987, p. 1.

O'Brien, Karen, "OWMC to be the issue, says Gladys Huffman," *West Lincoln Review*, 7 January 1987.

"Lincoln rep fears council will be "sold' on OWMC plan," *West Lincoln Review*, 8 January 1987.

O'Brien, Paul, "Chant certain OWMC has "the best site possible"," *Guardian Express*, 17 January, 1987, p. 9

Forsyth, Paul, "OWMC releases 22-volume EA document Next step: hearings Hopefully by year's end: Chant," Niagara South Shopping News, 20 January 1987.

"Snowed under with words," *Lincoln Post Express*, 21 January 1987.

"Start-up date unknown," *Niagara South Shopping News*, 20 January 1987.

Chant, D. A., "Text of remarks by OWMC chairman Dr. Chant," *The Standard*, 16 January 1987, p. 7.

Chapter 11

"Spicing up the draft E.A.," *West Lincoln Review*, 28 January 1987.

Draper, Doug, "Accelerated hearing plans anger OWMC site foes," *The Standard*, 10 January 1987, p. 9.

Draper, Doug, "Let the hearings start: Chant," *The Standard*, 16 January 1987, p.1.

Draper, Doug, "Dr. Chant's toxic waste report defies province," *The Standard*, 22 January 1987.

O'Brien, Paul, "Dick Accuses OWMC Of Stalling," *Guardian Express*, 17 January 1987, p.1.

Gracey, Clifford B., Gracey, Marilyn J., "Citizens should be wary of staunch assurances from West Germany," *Lincoln Post Express*, 4 February 1987, p. 5.

"Air monitoring continues at OWMC proposed site," *West Lincoln Review*, 11 February 1987.

Gracey, Clifford B., Gracey, Marilyn J., "Deliver us, Lord, from OWMC," *West Lincoln Review*, 4 February 1987.

II Chronicles 20:12; 6; 15-17; 20; 21.

II Chronicles 20:12.

I Samuel 3:13

Chapter 12

Draper, Doug, "PCB threat closes reservoir," *The Standard*, 13 February 1987, p. 1.

Draper, Doug, "Toxin reaches Smithville well," *The Standard*, 23 February 1989, p. 1.

Draper, Doug, "PCBs—Smithville's dilemma for nearly a decade," *The Standard*, 27 February 1987, p. 14.

Draper, Doug, "OWMC consultant says criticism unjustified," *The Standard*, 19 March 1987, p. 9.

Forsyth, "No leachate for minimum of 800 years."

Draper, Doug, "OWMC expert's credibility questioned at forum," *The Standard*, 16 February 1987.

O'Brien, Paul, "OWMC Distances Itself From PCB Spill," *Guardian Express*, 18 February 1987, p. 1.

"Residents present petition, demand council support," *West Lincoln Review*, 18 February 1987, p. 1.

Nelson, Norman, "A honeymoon with the voters," *Pelham Herald*, 17 February 1987.

Dykstra, John, "No safe dump in the world," *Pelham Herald*,

25 February 1987, p. 17.
O'Brien, Karen, "Premier supports OWMC," *West Lincoln Review*, 4 March 1987.
II Chronicles 20:6; 30.

Chapter 13
Gracey, Clifford B., Gracey, Marilyn J., "Stand fast and fight hard," *Guardian Express*, 18 March 1987.
"Council will join trek," *West Lincoln Review*, 4 March 1987.
"Citizens to lobby Region," *West Lincoln Review*, 18 March 1987.
"Citizens want Region's support in OWMC protest," *The Standard*, 20 March 1987, p. 9.
Environment Canada, "Storm Warning".
Gracey, Clifford B., Gracey, Marilyn J., Letter, 5 March 1987, p. 2.
Environment Canada, "Storm Warning".

Chapter 14
II Corinthians 4:8.
II Corinthians 7:5.
Editorial, "Who said, Get on with it?", *West Lincoln Review*, 25 March 1987.
"OWMC two-stage hearing proposal vetoed," *West Lincoln Review*, 11 March 1987.
Gracey, Clifford B., Gracey Marilyn J., "Strings attached to Intervenor funding," *Pelham Herald*, 22 April 1987.
Public Notice, "Township of West Lincoln Toxic Waste Committee Notice of Public Meeting," *Pelham Herald*, 25 March 1987, p. 12.
McCallum, David R., "Public meeting to be held April 8," *Pelham Herald*, 25 March 1987, page 17.
"Few residents greet Township consultants," West Lincoln Review, 15 April 1987, p. 1.

Chapter 15
D.A. Chant, "Questionnaire Questionable," *West Lincoln Review*, 15 April 1987.
"3 say "Get on with it' in OWMC survey," *West Lincoln Review*, 25 March 1987, p. 1
Gracey, Clifford B., Gracey, Marilyn J., "We hold land in trust for future generations," *Pelham Herald*, 13 May 1987.
Gracey, Clifford B., Gracey, Marilyn J., "Entrepreneurs will use us, leave," *West Lincoln Review*, 20 May 1987.
McEwen, Judy, "OWMC "excellent for business' says new Chamber president," *West Lincoln Review*, 29 April 1987.
"Help for stressed politicians," *West Lincoln Review*, 29 April 1987.
Nelson, Norman, "A formidable battle against Dr. Chant," *Pelham Herald*, 29 April 1987, p.4.
Nelson, Norman, "Delegation against the OWMC," *Pelham Herald*, 29 April 1987, p.4.
Isaiah 1:18.

Chapter 16
"Consultant should explain remarks," *West Lincoln Review*, 6 May 1987, p.1.
"Residents grill consultant," *West Lincoln Review*, 13 May 1987, p.1.
"Council to support fight against toxic waste site," *Guardian Express*, 25 April 1987, p. 17.
"Lincoln Liberals are just wild about Harry," *West Lincoln Review*, 20 May 1987, p. 1.

Chapter 17
Chant, D.A., "Can't wish problems away," *Pelham Herald*, 6 May 1987.
Gracey, Clifford B., Gracey, Marilyn J., "OWMC uses statistics that suit," *West Lincoln Review*, 27 May 1987.
OWMC, "Environmental Assessment," p. 119.
"Waste Quantities Study," p. 59.

Gracey, Clifford B., Gracey, Marilyn J., "Worried about spills," *The Tribune*, 13 June 1987.

Draper, Doug, "Cutting hazardous waste saved industry $400 million," *The Standard*, 29 May 1987.

"Mayor and residents at Queen's Park today," *West Lincoln Review*, 3 June 1987, p. 1.

Dykstra, John, & Gracey, Marilyn, "Totally opposed to OWMC dump," *Pelham Herald*, 27 May 1987.

Forsyth, Paul, ´OWMC opponents will get their audience with Peterson, Bradley," *The Tribune*, 21 May 1987.

OWMC opponents set up meeting with premier," *Niagara Farmers' Monthly*, June 1987, p.10.

"Local group has a date with Peterson," *Pelham Herald*, 27 May 1987, p. 1.

"Ontario's bread basket threatened," *The Tribune*, 21 May 1987.

Draper, Doug, "Auto workers join convoy to protest disposal facility," *The Standard*, 2 June 1987.

Forsyth, Paul, "Anti-waste site convoy heading to Queen's Park," *The Tribune*, 3 June 1987.

Chapter 18

Photo, *West Lincoln Review*, 10 June, 1987.

"West Lincoln Residents & Supporters Protest OWMC Proposal at Queen's Park – June 3," *Niagara Farmer's Monthly*, July 1987, p. 18.

Photo, Nelson, Norman, "A trip to Queen's Park," *Pelham Herald*, 10 June 1987, p. 18.

Photo, Nelson, Norman, "A trip to Queen's Park," *Pelham Herald*, 10 June 1987, p. 18.

Photo, Argue, Kevin, *The Spectator*, 4 June 1987, p. D1.

"West Lincoln Residents & Supporters Protest OWMC Proposal at Queen's Park – June 3," *Niagara Farmer's Monthly*, July 1987, p. 18.

Tait, Eleanor, "Queen's Park rally eases residents' toxic waste fears," *The Spectator*, 4 June 1987, p. D1.

Forsyth, Paul, "Opposition slams toxic waste plan," *The Tribune*, 4 June 1987, p. 5.

"Couldn't give land away," *The Tribune*, 4 June 1987, p.6.

Draper, Doug, "Residents carry waste protest to Queen's Park," *The Standard*, 4 June 1987, p. 10.

Forsyth, Paul, "Liberals unwilling, unable to deal with toxic waste problem," *The Tribune*, 4 June 1987.

"Couldn't give land..." *The Tribune*, 1987.

Nelson, Norman, "A trip to Queen's Park," *Pelham Herald*, 10 June 1987, p. 18.

"Choose other alternatives, says West Lincoln mayor," *The Tribune*, 5 June 1987.

Nelson, Norman, "Venting frustration at Queen's Park," *Pelham Herald*, 10 June 1987, p. 1.

"Stop game of "environmental roulette," *The Tribune*, 5 June 1987.

Nelson, Norman, "Venting frustration at Queen's Park," *Pelham Herald*, 10 June 1987, p. 1.

Forsyth, Paul, "Opposition support hailed as victory," *The Tribune*, 4 June 1987, p.5.

Nelson, Norman, "A trip to Queen's Park," *Pelham Herald*, 10 June 1987, p. 18.

Janes, Gracia, Poem, "Let us walk with a light foot upon the land." (By Permission—March 2, 2005)

Coffman, Margery, "A Prayer." (by permission—February 23, 2005)

Draper, Doug, "Residents carry waste protest to Queen's Park," *The Standard*, 4 June 1987, p. 10.

Forsyth, Paul, "Uniform causes problem for rally participant," *The Tribune*, 4 June 1987.

Draper, Doug, "Residents carry waste protest to Queen's Park," *The Standard*, 4 June 1987, p. 10.

"Bradley assures public hearings will be fair," *The Tribune*, 4 June 1987.

Draper, Doug, "Residents carry waste protest to Queen's Park," *The Standard*, 4 June 1987, p. 10.

Nelson, Norman, "A trip to Queen's Park," *Pelham Herald*, 10 June 1987, p. 18.

Tait, Eleanor, "Queen's Park rally eases residents' toxic waste fears," *The Spectator*, 4 June 1987, p. D1.

"Bradley assures public hearings will be fair," *The Tribune*, 4 June 1987.

Chapter 19

Draper, Doug, "Lawyer says provincial funding too little for toxic waste hearings," *The Standard*, 26 June 1987.

Gracey, Marilyn, personal notes re: Intervenor Funding, 10 June 1987.

Chapter 20

Sadleir, Dick, "Lincoln fire chief worried about chemical spills on route through urban Vineland or Beamsville," *The Standard*, 2 January 1987, p. 28.

Gracey, Clifford B., Gracey, Marilyn J., "OWMC makes big plans," *Niagara Falls Review*, 30 June 1987.

Mitchell, John, "Careful study, planning lead to choice of waste disposal site," *The Shoppers Guide*, Dunnville Chronicle, 3 June 1987

Nelson, Norman, "Chant is doing a great job," *Pelham Herald*, 24 June 1987.

Draper, Doug, "OWMC invites Niagara politicians to tour European toxic waste plants," *The Standard*, 24 June 1987, p. 4.

"The Toxic Tourist Trade," *West Lincoln Review*, 12 August 1987.

"At least two to take OWMC trip," *West Lincoln Review*, 12 August 1987, p. 2.

"Treasurer joins trip to waste facilities," *West Lincoln Review*, 2 September 1987, p. 2.

Chant, D. A., Letter to John Jackson re: no money for Coalition members to go on a trip to Europe, 17 November 1987.

Gracey, Clifford B., Gracey, Marilyn J., "Expansion inevitable?," *The Tribune*, 15 July 1987.

OWMC drawing, "Perspective of OWMC Facility," OWMC and the West Lincoln Community: Managing Change, July 1987, p. 5.

Micak, Jim, project manager environmental projects OWMC, "3_ acres a year for burial," *West Lincoln Review*, 26 August 1987.

Gracey, Clifford B., Gracey, Marilyn J., "The famous 'half twist'," *Pelham Herald*, 7 October 1987.

"Waste site is nuisance to 56 homes report says," *The Spectator*, 11 July 1987.

OWMC, Map, "Basis for OWMC's Nuisance Impact Zone.

Chapter 21

"Noise, dust will plague residents near OWMC site," *West Lincoln Review*, 15 July 1987, p. 2.

Nelson, Norman, "OWMC recommends a 'nuisance zone'," *Pelham Herald*, 15 July, 1987, p. 1.

McCallum, David R., "Property values down; taxes up," Pelham Herald, 15 July 1987.

"NDP led by Fonthill resident," *Pelham Herald*, August 1987, p.1.

Draper, Doug, "Waste site neighbors excluded," The Standard, 15 August 1987.

McEwen, Judy, "Planeload of politicos will head for Europe," *West Lincoln Review*, 19 August 1987, p. 1.

"Toxic tour travelogue," *West Lincoln Review*, 26 August 1987

"Pelissero says he's the man," *Pelham Herald*, August 1987, p. 1.

Chapter 22

"Citizens' float places first," *West Lincoln Review*, September 16, 1987.

Romans 8:28

Forsyth, Paul, "Plenty of problems at European facilities," *The Tribune*, 2 October 1987, p. 6.

Draper, Doug, "Group "favorably impressed' by toxic waste incinerators," *The Standard*, 23 September 1987.

II Chronicles 20:6
Draper, Doug, "A long-distance look," *The Standard*, 26 September 1987, p. 15.

Chapter 23
Colyn, Allard, (Mayor), "Press Release on European Hazardous Waste Tour," 28 September 1987.
"Plenty of problems..." *The Tribune*, 1987.
Gracey, Marilyn, Personal notes re: Mayor's Press Conference, 28 September 1987.
Draper, Doug, "Stance remains after toxic tour," *The Standard*, 30 September 1987, p. 9.
Tait, Eleanor, "European toxic plants don't impress mayor," *The Spectator*, 30 September 1987.
Chant, D. A., "Myths' compel Chant to reply," *West Lincoln Review*, 12 August 1987.
Gracey, Clifford B., Gracey, Marilyn J., "Where is Premier's logic?", *Niagara Falls Review*, 7 October 1987.
"Peterson promises not to sell the farm," *The Tribune*, 13 August 1987.
"Peterson confirms stand on OWMC," *The Guardian*, 26 August 1987.
Draper, Doug, "West German farmers feel waste sites a threat," *The Standard*, 9 September 1987.
Chant, D. A., "OWMC prefers "proven' technology," *Pelham Herald*, 27 May 1987.
Draper, Doug, "Tricil kiln may compete with OWMC," *The Standard*, 6 October 1987, p. 1.
Public Notice: "Township of West Lincoln Toxic Waste Committee," October 1987.
West Lincoln, The Township of; Municipality of Niagara, The Regional, "Joint Review of OWMC's Waste Quantities Study, 21 October 1988.
Draper, Doug, "Property values drop near proposed OWMC site," *The Standard*, 10 October 1987, p. 1.

Chapter 24

The Standard, September 12, 1987.

Gracey, Clifford B., Gracey, Marilyn J., "OWMC free to change the rules," *West Lincoln Review*, 4 November 1987.

Draper, Doug, "OWMC scientist says health risk 'acceptable'," *The Standard*, 16 October 1987.

Gracey, Clifford B., Gracey, Marilyn J., *West Lincoln Review*, 30 December 1987.

Jackson, John, "Waste Watch," Niagara Farmers' Monthly, November 1987, p. 16.

Chant, D. A., "Chant quoted out of context," *West Lincoln Review*, 2 December 1987.

Gracey, Clifford B., Gracey, Marilyn J., "OWMC's famous 'half twist'," *Pelham Herald*, 3 February 1988.

May, Shirley Ruth, "Fiddler tunes up his memories," *The Tribune*, 18 November 1987, p. 12.

"Consultants to air findings," *West Lincoln Review*, 2 December 1987.

"Be wary of duplication," *The Tribune*, 5 December 1987.

"Regional Council looking for ways to speed up OWMC hearings," *The Times-Review*, 8 December 1987, p. 2.

O'Brien, Paul, "Is the OWMC battle with Region a consultant's dream?," *Guardian Express*, 12 December 1987, p. 14.

O'Brien, Karen, "Region wants quick EA hearings," *West Lincoln Review*, 9 December 1987, p. 1.

Gerritsma, Joop, "OWMC process too slow: Region," *The Tribune*, 5 December 1987.

Dick, D. N., Letter to The Honourable James Bradley re: speed process, 29 December 1987.

Hallihan, Paul, "Spirit of willing volunteers "greatest gift of all' for family," *The Standard*, 21 December 1987, p. 1.

"Manual helps reduce wastes," *Pelham Herald*," 29 December 1987.

"Abuse, growth in 1988," *West Lincoln Review*, 30 December 1987.

Hallihan, Paul, "Spirit of willing volunteers "greatest gift of all' for family," *The Standard*, 21 December 1987, p. 1.

Manual helps reduce wastes," *Pelham Herald*, 29 December 1987.

"Abuse, growth in 1988," *West Lincoln Review*, 30 December 1987.

Chapter 25

May, Shirley Ruth, "Chemical-free gardening helps couple fight son's allergies," *The Tribune*, 7 January 1988.

"OWMC facility would take pressure off industry, government: Jackson," *The Tribune*, 13 November 1987, p. 6.

Forsyth, Paul, "Build small regional waste plants – Jackson," *The Tribune*, 4 March 1988, p. 6.

Chant, D. A., "Formula for Disaster," *The Tribune*, 9 December 1987.

"Citizens' groups developing new ways to handle wastes effectively: Jackson," *The Tribune*, 10 December 1987.

Gracey, Clifford B., Gracey, Marilyn J., "Chant's position doesn't convince the sensible," Times-Review, 12 April 1988.

Draper, Doug, "OWMC ready for showdown," *The Standard*, 9 January 1988.

"Deep Clay not needed for dump – consultant," *West Lincoln Review*, 10 February 1988.

OWMC drawing, "perspective of OWMC's proposed industrial waste treatment & disposal facility," OWMC Exchange, January 1987, p. 5.

Chapter 26

OWMC, Background Material for the Ontario Waste Management Corporation Phase 4B Public Consultation Site Assessment Meetings, Appendix F, Comparative Evaluation of Facility Refinement Options, Table G.1, July 1987, p. 19

Draper, Doug, "Above-ground storage of waste 'safer' than burial," *The Standard*, 5 March 1988.

Draper, Doug, "Compensation area too small, critics charge," *The Standard*, 8 march 1988.

Draper, Doug, "Waste plant plagues residents: OWMC foes," *The Standard*, 16 March 1988.

Jackson, John, "OWMC no better than Tricil," *West Lincoln Review*, 16 March 1988.

Chapter 27

"Ontario first, then U.S. waste—OWMC," *West Lincoln Review*, 27 January 1988.

"NIMBY term indefensible," *West Lincoln Review*, 27 January 1988.

May, Shirley Ruth, "Lessons not necessary for Fenwick cartoonist," *The Tribune*, 27 February 1988, p. 6.

Draper, Doug, "Treatment plant state-of-the-art, Chant argues," *The Standard*, 1 March 1988, p. 10.

"OWMC chemicals add to traffic," Lincoln Post Express, 23 March 1988.

Gracey, Marilyn J., Personal Notes re: Mock hearing, 9 March 1988.

Patterson, Theresa, "Larger OWMC compensation area backed by Lincoln town council," *Lincoln Post Express*, 23 March 1988.

Nelson, Norman, "Selling Fenwick to the OWMC?," *Pelham Herald*, 30 March 1988.

Roik, Richard, "OWMC told to buy Fenwick, *Guardian Express*, 23 March 1988.

"Selling Fenwick..." *Pelham Herald*.

"Larger OWMC compensation area..." *Lincoln Post Express*.

Jackson, John, News Release.

Wilson, Patricia, "Wainfleet residents explain fears of OWMC," Dunnville Chronicle, 13 April 1988.

Lees, Janet, "Waste site planning faulty, township's consultants report," *The Spectator*, 13 April 1988.

Draper, Doug, "OWMC giving us short shrift, says West Lincoln," *West Lincoln Review*, 13 April 1988.

OWMC—The "Wasted" Years: The Early Days

Chapter 28

"Thorold mayor wants toxic waste facility," *The Spectator*, 22 April 1988.

Herod, Doug, "Thorold mayor would welcome OWMC plant," *The Standard*, 22 April 1988, p. 9.

Schilstra, John (Alderman), Letter re: presentation to Region of Niagara, 25 April 1988.

"Thorold would welcome waste site," *The Spectator*, 22 April 1988.

Gerritsma, Joop, "Thorold mayor disappointed city not considered for dump," *The Tribune*, 22 April 1988.

O'Brien, Paul, "OWMC welcome to build toxic waste plant in Thorold," *Guardian Express*, 23 April, 1988.

Carfagnini, Paul, "Thorold councilor "deeply shocked' Longo welcomed toxic waste facility," *The Tribune*, 6 May 1988.

Krick, Ella, Letter of encouragement, 25 April 1988.

Draper, Doug, "Battle for compensation requires facts—Region," *The Standard*, 11 May 1988.

"OWMC salaries," *West Lincoln Review*, 11 May 1988, continued from p. 1.

Chapter 29

Citizens, Concerned, "Queen's Park, Presentation: Ministry of the Environment, 9 June 1988.

May, Shirley Ruth, "Waste-site neighbours feel like 'pawns'," *The Tribune*, 10 June 1988.

Lees, Janet, "Township should pull out of waste site battle: Mayor," *The Spectator*, 9 June 1988, p. C5.

Dykstra, John, "OWMC does not consider farmers," *West Lincoln Review*, 22 June 1988.

Garr, Mary Lou, "You can't please everyone," *West Lincoln Review*, 22 June 1988.

Chapter 30

OWMC Map, Fig. 1 Approximate Rye Grass locations

Daniels, Leslie, "OWMC replies to citizens' groups," *Pelham*

Herald, 13 July 1988.

Dykstra, John, "Better way say Concerned Citizens group," *Pelham Herald*, 3 August 1988.

Daniels, Leslie, Letter requesting position papers, 14 July 1988.

Dykstra, John, "Better way..."

"Chant on International Great Lakes commission," *West Lincoln Review*, 13 May, 1987, p. 11.

"Chant re-appointed," *The Tribune*, 13 July 1987.

"OWMC boss joins Order of Canada," *The Standard*, 15 July 1988, p. 10.

– awarded Thurs. July 14, 1988.

Sexton, Rosemary, "Peterson gives dinner for Michigan Governor," Globe and Mail, 21 April 1988, p. A2.

Chapter 31

Lemon, Edgar, "Fleecing the taxpayer," *The Standard*, 4 July 1988.

II Chronicles 20:12.

"Share it with the world," Pelham Herald, 30 March 1988, p. 1.

Bergsma, Marlene, "Council turns down grant for group opposing OWMC" *The Standard*, 26 July 1988, p. 10.

Gracey, Marilyn J., Personal notes, Funding Hearing, 3 & 4 August 1988.

Chapter 32

Jackson, John, Letter to Dr. Chant re: funding for Coalition, 22 August 1988.

Scott, Michael, Letter to John Jackson re: no funding for Coalition, 30 August 1988.

Daniels, Leslie, Letter to Residents re: excavating test pits, 1 September 1988.

Gracey, Marilyn, visit to test pits based on personal notes, 8 September 1988.

Draper, Doug, "OWMC critics' demands 'astonishing'," *The Standard*, 15 September 1988.

"No cash yet for OWMC foes," *West Lincoln Review*, 21 September 1988.

Gracey, Marilyn, Personal notes at Legion Hall, Smithville, 14 September 1988.

Martin, Karen, "Health department supports monitoring of waste facility," *Guardian Express*, 28 September 1988, p. 17.

"Coalition grant," *Pelham Herald*, 28 September 1988.

"Destroy waste at origin point," *The Tribune*, 7 September 1988, p. 6.

Chapter 33

Bradley, Lynda, "OWMC has failed to respond to public," *The Standard*, 25 August, 1988, p. 3.

Chapter 34

Gracey, Clifford B., Gracey, Marilyn J., "OWMC opponents not 'fleecers'," *West Lincoln Review*, 26 October 1988.

Gracey, Marilyn, Personal Notes re: Leslie Daniels, Brock University, 11 October 1988.

"OWMC to Honour Successful Waste Reduction," *The Shoppers Guide*, 11 October 1988, p. 8.

"Still no cash for hearings," *West Lincoln Review*, 19 October 1988, p. 1.

Draper, Doug, "OWMC foes clash over cost-cutting proposal," *The Standard*, 14 October 1988, p. 9.

Chapter 35

West Lincoln, The Township of, & Niagara, The Regional Municipality of, "Joint review of OWMC's Waste Quantities Study, 3.3 Experience in Other Jurisdictions," 21 October 1988, p. 3-4.

Gracey, Marilyn, Personal Notes taken from taped Interview with Dr. Chant & John Dykstra on radio station CBC's *Radio Noon* with Donna Tranquada, December 2, 1988.

Gracey, Marilyn, Personal Notes taken from taped Interview with John Dykstra on radio station CKOC. December 1988.

Giovannone, Frank, "German newspapers shed doubt on

OWMC's "state of art' claim," *West Lincoln Review*, 9 July 1986.

"Four processes of waste treatment involved," *Niagara Falls Review*, 10 December 1988.

Jackson, John, Letter re: Coalition needs money to pay bills, 12 December 1988.

McKay, Lloyd, "Township has no clout at Region," *West Lincoln Review*, 14 December 1988.

West Lincoln Review, 14 December 1988.

Jackson, John, Letter re: Coalition received Intervenor funding, 17 December 1988.

Gracey, Clifford B., Gracey, Marilyn J., "Wastes are imported," *The Tribune*, 14 January 1989, p. 4.

"Import record is set for hazardous wastes," *The Standard*, 13 September 1988.

"Jugs lose out," Farm and Country, 15 November 1988.

Coalition Flyer, video, "The Turnaround Decade."

Chapter 36

"Township gets scraps in funding allocation," *West Lincoln Review*, 7 December 1988, p. 1.

McEwen, Judy, "Opponents may get $3.8-M," *West Lincoln Review*, 21 December 1988, p. 2.

"Funding is unfair to West Lincoln," *Niagara Falls Review*, 20 January 1989.

"OWMC workshop for council," *West Lincoln Review*, 11 January 1989.

Bogacz, Rick, "Region's OWMC fight in jeopardy," *The Standard*, 19 January 1989, p. 1.

Curran, Sheldon, "Ontario has abandoned region over OWMC plan: Dick," *Niagara Falls Review*, 20 January 1989.

Gerritsma, Joop, "Niagara 'dumped on'," *The Evening Tribune*, 20 January 1989, p.1.

Gerritsma, Joop, "Funding news 'hit like a thunder clap'," *The Tribune*, 21 January 1989, p. 3.

Arnold, Steve, "Ontario trying to cripple us say toxic-waste opponents," *The Spectator*, 20 January 1989, p. C3.

"Plenty to chew on," *The Evening Tribune*, 24 January 1989, p. 4.

Arnold, Steve, "Grits rapped for their stand on toxic waste," *The Spectator*, 23 January 1989.

Curran, Sheldon, "Less-expensive lawyers are available to fight OWMC, Bradley tells region," *Niagara Falls Review*, 23 January 1989, p. 3.

Forsyth, Paul, "Ontario "unlikely' to raise Intervenor funds," *The Standard*, 23 January 1989, p. 9.

"MPPs take bruising for Region's shortfall," *The Standard*, 23 January 1989.

"Tempers flare during meeting," *The Standard*, 23 January 1989.

"Waste of time' Riddell says of bi-level parley," *The Tribune*, 24 January 1989, p. 3.

O'Brien, Paul, "Region fully committed to fight with the OWMC," *Guardian Express*, 1 February 1989, p. 9.

"Legal stock futures," *West Lincoln Review*, 25 January 1989.

McEwen, Judy, "Lawyers to bail out unless fees raised," *West Lincoln Review*, 25 January 1989, p. 1.

Chapter 37

Gracey, Clifford B., Gracey, Marilyn J., Letter to the editor re: nuisance impact zones, 12 January 1989.

Gracey, Clifford B., Gracey, Marilyn J., Letter to the editor re: "When is a buffer zone not a buffer zone?, 25 January 1989.

Gracey, Clifford B., Gracey, Marilyn J., "Hold OWMC hearings in West Lincoln," *Niagara Falls Review*, 27 January 1989, p. 6.

Draper, Doug, "Miserly legal fees could thwart thorough OWMC review," *The Standard*, 27 January 1989.

Gracey, Clifford B., Gracey, Marilyn J., "Unequal funding in facility fight," *Guardian Express*, 17 February 1989, p. 4.

Draper, Doug, "Chant defends funding—Best deal ever for Intervenors," *The Standard*, 27 January 1989, p. 9.

"Hearing Board Appointed," Update – Ontario Waste Management Corporation, Volume 4/ Number 2, March/April 1989.

Draper, Doug, "Ontario embraces outspoken federal scientist," *The Standard*, p. 9.

Gracey, Marilyn J., Personal notes re: Preliminary Hearing #1, 18 January 1989.

Alaimo, Carol, "Dick holds out hope for funding," *The Standard*, 9 February 1989.

Alaimo, Carol, "Showdown looms in OWMC feud," *The Standard*, 17 February 1989.

"An odd compromise," *West Lincoln Review*, 15 February 1989.

Draper, Doug, "West Lincoln pledges to keep fighting OWMC," *The Standard*, 4 February 1989, p. 9.

Lees, Janet, "Rift widens in fight over waste plant," *The Spectator*, 10 February 1989.

"Coalition prefers united front," *The Spectator*, 14 February 1988.

"Mayors off on western toxic tour," *Niagara Falls Review*, 9 February 1989.

McEwen, Judy, "Swan Hills people like waste facility," *West Lincoln Review*, 1 February 1989, p. 1.

Rooney, Bob, "Roses for some...thorns to others," *The Tribune*, 21 January 1989, p. 4.

Gracey, Clifford B., Gracey, Marilyn J., "Offended by columnist," *The Tribune*, 18 February 1989, p.5.

Draper, Doug, "U.S. expert praises OWMC plans," *The Standard*, 21 February 1989, p.1.

Chapter 38

"Study finds OWMC detail scant," *West Lincoln Review*, 22 February 1989.

Gracey, Clifford B., Gracey, Marilyn J., "Risk to a few is risk to all," *West Lincoln Review*, 15 March 1989.

Lees, Janet, "Video claims OWMC outdated," *The Spectator*, 14 February 1989.

"The Regional Municipality of Niagara Council Session 5-89 order of Business," 16 February 1989.

"The Regional Municipality of Niagara OWMC Steering Committee Agenda 5-89," 21 February 1989.

Dykstra, John D., "Presentation to West Lincoln Council," 20 February 1989.

Forsyth, Paul, "OWMC review set for Oakville," *The Standard*, 31 May 1989.

"Cheaper location sought," *West Lincoln Review*, 22 February 1989.

"Niagara venues for hearings on toxic waste site," The Tribune, 22 February 1989, p. 3.

"Two hearings," *The Tribune*, 4 March 1989, p. 3.

"EA hearings cheaper if near Toronto?" West Lincoln Review, 8 March 1989.

Gracey, Clifford B., Gracey, Marilyn J., letter to Dr. Kingham re: hearing location, 8 March 1989.

Dykstra, John, "Presentation to West Lincoln Council," 6 March 1989.

Draper, Doug, "West Lincoln trouble-shooter," *The Standard*, Spectrum, 13 February 1989, p. 13.

Chapter 39

Gracey, Marilyn J., Personal notes re: Day 1 — Preliminary Hearing #2, 28 March 1989.

Gracey, Marilyn J., Personal notes re: Day 2 — Preliminary Hearing #2, 29 March 1989.

"Cheaper location..." Review, 1989.

Gracey, Marilyn J., Personal notes re: Day 1 — Preliminary Hearing #2, 30 March 1989.

Lees, Janet, "Toxic-waste fighters not surprised by vote," *The Spectator*, 7 April 1989.

Packham, Mayor Joan, Letter to Jim Bradley, Minister of the Environment re: industrial development of farmlands in immediate area of the proposed OWMC facility, 11 April 1989.

Gracey, Clifford B., Gracey, Marilyn J., "E.A.—a forum for testing," 26 April 1989.

"John Jackson elected president of Great Lakes United,"

Niagara Farmers' Monthly, June 1989.

Berry, Shawn, "Potters launch lakes crusade," *The Standard*, 17 May 1989, p. 8.

Leach, Willa, Letter, *West Lincoln Review*, 10 May 1989.

Chapter 40

Kingham, D. J., Thomson, G. I., & Doherty, B. L., "Hearing Locations – Decision and Reasons," 26 May 1989.

Lees, Janet, "We feel abandoned," *The Spectator*, 1 June 1989, p. C1.

Forsyth, Paul, "OWMC review set for Oakville," *The Standard*, 31 May 1989.

Gerritsma, Joop, "Region's stand: Hold OWMC hearings here," *The Tribune*, 2 June 1989, p. 3.

Gracey, Clifford B., Gracey, Marilyn J., "Says OWMC out of step," *The Tribune*, 14 June 1989, p. 5.

Dechter, Richard, "Water Quality—The Public's Perception," Water and Pollution Control, October/November 1988.

Alaimo, Carol, "$116,500 request West Lincoln closer to loan," *The Standard*, 8 June 1989, p. 9.

"$116,500 to fight toxic waste is urged on regional council," *Niagara Falls Review*, 9 June 1989.

Sherwin, John, "West Lincoln gets $116,500 to fight toxic waste plant," *The Tribune*, 16 June 1989, p. 6.

Chapter 41

Gracey, Clifford B., Gracey, Marilyn J., "False advertising," *The Standard*, 23 June 1989, p. 3.

Daniels, Leslie, "OWMC's rebuttal to letter," *Guardian Express*, 26 July 1989, p. 5.

Gracey, Clifford B., Gracey, Marilyn J., letter #54 – re: Ms. Daniels' choice of articles in OWMC Update re: "proposed", 18 July 1989.

"Monitoring Programs Will Help Protect Environment," Update, March/April 1988, Volume 3 / Number 2, p. 1.

(Photo) "Sidewalk Marketplace," *West Lincoln Review*, 28

June 1989, p. 10.

"Township avoids debt in hiring new lawyers," *West Lincoln Review,* 28 June 1989, p.1.

Gracey, Marilyn J., Personal notes re: #3 Preliminary Hearing, 4 July 1989.

Hodgson, Douglas, Letter to Ian Blue, 24 June, 1989.

Chapter 42

Heule, Nandy, "Dispute delays refunds for Region's waste fight," *The Standard,* 26 July 1989, p. 9.

Hansard, 26 June, 1986, p. R-95.

Chapter 43

"An invitation to comment on the environmental assessment for the proposed Ontario Waste management Corporation hazardous waste treatment system...Notice of Completion of Review," *The Standard,* 18 September 1989, p. 36.

Draper, Doug, "The deal has been cooked," *The Standard,* 9 September 1989, p. 1.

Smith, Shaun, "Kormos joins the fight," *Guardian Express,* 27 September 1989, p. 14.

"Keep on fighting, he tells residents," *The Tribune,* 27 September 1989, p. 10.

Photo, "Wastebusters," *The Tribune,* 27 September 1989, p.9.

Draper, Doug, "Region wants delay in OWMC hearings," *The Standard,* 27 September 1989.

Curran, Sheldon, "OWMC report deemed inadequate, hearing is set back longer," *Niagara Falls Review,* 29 September 1989, p. 5.

Farr & Associates Reporting, Inc., Transcript for Preliminary Hearing, 3 October 1989, pp.851, 852, 987,988, 1006-1012.

Farr, Transcript for Preliminary Hearing, 4 October 1989, pp. 1077-1080, 1154-1156.

Gracey, Marilyn J., Personal notes re: Preliminary Hearings, 3&4 October 1989.

Gracey, Clifford B., Gracey, Marilyn J., Letter: "Not permitted to speak on own behalf", *Guardian Express,* 4 November 1989.

Chapter 44

Smart, Scott, Welch's – Memo, 11 July 1989.

Robertson, Marianne, letter from Mott's Canada to Michael Scott, 25 July 1989.

Bogacz, Rick, "OWMC hearings set to go but delays feared," *The Standard*, 12 October 1989.

"EA hearings start Nov. 23," *West Lincoln Review*, 11 October 1989.

"Office of Consolidated Hearings Notice of Public Hearing," *West Lincoln Review*, 25 October 1989.

Draper, Doug, "OWMC doesn't keep promises, Region told," *The Standard*, 25 October 1989, p. 9.

Gracey, Clifford B., Gracey, Marilyn J., Letter: "Who will set the record straight?" *The Times-Review*, 23 January 1990.

Currana, Sheldon, "Region's lawyer cries foul over late reports by OWMC," *Niagara Falls Review*, 25 October 1989.

Farr, Transcript for Preliminary Hearing, 4 October 1989, pp. 1079, 1080.

Chapter 45

Chant, D. A., Letter to Gary Posen, Deputy Minister of the Environment re: subsidy and regulations, 10 January 1989.

Jackson, John, Deposed 6 November 1989.

Burgess, Blair, "OWMC seeks tax cash but no market shown," *West Lincoln Review*, 8 November 1989.

Forsyth, Paul, "OWMC seeks tax subsidies," *The Standard*, 8 November 1989, p. 1.

Gerritsma, Joop, "Will toxic waste site be a "white elephant'?, *The Tribune*, 8 November 1989

Avey, Ken, "OWMC rebuts charges," *The Tribune*, 9 November 1989, p.3.

Smith, Shaun, "Opponents Fear OWMC Monopoly," *Guardian Express*, 18 November 1989, p. 15.

McEwen, Judy, "OWMC foes pledge solidarity, *West Lincoln Review*, 15 November 1989.

Gracey, Marilyn J., Personal notes re: Township public meeting,

9 November 1989.

Packham, Mayor Joan, Letter to Donald Chant re: travel expenses for members of the Coalition, 25 October 1989.

Chapter 46

Gracey, Clifford B., Gracey, Marilyn J., Letter re: best available technology, 20 November 1989.

"Design Changes Improve Emissions," *Update*, March/April 1989, p. 2.

Bradley, Jim, Minister of the Environment, Letter to John Jackson re: funding under the Intervenor Funding Project Act., 21 November 1989.

Draper, Doug, "Hearings into OWMC plant ready to begin," *The Standard*, 22 November 1989.

McInnes, Craig, "Lines drawn as waste plant hearings open," *The Globe and Mail*, 23 November 1989.

Nethercott, Diana, Photographer, "Helen Kszans' farm is on planned site of a waste management plant," *The Globe and Mail*, 23 November 1989.

Gracey, Marilyn J., Personal notes re: Preliminary Hearings, 23 &24 November 1989.

Gerritsma, Joop, "OWMC steps-up feud with Region," *The Tribune*, 25 October 1989, p. 3.

Smith, Shaun, "What person dumps his toxic garbage in his garden?" *Guardian Express*, 29 November 1989

"New foes of OWMC seek $1-M," *West Lincoln Review*, 3 January 1989.

"Trough reopens," *West Lincoln Review*, 6 December 1989.

Farr & Associates Reporting, Inc., Transcript for Preliminary Hearing, 1 December 1989, pp.172, 377-383, 411.

Gracey, Marilyn J., Personal notes re: Preliminary Hearing, 1 December 1989.

Gerritsma, Joop, "They're abusing the hearing process," *The Tribune*, 25 November 1989, p. 3.

Kelso, Brigid, "Toxic waste site is too close to fruitland," *Ontario Farmer*, Western edition, 29 November 1989.

Gracey, Clifford B., Gracey, Marilyn J., "Site is prime fruit land," *Ontario Farmer,* Western edition, 13 December 1989.

Haist, Cindy, Letter from Niagara Peninsula Fruit & Vegetable Growers' Association re: resolution to support the Coalition's strategy for waste management, 28 December 1989.

Jackson, John, Letter to Coalition supporters re: financial support for public education, 13 December 1989.

"PCB hearings begin in Smithville," *West Lincoln Review,* 20 December 1989, p. 8.

Chapter 47

Gracey, Clifford B., Gracey, Marilyn J., Letter to editor, 22 January 1990.

Volume III – Alternative Methods Site Selection Appendices Part One Table E.5, November 1988, p. E5-25, E5-26.

Scott, Angus, "The Fighters," *Dunnville Chronicle,* 24 January 1990.

Gerritsma, Joop, "Maybe it's only $1, but Region considers it a victory," *The Tribune,* 24 January 1990, p. 6.

"Protesters plan opposition rally," Pelham Herald, 24 January 1990.

"Residents plan march to protest against OWMC plan," *West Lincoln Review,* 17 January 1990.

Globe and Mail, 23 November 1989.

Burgess, Blair, "OWMC officials undisturbed about planned protest rally," *West Lincoln Review,* 31 January 1990.

Draper, Doug, "Chant denies OWMC seeking subsidies, monopoly," *The Standard,* 2 February 1990.

Draper, Doug, "$80 million later, OWMC proposal to get public test," *The Standard,* 2 February 1990.

"Critics make no apologies for NIMBY syndrome," *The Standard,* 2 February 1990.

Draper, Doug, "A glimmer of hope," *The Standard,* 3 February 1990, p. 13.

Draper, Doug, "The German Experience," *The Standard,* 3 February 1990.

OWMC—The "Wasted" Years: The Early Days

Chapter 48

Photo: "Niagara Protests OWMC Proceedings," *Niagara Farmers' Monthly*, March 1990.

Photo: "Protest sign: Dick, Sojka, Smith and Jackson at unveiling," *Guardian Express*, 10 February 1990.

Barevich, Pat, "400 march to protest toxic waste proposal," *The Tribune*, 6 February 1990.

Smith, Shaun, "Marchers protest as hearings begin," *Guardian Express*, 7 February 1990.

Hallihan, Paul, "'David and Goliath' battle starts as Niagara, OWMC square off," *The Spectator*, 6 February 1990.

Smith, Shaun, "OWMC opponents encouraged by turnout at demonstration," *Guardian Express*, 10 February 1990.

Draper, Doug, "OWMC cuts witness list to comply with order," *The Standard*, 10 January 1990, p. 8.

Abbot Professionals Reporting & Transcription Ltd., Transcript: "The Joint Board The Consolidated Hearings Act, 1981, re: Ontario Waste Management Corp.," 5 February 1990, p. 22.

Autographs

Autographs

www.ingramcontent.com/pod-product-compliance
Lightning Source LLC
Chambersburg PA
CBHW020719180526
45163CB00001B/37